ELASTIC-PLASTIC PROBLEMS

ASME PRESS TRANSLATIONS

Elastic-Plastic Problems
by B.D. Annin and G.P. Cherepanov

Synchronization in Science and Technology
by I. I. Blekhman

Prediction of Lifetime for Machines and Structures
by V. V. Bolotin

Friction Units of Machines
by I. V. Kragelsky

Accuracy of Machine Tools
by D.N. Reshetov and V. T. Portman

ELASTIC-PLASTIC PROBLEMS

B.D. Annin and G.P. Cherepanov

Academy of Sciences of the USSR, Siberian Branch
M.A. Lavrentiev Institute of Hydrodynamics

Translated from the Russian by Jamil I. Ghojel
Translation edited by V.A. Zilberstein

ASME PRESS NEW YORK 1988

The Russian edition of this book was published by Nauka Publishers, Siberian Branch, Novosibirsk in 1983.

The book deals with elastic-plastic problems other than one-dimensional. It covers shear, torsion, plane strain, plane stress, three-dimensional problems, and other related problems. The more important analytical solutions are given and some numerical results are presented. The book is intended for scientists specializing in the mechanics of deformation of solids and for engineers engaged in strength analysis.

Library of Congress Cataloging-in-Publication Data

Annin, B.D. (Boris Dmitrievich)
Elastic-plastic problems.

Translation of: Uprugo-plasticheskaia zadacha.
Bibliography: p.
1. Elasticity. 2. Plasticity. I. Cherepanov, Gennadiĭ Petrovich. II. Title.
QA931.A613 1988 531′.3823′015 88-70032
ISBN 0-7918-0000-8

CONTENTS

1. INTRODUCTION

All solid bodies, when deformed, can exhibit elastic, plastic, or viscous properties. The latter characterize a change in deformation of the body with time at constant load. Depending on temperature, load level, time, and loading rate, these properties are exhibited to various degrees. At temperatures well below the melting point, viscous properties for the majority of structural materials can be ignored.

Plasticity is the property of solids to acquire a permanent set that remain unchanged under constant loads. At sufficiently high load levels any solid can acquire a permanent set.

The mathematical theory of plasticity, conceived more than 100 years ago by Saint-Venant and Lévy, has become a diverse discipline serving highly differing branches of human practice. Of the vast number of publications, we note only some references [5–12, 14, 17–31] that give a sufficient idea about the theoretical and applied results in this field. The most complicated and least investigated problems of the mathematical theory of plasticity are elastic–plastic problems other than one-dimensional. These problems become important in the analysis of stresses and strains in the vicinity of various grooves, holes, angles, cracks, and so on. The complexity of these problems is due to the shape and dimensions of the plastic region, which are not known beforehand and must be determined as a part of the solution process.

This book presents the fundamental equations of the theory of elasticity and theory of plasticity needed for the solution of the elastic–plastic problems, which are considered in the following chapters. Attention will also be given to the general properties of the presented system of equations.

§ 1. ELASTIC–PLASTIC MEDIUM

Let us substitute the real solid body with a continuum and introduce a number of additional assumptions characterizing its elastic–plastic behavior.

Let x_1, x_2, x_3 denote the Cartesian coordinates. Components of the stress tensor, strain tensor, and displacement vector are designated, respectively, σ_{ij}, ε_{ij}, and u_i, $i, j = 1, 2, 3$.

Consider an elementary volume $dx_1\,dx_2\,dx_3$ loaded on the surface by stresses σ_{ij}, $i, j = 1, 2, 3$, as a "black box" with an input stress σ_{ij} and an output strain ε_{ij}.[1] We shall assume that the state of the elastic–plastic medium at each point is described by σ_{ij}, ε_{ij}, and temperature T. Thus, infinitesimally small increments of the output values $d\varepsilon_{ij}$ can be written in terms of the

[1] The strains of a solid body are considered small everywhere in this book.

corresponding increments $d\sigma_{mm}$ and dT in the following form[2]:

$$d\varepsilon_{ij} = A_{ijmn}\, d\sigma_{mn} + C_{ij}\, dT$$

or (1.1.1)

$$d\sigma_{ij} = A^*_{ijmn}\, d\varepsilon_{mn} + c^*_{ij}\, dT.$$

Here, the coefficients A_{ijmn} and C_{ij} can be dependent only on the instantaneous values of the stresses, strains, temperature, deformation history, and the direction of the increment vector $(d\sigma_{ij}, dT)$ in space (σ_{ij}, T). Deformation history is determined by a curve (or path) in space $(\sigma_{ij}, \varepsilon_{ij}, T)$. Since $\varepsilon_{ij} = \varepsilon_{ji}$ and $\sigma_{ij} = \sigma_{ji}$ (instantaneous stresses are assumed absent), tensor C_{ij} must be symmetrical and tensor A_{ijmn} must be symmetrical with respect to the first two indices and the last two indices. For this case, the following assumptions were made:

a) The elastic–plastic body is homogeneous (as a result, A_{ijmn} and C_{ij} do not explicitly depend on the coordinates x_1, x_2, x_3 of a point in the body).

b) The rheological properties of the material within the elementary volume under consideration do not change with time; therefore, the coefficients A_{ijmn} and C_{ij} do not explicitly depend on time t.

c) The effect of the derivatives of any order, with respect to x_1, x_2, x_3, of parameters ε_{ij}, σ_{ij}, and T on the governing equations (1.1.1) is negligibly small.

These simplifying assumptions are quite admissible for a wide class of real cases.

The study of the character of a system's response to an external perturbation is of special interest. The response of our system (elementary volume) is given in terms of the strains ε_{ij}, while the external perturbations include the load σ_{ij} and temperature T on the surface of the elementary volume. In the elementary volume of the elastic–plastic material, the response to an instantaneous external perturbation occurs momentarily and generally stays constant if σ_{ij} and T do not change. Therefore, the elastic–plastic body can be classified as a hereditary system with an instantaneous response.

Let us assume that the external perturbation vanishes with time. The total response of an arbitrary system to the vanishing external perturbation can be described as the sum of the reversible reaction ε^e_{ij} (which also vanishes) and irreversible, or residual, reaction ε^p_{ij} (which characterizes the "memory" of the system about the vanishing external perturbation). Consequently,

$$\varepsilon_{ij} = \varepsilon^e_{ij} + \varepsilon^p_{ij}. \tag{1.1.2}$$

Thus, the behavior of an anisotropic elastic–plastic body is fully described by 36 independent functionals A_{ijmn} and 6 functionals C_{ij}. When the reciprocity condition is satisfied, the number of independent functionals A_{ijmn} reduces

[2] Summation is performed with respect to repeating indexes.

to 18. It can be shown that in the case of an isotropic elastic–plastic body, the coefficients A_{ijmn} contain only two independent functionals, and C_{ij} becomes a spherical tensor with identical diagonal elements, i.e., it has only one independent functional. These functionals can, in principle, be determined experimentally; however, this is impractical due to the large number of variables and the enormous volume of experimental work. Therefore, the "black box" method can practically be used only for systems having one input and one output. Let, for example, the elastic–plastic body be subjected to tension by force P (input) resulting in relative displacement u (output) at two characteristic points. The only functional, describing the behavior of this system (i.e., the evolution of u as a function of P), is determined empirically from a series of loading–unloading tests at various values of P.

The most rational method of investigation, i.e., the determination of the unknown functionals in (1.1.1), consists of constructing a heuristic general model (theory) which must describe the available set of empirical data for the various simplest systems (obtained by the "black box" method).

Let us assume, for example, that the process is isothermal, and the body is isotropic (anisotropy, which usually results from plastic deformation, is ignored). In this case, only two functionals f and g are left in the general relations (1.1.1), and the relations can be written in Cartesian coordinates as

$$d\varepsilon_{11} = f d\sigma_{11} - g d(\sigma_{22} + \sigma_{33}),$$

$$d\varepsilon_{12} = (f + g)\, d\sigma_{12}, \qquad \begin{matrix} & 1 & \\ \nearrow & & \searrow \\ 3 & \leftarrow & 2 \end{matrix}. \tag{1.1.3}$$

The symbol $\begin{matrix} & 1 & \\ \nearrow & & \searrow \\ 3 & \leftarrow & 2 \end{matrix}$ indicates circular substitution of indices. Additional assumptions lead to further simplifications of the governing equations (1.1.3). For example, let us also assume that the volumetric deformation is elastic, i.e.,

$$d(\varepsilon_{11} + \varepsilon_{22} + \varepsilon_{33}) = d(\sigma_{11} + \sigma_{22} + \sigma_{33})/h, \tag{1.1.4}$$

where h is a material constant. Summing the first three relations (1.1.3), we obtain

$$g = (f - h^{-1})/2. \tag{1.1.5}$$

Therefore, functional f remains indeterminate. In the general isotropic case, f can depend only on the invariants of the tensors of stresses and strains and on the path of deformation. When f depends only on one invariant (for example, on the second invariant of the stress deviator), functional f can be determined directly by the "black box" method from a series of tests on uniaxial tension–compression of a rod.

In the following two sections we shall present the most common versions of the modern theory of elastic–plastic media; the formulation of this theory is based on the use of thermodynamic laws and principles.

§ 2. FUNDAMENTAL EQUATIONS OF THE THEORY OF ELASTICITY

Henceforth, we are going to consider isotropic bodies and isothermal processes. In this case, the second term on the right-hand side of (1.1.1) vanishes and the coefficient A_{ijmn} includes only two independent functionals.

A body with completely reversible response is called elastic; for an elastic body, the strains are uniquely related to stresses and the relation (1.1.1) is integrable.

1. Linear elastic body

Let the coefficients A_{ijmn} be constants in some region of the stress space. Such a body is called linear elastic; the corresponding region in the stress space is called the elastic region. By integrating (1.1.3), the resulting relation for the linear elastic body can be written as

$$\varepsilon_{11} = \frac{1}{E}\left[\sigma_{11} - \nu(\sigma_{22} + \sigma_{33})\right] + \varepsilon_{11}^{0},$$
$$\varepsilon_{12} = \frac{1+\nu}{E}\sigma_{12} + \varepsilon_{12}^{0}, \qquad \begin{matrix} & 1 & \\ \nearrow & & \searrow \\ 3 & \leftarrow & 2 \end{matrix}. \tag{1.2.1}$$

For the determination of the constants E (Young's modulus) and ν (Poisson's ratio), it is sufficient to conduct one test with a bar of rectangular section under tension and measure the longitudinal and transverse elongations. The values of ε_{ij}^{0} represent, generally speaking, coordinate functions characterizing the "memory" of the system about its past (before the system entered the elastic region under consideration). For the sake of simplicity, the memory part of the system will be ignored here so that all ε_{ij}^{0} are equal to zero.

Therefore, (1.2.1) yields the equation

$$\varepsilon_{ij} = \frac{\partial \hat{U}}{\partial \sigma_{ij}}. \tag{1.2.2}$$

Here

$$\hat{U} = \frac{1}{2E}(\sigma_{11} + \sigma_{22} + \sigma_{33})^2 + \frac{1}{2G}\left[\frac{\sigma_{12}^2}{2} + \frac{\sigma_{21}^2}{2} + \frac{\sigma_{13}^2}{2} + \frac{\sigma_{31}^2}{2} + \frac{\sigma_{23}^2}{2} + \frac{\sigma_{31}^2}{2} - \sigma_{11}\sigma_{22} - \sigma_{11}\sigma_{33} - \sigma_{22}\sigma_{33}\right], \tag{1.2.3}$$

$G = E/2(1+\nu)$ is the shear modulus.

For any continuous medium, the following relations must hold:
a) *equilibrium equations*

$$\frac{\partial \sigma_{11}}{\partial x_1} + \frac{\partial \sigma_{12}}{\partial x_2} + \frac{\partial \sigma_{13}}{\partial x_3} = 0,$$

$$\frac{\partial \sigma_{12}}{\partial x_1} + \frac{\partial \sigma_{22}}{\partial x_2} + \frac{\partial \sigma_{23}}{\partial x_3} = 0, \qquad (1.2.4)$$

$$\frac{\partial \sigma_{13}}{\partial x_1} + \frac{\partial \sigma_{23}}{\partial x_2} + \frac{\partial \sigma_{33}}{\partial x_3} = 0$$

(quasistatic process without external mass forces);
b) *Cauchy's equations*

$$\varepsilon_{ij} = \frac{1}{2}\left[\frac{\partial u_i}{\partial x_j} + \frac{\partial u_j}{\partial x_i}\right] \qquad (1.2.5)$$

(small strains). Here (u_1, u_2, u_3) is the displacement vector. Displacements and strains are reckoned from the initial unstressed state of the body. Residual stresses and strains must also satisfy relations (1.2.4) and (1.2.5).

The system of equations (1.2.1), (1.2.4), and (1.2.5) can be defined as closed. The majority of solids encountered in nature satisfy these equations under sufficiently small loads.

By excluding the displacement vector from equations (1.2.5), we obtain the six Saint-Venant compatibility equations

$$\frac{\partial^2 \varepsilon_{11}}{\partial x_2^2} + \frac{\partial^2 \varepsilon_{23}}{\partial x_1^2} - 2\frac{\partial^2 \varepsilon_{12}}{\partial x_1 \partial x_2} = 0,$$

$$\frac{\partial^2 \varepsilon_{11}}{\partial x_2 \partial x_3} + \frac{\partial}{\partial x_1}\left[\frac{\partial \varepsilon_{23}}{\partial x_1} + \frac{\partial \varepsilon_{13}}{\partial x_2} - \frac{\partial \varepsilon_{12}}{\partial x_3}\right] = 0, \qquad \begin{matrix} & 1 & \\ \nearrow & & \searrow \\ 3 & \leftarrow & 2 \end{matrix} . \qquad (1.2.6)$$

By substituting the strains expressed in terms of stresses (1.2.2) into equations (1.2.6), we find

$$\frac{\partial^2}{\partial x_2^2}\left(\frac{\partial \hat{U}}{\partial \sigma_{11}}\right) + \frac{\partial^2}{\partial x_1^2}\left(\frac{\partial \hat{U}}{\partial \sigma_{22}}\right) - 2\frac{\partial^2}{\partial x_1 \partial x_2}\left(\frac{\partial \hat{U}}{\partial \sigma_{12}}\right) = 0,$$

$$\frac{\partial^2}{\partial x_2 \partial x_3}\left(\frac{\partial \hat{U}}{\partial \sigma_{11}}\right) + \frac{\partial}{\partial x_1}\left[\frac{\partial}{\partial x_1}\left(\frac{\partial \hat{U}}{\partial \sigma_{23}}\right) - \frac{\partial}{\partial x_2}\left(\frac{\partial \hat{U}}{\partial \sigma_{13}}\right) - \frac{\partial}{\partial x_3}\left(\frac{\partial \hat{U}}{\partial \sigma_{12}}\right)\right] = 0, \qquad \begin{matrix} & 1 & \\ \nearrow & & \searrow \\ 3 & \leftarrow & 2 \end{matrix} . \qquad (1.2.7)$$

Equations (1.2.4) and (1.2.7), together with the boundary conditions in terms of stresses (first basic problem), uniquely determine the stresses.

2. Nonlinear elastic body

The state of an elastic body is completely described by strains. According to the first and second laws of thermodynamics for reversible systems, there exists the elastic potential[3] $U(\sigma_{ij})$ such that[4]

$$dU = \varepsilon_{ij}\, d\sigma_{ij}. \tag{1.2.8}$$

Thus,

$$\varepsilon_{ij} = \frac{\partial U}{\partial \sigma_{ij}}. \tag{1.2.9}$$

If function $U(\sigma_{ij})$ is known from experience or is initially assumed from physical considerations, equations (1.2.4), (1.2.5), and (1.2.9) form a closed system of equations for a nonlinear elastic body. In this case, the coefficients A_{ijmn} in the general equations (1.1.1) are unique stress functions. If function $U(\sigma_{ij})$ transforms into a quadratic form, the body will be linear elastic and vice versa. For the sake of fairness, it is worth noting that in the other cases it becomes practically impossible to determine a function of six variables from tests alone. Therefore, it is always necessary to resort to certain assumptions, which are justified a posteriori.

For an isotropic body (which is always somewhat idealized), potential U depends only on three variables. The latter can be represented by any three independent invariants of the stress tensor. The most common and simplest version of the nonlinear theory of elasticity is based on the following additional assumption: Potential U depends only on the second invariant I of the stress deviator $s_{ij} = \sigma_{ij} - \sigma\delta_{ij}$, where

$$3\sigma = \sigma_{ij}\delta_{ij},$$

$$I = \tfrac{1}{2}s_{ij}s_{ij} = \tfrac{1}{6}\Big[(\sigma_{11}-\sigma_{22})^2 + (\sigma_{22}-\sigma_{33})^2 + (\sigma_{11}-\sigma_{33})^2 \tag{1.2.10}$$

$$+\,3\big(\sigma_{12}^2 + \sigma_{21}^2 + \sigma_{13}^2 + \sigma_{31}^2 + \sigma_{12}^2 + \sigma_{21}^2\big)\Big]$$

(δ_{ij} is the Kronecker delta). In this case the relation between the stresses and

[3] For the isothermal process under consideration, this potential is the Gibbs energy.

[4] Here, all σ_{ij} must be considered independent.

strains according to (1.2.9) take the form

$$\varepsilon_{11} = \frac{1}{3}\frac{\partial U}{\partial I}(2\sigma_{11} - \sigma_{22} - \sigma_{33}),$$

$$\varepsilon_{12} = \frac{\partial U}{\partial I}\sigma_{12}, \qquad \begin{matrix} & 1 & \\ \nearrow & & \searrow \\ 3 & \leftarrow & 2 \end{matrix}. \tag{1.2.11}$$

Here, $\varepsilon_{11} + \varepsilon_{22} + \varepsilon_{33} = 0$, i.e., the body is incompressible. The function $U(I)$ is easily determined from a single test of a rod under tension–compression.

As experiments show, equations (1.2.11) describe the plastic deformation of many metals satisfactorily if the external loads vary proportionally to the same parameter. In the test with proportional loading, the behavior of the elastic–plastic body as viewed by an observer, who monitors stresses and strains only, always coincides with the behavior of some imaginary nonlinear elastic body. The difference can be detected only by conducting additional observations (measurement of heat released due to the deformation of the body, x-ray diffraction analysis, examination of metallographic specimens, etc.). This observation is also valid for any fixed loading path (and not only for proportional loading).

For a nonlinear elastic body with the volumetric strain $e = \varepsilon_{ij}\delta_{ij}/3$ being a linear function of the mean stress σ, the relations between the stresses and strains are

$$\varepsilon_{ij} = \frac{\sigma}{3K}\delta_{ij} + \frac{1}{2}s_{ij}\frac{\Gamma}{T(\Gamma)},$$

$$T^2 = I, \qquad \Gamma^2 = 2(\varepsilon_{ij} - e\delta_{ij})(\varepsilon_{ij} - \delta_{ij}), \qquad T \geq 0, \qquad \Gamma \geq 0. \tag{1.2.12}$$

Function $T = T(\Gamma)$ here is determined from tests on pure shear; K is the bulk modulus. These relations are called the *deformation theory of plasticity under loading.*

§ 3. ASSOCIATED LAW OF PLASTIC FLOW

We shall present the basic equations of the theory of an ideal elastic–plastic body, which is defined by the following description.

In the six-dimensional stress space σ_{ij}, there exists a fixed surface (called *plasticity condition* or *yield surface*)

$$f(\sigma_{11}, \sigma_{22}, \sigma_{33}, \sigma_{12}, \sigma_{13}, \sigma_{23}) - k = 0 \tag{1.3.1}$$

(k is a material constant). This surface divides the entire stress space into two regions; only one of them (elastic region D_0) can be reached by the stressed state at any point within the continuum. The body is assumed to be continuous and isotropic; therefore, function $f(\sigma_{ij})$ does not depend on the choice of the coordinate system. Additionally, the function is the same for all the points within the continuum. Therefore, function $f(\sigma_{ij})$ can essentially depend only on any three independent invariants of the stress tensor.

When the stressed state lies in region D_0, the body behaves like an elastic one; when the stressed state lies on the yield surface, plastic (irreversible) deformation also occurs. The states outside the region are not attainable. The loading condition in the plastic state signifies the movement of a representative point in the stress state along the yield surface. From physical considerations, the yield surface must be convex and must contain the origin of the coordinates.

The law of conservation of energy for an elementary volume of the elastic–plastic medium can be written in the following form:

$$\sigma_{ij}\,d\varepsilon_{ij}^{e} + \sigma_{ij}\,d\varepsilon_{ij}^{p} = dA^{e} + dA^{p}, \qquad \varepsilon_{ij} = \varepsilon_{ij}^{e} + \varepsilon_{ij}^{p}. \tag{1.3.2}$$

Here, ε_{ij}^{e} and ε_{ij}^{p} are the elastic and plastic components of strain, respectively, and A^{e} and Q^{p} are the corresponding reversible and irreversible components of the total work. The value A^{e} is a function of state and corresponds to an isothermal thermodynamic process (Helmholtz energy); the value A^{p} is the dissipation energy. It is assumed that A^{p} is dependent on ε_{ij}^{p} only and independent of ε_{ij}^{e}; A^{e}, on the other hand, is dependent only on ε_{ij}^{e} and is independent of ε_{ij}^{p}. As a consequence to this assumption,

$$\sigma_{ij}\,d\varepsilon_{ij}^{e} = dA^{e}, \qquad \sigma_{ij}\,d\varepsilon_{ij}^{p} = dA^{p}. \tag{1.3.3}$$

Since there is a one-to-one correspondence between ε_{ij}^{e} and σ_{ij}, the following dependence [cf. (1.2.2)] is obtained from the first relation of (1.3.3):

$$\varepsilon_{ij}^{e} = \frac{\partial U}{\partial \sigma_{ij}}, \qquad U(\sigma_{ij}) = \sigma_{ij}\varepsilon_{ij}^{e} - A^{e}, \tag{1.3.4}$$

where U is the elastic potential (Gibbs energy).

In the most common and important case of isotropic linear elasticity, the relations (1.3.4) coincide with equations (1.2.2) with the corresponding substitution of ε_{ij}^{e} for ε_{ij}.

Let us apply the following assumption (**the maximum principle**[5]): The increment of irreversible work dA^{p} is maximal relative to *the stresses* σ_{ij}, *which are allowed by the plasticity condition.*

The condition of an analytical extremum based on (1.3.3) and (1.3.1) can be expressed in the form

$$\frac{\partial \Psi}{\partial \sigma_{ij}} = 0, \qquad \Psi = \sigma_{ij}\,d\varepsilon_{ij}^{p} - \lambda\left[f(\sigma_{ij}) - k\right], \tag{1.3.5}$$

where λ is the Lagrange undetermined multiplier.

[5] This principle is also called the Mises principle in honor of the German scientist.

From equation (1.3.5), the associated law of plasticity results in

$$d\varepsilon_{ij}^{p} = \lambda \frac{\partial f}{\partial \sigma_{ij}}, \tag{1.3.6}$$

which allows determination of the increments of the components of plastic strain, utilizing the plasticity condition, in terms of a new unknown value λ which must be determined.

The associated law of plasticity (1.3.6) can be formulated as follows: The vector of the plastic strain increment is perpendicular to the yield surface at the point in the stress space.

Equations (1.2.4), (1.2.5), (1.3.1), (1.3.4), and (1.3.6) form a closed system of equations that describes the stresses and strains in plastic zones of the ideal elastic–plastic body.

Equation (1.3.6) is true only for smooth parts of the yield surface. At singular points of the yield surface, the maximum principle allows obtaining the associated law of plasticity by the same Lagrange method in the following form:

On an edge that represents the intersection line of two smooth surfaces

$$f_1(\sigma_{ij}) - k_1 = 0, \qquad f_2(\sigma_{ij}) - k_2 = 0, \tag{1.3.7}$$

we get

$$d\varepsilon_{ij}^{p} = \lambda_1 \frac{\partial f_1}{\partial \sigma_{ij}} + \lambda_2 \frac{\partial f_2}{\partial \sigma_{ij}} \tag{1.3.8}$$

(λ_1 and λ_2 are two new unknown quantities).

On a corner point that represents the intersection point of three smooth surfaces

$$f_1(\sigma_{ij}) - k_1 = 0, \qquad f_2(\sigma_{ij}) - k_2 = 0, \qquad f_3(\sigma_{ij}) - k_3 = 0, \tag{1.3.9}$$

we get

$$d\varepsilon_{ij}^{p} = \lambda_1 \frac{\partial f_1}{\partial \sigma_{ij}} + \lambda_2 \frac{\partial f_2}{\partial \sigma_{ij}} + \lambda_3 \frac{\partial f_3}{\partial \sigma_{ij}} \tag{1.3.10}$$

(λ_1, λ_2, and λ_3 are the three new unknown quantities).

Thus, for singular points of the yield surface, a closed system of equations is also obtained.

The most commonly used plasticity conditions are:

Tresca–Saint-Venant condition

$$\tau_{max} = \tfrac{1}{2}\max\{|\sigma_1 - \sigma_2|, |\sigma_2 - \sigma_3|, |\sigma_3 - \sigma_1|\} = k, \tag{1.3.11}$$

where τ_{max} is the maximum shear stress and $\sigma_1, \sigma_2, \sigma_3$ are principal stresses.

Guber–Mises condition

$$I = (\sigma_1 - \sigma_2)^2 + (\sigma_2 - \sigma_3)^2 + (\sigma_3 - \sigma_1)^2 = 6k^2. \tag{1.3.12}$$

These conditions describe satisfactorily the onset of shear-type plastic deformations in metals and polymers.

In granular materials and rocks, the initiation of plastic deformations is best described by the *Schleiher–Mohr condition*

$$(\sigma_1 - \sigma_2)^2 + (\sigma_2 - \sigma_3)^2 + (\sigma_3 - \sigma_1)^2 = g^2(\sigma), \tag{1.3.13}$$

where $g(\sigma)$ is a function of the mean stress. For example, $g(\sigma) = \sqrt{6}\,k(1 - e^{(\sigma-\sigma_0)/k})$; σ_0, k are positive constants [1].

For a particular problem, it is expedient to use the general system of the equations and take into account the properties of symmetry inherent in the given problem. In the case of a yield surface with singular lines [for example, condition (1.3.11)], plastic zones of different types can exist with various governing equations. For analytical yield surfaces, the general equations can be written in a simpler form. Let us, for example, present the general system of the governing equations in the plastic region in the case of Guber–Mises condition (1.3.12). This system is obtained based on equations (1.3.4), (1.3.6), (1.3.1), and (1.3.2) in the following form:

$$\begin{gathered}
de_{ij} = s_{ij} \cdot \lambda + ds_{ij}/(2G), \\
de = E^{-1}(1 - 2\nu)\, d\sigma, \qquad s_{ij}s_{ij} = 2\tau_s^2, \\
e = \varepsilon_{ij}\delta_{ij}, \qquad e_{ij} = \varepsilon_{ij} - e\delta_{ij}, \\
\lambda = \begin{cases} 0, & \text{if } s_{ij}s_{ij} < 2\tau_s^2, \\ 0, & \text{if } s_{ij}s_{ij} = 2\tau_s^2 \text{ and } s_{ij}\,ds_{ij} < 0, \end{cases} \\
\lambda \geq 0, \quad \text{if } s_{ij}s_{ij} = 2\tau_s^2 \text{ and } s_{ij}\,ds_{ij} = 0.
\end{gathered} \tag{1.3.14}$$

Here, τ_s is the yield limit under pure shear, ν is the Poisson's ratio $G = E \times 2^{-1}(1 + \nu)^{-1}$, and E is Young's modulus.

Equations (1.3.14) were proposed by Reuss in 1930 and, for a plane problem, by Prandtl in 1924. Therefore, they are called Prandtl–Reuss equations.

The boundary conditions must be added to the general equations. We present the most typical boundary value problems.

First problem—one normal and two shear stresses are specified at the boundary of the body.

Second problem—displacement vector is specified at the boundary of the body.

Contact problem—along a region of the boundary, the load is given, as in the first problem, while outside of the region the displacement vector component normal to the boundary and two conditions of dry friction are given. These conditions relate the tangential loads to the normal stress on the contact area (in particular, when friction is neglected, the tangential loads are equal to zero).

It should be noted that other formulations of proper boundary value problems are possible.

All tensor components of stresses and strains are assumed to be continuous on the internal surface separating elastic and plastic zones and also, possibly, on the interface between various plastic zones. For specific problems, these conditions of continuity serve as boundary conditions due to the fact that the differential equations of the problem for each of these zones will be different.

The expected degree of smoothness of the required solution is essential. The fact of the matter is that when the plastic state is reached initially at one point, further development of plastic deformation can proceed either continuously with the formation of three-dimensional zones or with the formation of discontinuity lines. It is also possible for plastic deformation to develop entirely along some discontinuity surfaces. A similar problem arises in transonic aerodynamics in connection with the problem of the initiation of a shock wave in a local supersonic zone. There is no simple answer to this question yet; therefore, selection of an appropriate solution must be based on experimental data. Hereafter in the book, the exact required solutions for stresses, strains, and displacements are assumed continuous everywhere in the elastic and plastic zones. The solutions with plastic discontinuity lines, which are also presented in some cases, are approximate.

The uniqueness of the continuous solution of the elastic–plastic problem was shown by Melan in 1939 (see [12]), and the questions of the existence of solutions are partially studied in [3, 12]. It should be noted that the use of the nonassociated law in some cases might lead to incorrect problems. An experimental study of possible deviations from the associated law of plasticity is of great interest regarding the elucidation of the boundaries within which the maximum principle is applicable. The latter is widely used in modern thermodynamics, though, undoubtedly, to a lesser extent than the first and second laws of thermodynamics. The above question remains open at the present time.

The integral generalization of the Mises maximum principle is the principle of selection which was first formulated in 1974 [35]. When the volume of the plastic region is not equal to zero, the principle of selection leads to the associated law of plasticity the same way as the maximum principle does. In the case of slip lines, when the volume of the plastic zone is equal to zero, it leads to new results and allows, in particular, solution of the problem of nonuniqueness in the theory of plasticity. This problem results from various topological structures of possible solutions (we should note that Melan's proof of the uniqueness of the solution of the elastic–plastic problem is valid only when the volume occupied by the plastic zone is not equal to zero). The principle of selection states: The rate of energy dissipation for the entire body is maximal [35]. For more details, the reader is referred to monograph [36].

The incremental theory of plasticity, which leads to the associated law, is the most general and substantiated theory. However, other deformation theories, equivalent to the nonlinear theory of elasticity, are also valid in their own right, since they provide a good description of certain classes of loading for some types of materials. One of these approaches is discussed in the following section.

§ 4. HAAR–KARMAN PRINCIPLE

Consider a body V bounded by a sufficiently smooth surface S. Let displacements

$$u_i = u_i^*, \qquad i = 1,2,3, \tag{1.4.1}$$

be given on the part S_u of the surface and the external forces

$$\sigma_{ij} n_j = T_i^*, \qquad i = 1,2,3, \tag{1.4.2}$$

on the part S_T of the surface. Here, u_i^*, T_i^* are specified functions of points on the surfaces S_u and S_T, respectively, and (n_1, n_2, n_3) is the vector of the external normals to surface S_T, $S_T = S \setminus S_u$. Let the transition of the body to the plastic state be determined by the plasticity condition

$$f(\sigma_{ij}) = k, \tag{1.4.3}$$

where k is a constant; therefore, the elastic region is determined in the stress space by inequality

$$f(\sigma_{ij}) < k. \tag{1.4.4}$$

Thus, at any point of body V, the following condition is satisfied:

$$f(\sigma_{ij}) \le k. \tag{1.4.5}$$

Let us assume that function $f(\sigma_{ij})$ is continuous and convex relative to σ_{ij} such that in a six-dimensional stress space $(\sigma_{11}, \sigma_{22}, \sigma_{33}, \sigma_{12}, \sigma_{23}, \sigma_{13})$ the set, determined by inequality (1.4.5), is closed and convex.

The Haar–Karman principle postulates that, in an elastic–plastic state of body V, the stresses are such that they minimize the functional

$$Э(\sigma) = \int_V \hat{U}(\sigma)\, dV - \int_{S_u} \sigma_{ij} n_i u_j^*\, dS \tag{1.4.6}$$

over the stress states that satisfy conditions (1.4.2), (1.4.5) and the equilibrium equations (1.2.4), i.e., over the statically permissible stress fields. Here, $\hat{U}(\sigma)$ is determined by equality (1.2.3). The Lagrange multipliers of the equilibrium equations (1.2.4) in this case are taken as the components of the displacement vector.

It should be noted that when inequality (1.4.4) is satisfied at all points within the body, i.e., when the entire body deforms elastically, the Haar–Karman principle becomes the Castigliano principle in the linear theory of elasticity.

Let us assume that at some points equality (1.4.3) is valid and these points form region V^p. Let $V^e = V \setminus V^p$. Inequality (1.4.4) is true in region V^e. The necessary conditions for the minimum of the functional in region V^e can be obtained if an additional condition (1.2.4) is taken into account, while in region V^p condition (1.4.3) should also be added.

Consider two important cases.

1. Let us use the Mises plasticity condition (1.3.12) written as

$$\frac{1}{2G}\left\{\frac{1}{6}\left[(\sigma_{11}-\sigma_{22})^2+(\sigma_{11}-\sigma_{33})^2+(\sigma_{22}-\sigma_{33})^2\right]+\sigma_{12}^2+\sigma_{13}^2+\sigma_{23}^2\right\}$$
$$=\frac{1}{2}\frac{\tau_s^2}{G}. \tag{1.4.7}$$

Here, τ_s is the yield limit under pure shear. Let u_1, u_2, u_3 be the Lagrange multipliers of the equilibrium equations (1.2.4) and let φ stand for condition (1.4.7). In region V^e, we have $\varphi = 0$. In region V^p, the necessary condition for the minimum of functional $\text{Э}(\sigma)$ under the additional conditions (1.2.4) and (1.4.7) leads to

$$\frac{\partial u_1}{\partial x_1} = \frac{1+\varphi}{2G}\sigma_{11} - \frac{\sigma}{2G}\left(\varphi + \frac{3}{m+1}\right), \qquad \begin{matrix} & 1 & \\ \nearrow & & \searrow \\ 3 & \leftarrow & 2 \end{matrix},$$
$$\frac{\partial u_1}{\partial x_2} + \frac{\partial u_2}{\partial x_1} = \frac{1+\varphi}{G}\sigma_{12}, \qquad m = \nu^{-1}, \qquad \begin{matrix} & 1 & \\ \nearrow & & \searrow \\ 3 & \leftarrow & 2 \end{matrix}. \tag{1.4.8}$$

Here, ν is the Poisson's ratio. Conditions (1.4.7) and (1.4.8) are called **Hencky's relations**. From the minimum of $\text{Э}(\sigma)$, it follows that conditions (1.4.1) on S_u are valid.

2. Let us use the Tresca plasticity condition (1.3.11) written as

$$\tau_{\max}(\sigma) = \max_{\alpha,\beta} \sigma_{ij}\alpha_i\beta_j = \tau_s,$$
$$\alpha_1^2+\alpha_2^2+\alpha_3^2 = 1, \qquad \beta_1^2+\beta_2^2+\beta_3^2 = 1, \qquad \alpha_1\beta_1+\alpha_2\beta_2+\alpha_3\beta_3 = 0, \tag{1.4.9}$$

where τ_s is the yield limit under pure shear. We first show that the set of stress space R^6, determined by inequality $\tau_{\max}(\sigma) \le \tau_s$, will be convex. Let $\tau_{\max}(\sigma') \le \tau_s$, $\tau_{\max}(\sigma'') \le \tau_s$. Inequality

$$\tau_{\max}(t\sigma' + (1-t)\sigma'') \le \tau_s$$

for any $t \in [0,1]$ follows from

$$\left[t\sigma'_{ij} + (1-t)\sigma''_{ij}\right] \le t\max_{\alpha,\beta}\sigma'_{ij}\alpha_i\beta_j + (1-t)\max_{\alpha,\beta}\sigma''_{ij}\alpha_i\beta_j \le \tau_s.$$

The minimization of functional (1.4.6) under conditions (1.4.2), (1.2.4), and $\tau_{\max} \le \tau_s$ leads to the Haar–Karman relations which have a different analytical form depending on whether one or two differences of the principle stresses in (1.3.11) reach the value τ_s. Let, for example, in some region of the body

$$|\sigma_1 - \sigma_2| = 2\tau_s, \qquad |\sigma_1 - \sigma_2| < 2\tau_s, \qquad |\sigma_2 - \sigma_3| < 2\tau_s. \tag{1.4.10}$$

Such a state is called **semiplastic state**. It follows from the extremum condition that the main axes of the stress and strain tensors coincide in this case and the following relations exist:

$$\sigma_1 + \sigma_2 + \sigma_3 = \frac{E}{1-2\nu}(\varepsilon_1 + \varepsilon_2 + \varepsilon_3), \tag{1.4.11}$$

$$\sigma_2 = \frac{E\nu}{(1+\nu)(1-2\nu)}(\varepsilon_1 + \varepsilon_2 + \varepsilon_3) + \frac{E}{1+\nu}\varepsilon_2. \tag{1.4.12}$$

Here ε_1, ε_2, ε_3 are the principle strains.

Relation (1.4.11) signifies the elastic volume change, while relation (1.4.12) implies the elastic constraint with respect to the second principal direction. Using relations (1.4.10)–(1.4.12) we can determine the principal stresses from the principal strains. The dependence between the tensor of stresses and tensor of strains in this case will be quadratic.

Let, in some volume of a body,

$$|\sigma_1 - \sigma_3| = 2\tau_s, \qquad |\sigma_1 - \sigma_2| = 2\tau_s, \qquad |\sigma_2 - \sigma_3| < 2\tau_s. \tag{1.4.13}$$

This state is called the state of **total plasticity**.

From the extremum condition for $Э(\sigma)$, it follows that

$$\sigma_1 + \sigma_2 + \sigma_3 = \frac{E}{1-2\nu}(\varepsilon_1 + \varepsilon_2 + \varepsilon_3). \tag{1.4.14}$$

Relations (1.4.13) and (1.4.14) can be used to determine the principal stresses in the case of total plasticity.

Recently, the Haar–Karman theory has received further development in the works of Khristianovich and Shemiakin [31].

§ 5. ON THE METHODS OF SOLUTION OF ELASTIC–PLASTIC PROBLEMS

The methods that are used to solve elastic–plastic problems can be arbitrarily divided into two groups[6]: analytical and variational difference methods.

The analytical techniques used in solving elastic–plastic problems are mainly based on the methods of the theory of functions of complex variables [13] and on perturbation methods [7].

Of the variational difference methods, the finite-element method [3], which will be discussed now briefly, is most widely used presently to solve elastic–plastic problems.

[6] Henceforth we shall not address any experimental methods (coating, etching, microhardness measurements, etc.).

The finite-element method was originally developed for the solution of static problems in structural mechanics. The specific feature of the method is the ability to construct a discrete computational model based on variational principles without resorting to a system of differential equations and boundary conditions. The region occupied by the body is broken into finite elements of an arbitrary structure. Relations are established between the forces acting at the nodal points of each element and the displacements caused by these forces and finally the equilibrium equations of each element are written. The construction of the solution using the finite element method can be interpreted as an approximate method of solving the differential equations with boundary conditions corresponding to the mechanical problem, and also as solving the problem of minimization of the energy functional for the problem using the Ritz–Galerkin method.

The finite-element method will be considered here as a special case of the Ritz–Galerkin method with an appropriate selection of coordinate functions. These functions are determined in such a way that the final system of algebraic equations become rarefied (contains a large number of zeros). Let us consider an application of the method to solution of the one-sided variational problem. The latter includes the elastic–plastic problem based on the Haar–Karman principle as a particular case.

Let it be required to minimize the functional

$$I(u)=\tfrac{1}{2}(au,u)-(f,u) \tag{1.5.1}$$

on a closed and convex subset

$$G\subset H_a(\Omega). \tag{1.5.2}$$

Here, $H(\Omega)$ is the Hilbert space of the functions determined in the region of the three-dimensional (two-dimensional) space Ω, the brackets designate a scalar product, a is a linear positively determined operator $(au, u)\geq c(u, u)$, $c>0$, and $H_a(\Omega)$ is the functional space obtained by closing $H(\Omega)$ with respect to the norm $|u|=(au, u)^{1/2}$, $u, f\in H(\Omega)$.

We assume that the boundary of region Ω is a polyhedron (polygon in plane case).[7] Let us divide the region Ω into n finite elements with piecewise linear boundaries. These elements normally are tetrahedra, rectangular parallelepipeds (triangles, rectangles). We construct finite-dimensional approximation $H_a^n=\{u_n\}$ of space $H_a(\Omega)$. On each element, u_n is assumed to be a polynomial. At some points, called nodes, the values of the polynomials and their derivatives are given. These are called nodal parameters. The conditions of polynomial splicing at the nodes, located on common regions of element boundaries, are governed by the requirement that $u_n\in H_a(\Omega)$. Further, each of the functions u_n is represented in the form

$$u_n=\sum_{k=1}^{N} c_k\varphi_k, \tag{1.5.3}$$

[7] If it is not the case, mapping of Ω on region Ω^* bounded by a polyhedral surface must be constructed.

where N is the number of independent nodal parameters, and the coordinate functions φ_k are determined in such a manner that the value of one independent nodal parameter is equal to 1, while the other parameters are equal to zero. An approximate solution of the problem defined by (1.5.1) and (1.5.2) is sought in the form of (1.5.3), where the coefficients c_k are determined from the conditions of the minimum of the functions with N variables.

$$f(c_1, c_2, \dots, c_N) = J(u_n) = \frac{1}{2} \sum_{i=1}^{N} \sum_{j=1}^{N} a_{ij} c_i c_j - \sum_{k=1}^{N} c_k e_k,$$

$$a_{ij} = (a\varphi_i, \varphi_j), \qquad e_k = (f, \varphi_k),$$

on set

$$\left\{ (c_1, c_2, \dots, c_N) : u_n = \sum_{k=1}^{N} c_k \varphi_k \in G \cap H_a^n(\Omega) \right\}.$$

Matrix $\|a_{ij}\|$ is called the Gram matrix or the stiffness matrix.

For more detailed discussion of the finite-element method and its applications, the reader is referred to references [4, 14, 37]. We should note, in particular, the interesting book by Morozov and Nikishkov [14], a large part of which is devoted to numerical experiments on the evolution of plastic zones in the vicinity of cracks and slits. Here we will briefly discuss only those questions which are not sufficiently elucidated in reference [14].

2. COMPLEX SHEAR

§ 1. FUNDAMENTAL EQUATIONS. BRIEF REVIEW

The simplest combined stress state is complex shear, which is also called longitudinal shear, pure shear, and antiplane deformation. **Complex shear** is a stress state in an infinitely long cylindrical body that results from the action of uniform loads acting along the generator. Such a stress state also occurs under torsion when the region of interest is small in comparison with the characteristic dimensions of the contour under torsion.

Let the Cartesian coordinates x and y lie in a plane connecting the slit and let axis z be perpendicular to plane xOy. Under complex shear, displacements take place in the direction of z only, i.e., the fields of displacements and stresses in this case are such that

$$\begin{gathered} u = v = 0, \qquad w = w(x, y), \\ \sigma_x = \sigma_y = \sigma_z = \tau_{xy} = 0, \\ \tau_{xz} = \tau_{xz}(x, y), \qquad \tau_{yz} = \tau_{yz}(x, y). \end{gathered} \tag{2.1.1}$$

The equilibrium equations in this case are particularly simple and are reduced to one equation [13]

$$\frac{\partial \tau_{xz}}{\partial x} + \frac{\partial \tau_{yz}}{\partial y} = 0 \tag{2.1.2}$$

in terms of stresses or

$$\frac{\partial^2 w}{\partial x^2} + \frac{\partial^2 w}{\partial y^2} = 0$$

in terms of displacements (in the elastic region).

From (2.1.2) and Hooke's law $\tau_{xz} = G(\partial w/\partial x)$, $\tau_{yz} = G(\partial w/\partial y)$ (G is the shear modulus), it follows that stresses and displacements in the elastic region can be expressed by one analytical function $f(z)$ of the complex variable $z = x + iy$[1]

$$w = \operatorname{Re} f(z), \qquad \tau_{xz} + i\tau_{yz} = G\overline{f'(z)}. \tag{2.1.3}$$

[1]The use of the same letter z to denote a Cartesian coordinate and a complex variable can not, evidently, lead to confusion.

In the case of ideal plasticity, in addition to the equilibrium equation (2.1.2), the following relations [1, 30] apply in the plastic region

$$\tau_{xz}^2 + \tau_{yz}^2 = k^2 \quad \text{yield condition,}$$
$$\tau_{yz}\frac{\partial w}{\partial x} - \tau_{xz}\frac{\partial w}{\partial y} = 0 \quad \text{Hencky's equation.} \tag{2.1.4}$$

Here, $k = \tau_s$ according to the Mises condition, $k = 2\tau_s/\sqrt{3}$ according to the Tresca–Saint Venant condition, and τ_s is the yield limit under pure shear.

Stresses are determined independently from displacements if the loads on the boundary of the plastic region are given. Let us represent the stresses by the magnitude of the shear stress vector and angle θ between the slip line and the x axis, in the form

$$\tau_{xz} = -k\sin\theta, \qquad \tau_{yz} = k\cos\theta. \tag{2.1.5}$$

Function $\theta(x, y)$ satisfies the equilibrium equation

$$\cos\theta\frac{\partial\theta}{\partial x} + \sin\theta\frac{\partial\theta}{\partial y} = 0. \tag{2.1.6}$$

The characteristics of equation (2.1.6) are a family of straight lines $y = x\tan\theta + C$, $\theta = \text{const}$, that coincide with the slip lines and the orthogonal vector (τ_{xz}, τ_{yz}) at each point.

According to the Hencky relation (2.1.4), condition $w = \text{const}$ exists along the slip line.[2] Thus, the stress field in a plastic region is determined entirely by the shape of the plastic region boundary and by the boundary load.

On the boundary of elastic and plastic regions, the stresses and displacements are assumed continuous.

For a hardening material, the dependence between stresses and strains is described by function $\tau = \tau(\gamma)$, which links the modulus of stress τ and modulus of strain γ. Here, τ and γ are determined by the expressions

$$\tau^2 = \tau_{xz}^2 + \tau_{yz}^2, \qquad \gamma^2 = \gamma_{xz}^2 + \gamma_{yz}^2,$$
$$\gamma_{xz} = \frac{\partial w}{\partial x}, \qquad \gamma_{yz} = \frac{\partial w}{\partial y}. \tag{2.1.7}$$

Complex shear problems were first studied by Trefftz in 1922. He presented a solution to the problem of complex shear for an ideal elastic–plastic body having an L-shaped profile, and a similar problem for the boundary of a round opening [29, 30]. He used the hodograph plane and the methods of the theory of complex variable functions which are analogous to the methods of plane

[2]According to a more rigorous associated law of plasticity, all conclusions based on the Hencky equation remain unchanged if w is substituted by the velocity $\dot{w}$.

hydrodynamics of an ideal incompressible fluid. Subsequently, Hult and McClintock [4, 14, 15], Koskinen [23], and Rice [25, 26] obtained a solution for a half plane with an angular notch using the same method. A solution for strips of finite width, weakened by an angular notch was obtained in [26]. Field [22], Erdogan [21], and Kostrov and Nikitin [3] discussed problems of complex shear for bodies with cracks; it was assumed that the plastic zone is concentrated in an infinitely narrow layer on the extension of the crack.

Cherepanov [16] presented a method of solution by quadratures for problems of complex shear in ideal elastic–plastic bodies, for any contour formed by sections of straight and curved lines. A special case was also considered when the sections of straight lines were free from stresses and sections of curved arcs were arbitrarily loaded and entirely covered by the plastic zone. The solution of these problems is essentially based on the solution of a nonlinear boundary problem [17]. Notably, the solution of the elastic problem for bodies of a similar shape is not expressed in quadratures so that elastic–plastic problem is, in essence, simpler than a pure elastic problem. In reference [18], a method is given for finding a closed-form solution for a similar class of contact elastic–plastic problems of complex shear. Using this method, exact solutions were found by Sukhikh [12] and also in references [16, 18] for the following body configurations:

a) one crack at the boundary of the half plane;
b) a rigid punch acting on the free boundary of the half plane;
c) a groove with a rounded bottom;
d) a periodic system of cracks at the boundary of the half plane.

Plastic zones appearing in the half plane with a notch formed by two generally nonparallel straight sections connected by an arc were investigated in detail in reference [9].

Neuber [5] and Sokolovsky [11] discussed some problems for a work-hardening body under complex shear for specially selected analytical relations between stresses and strains approximating real stress–strain curves. It is important to note that in case of work-hardening, the equations in the problem of complex shear are analogous to the equations of plane flow of compressible ideal fluid, and that the adopted technique is similar to the Chaplygin method. This method was used by Neuber [5], Sokolovsky [11], and Dobrovolsky [2] to obtain exact solutions for some shapes of notches in the half plane and in a strip. Panc studied shearing of a prismatic body with symmetrical sharp notches using piecewise-linear "stress–strain" curves [24]. Rice [27] used the locus method to study exhaustively the problem for the half plane with an angular notch under an arbitrary work-hardening law. The perturbation theory of elastic–plastic deformations under complex shear had been discussed by Tuba [28]. As an example, a plane with a circular hole was analyzed. The qualitative features of the complex shear problem are discussed in references [10, 20].

§ 2. TREFFTZ PROBLEM

Consider an elastic–plastic problem for a region with a circular arc that is partially enveloped by a plastic zone. The problem becomes much more complicated for an arbitrary contour when only a part of it is covered by the plastic region. In this case, exact solutions can be found only for a limited class of problems. As an example of such a problem, consider a surface with "notches" in the shape of two sectors of the same circle having radius r_0. These notches are symmetrical relative to axes x and y (Fig. 2.1). At the origin, concentrated and opposite forces P are applied as shown in Fig. 2.1; stress $\tau_{yz} = \tau_\infty$ acts at infinity. Plastic regions develop at the arcs. Generally speaking, plastic zones are also formed near the origin. However, if we are interested in the development of plastic regions located at the arc, the effect of the plastic region located near the origin can be taken into account within the bounds of the theory of elasticity by introducing a concentrated force, provided this region is small compared to the radius of the circle.

Consider the right-hand half plane $\operatorname{Re} z > 0$; based on

$$w = \operatorname{Re} f(z), \qquad \tau_{xz} + i\tau_{yz} = G\overline{f'(z)},$$

$$f'(z) = (k/G)e^{-i\theta},$$

the following boundary conditions exist on the boundary of the elastic region:

$$\operatorname{Im}[zf'(z)] = 0 \quad \text{when } x = 0 \text{ and when } \theta = \pm\theta_0,\ r < r_0,$$

$$\operatorname{Re}[zf'(z)] = 0 \quad \text{when } r = r_0,\ |\theta| < \theta_0, \tag{2.2.1}$$

$$zf'(z) = \frac{ik}{G}|z| \quad \text{on an unknown boundary;}$$

$$f'(z) = -\left(\frac{i}{G}\right)\tau_\infty + O(z^{-2}) \quad \text{when } z \to \infty,$$

$$f'(z) = -\frac{2P}{G(\pi - 2\theta_0)}\frac{1}{z} + O(1) \quad \text{when } z \to 0. \tag{2.2.2}$$

Let us go over to the parametric plane of the complex variable $\zeta = \xi - i\eta$:

$$\zeta = zf'(z), \qquad z = \omega(\zeta). \tag{2.2.3}$$

Using the conformal transformation (2.2.3), the right-hand half plane $\operatorname{Re} z > 0$ on the basis of (2.2.1) becomes the region illustrated in Fig. 2.2. For

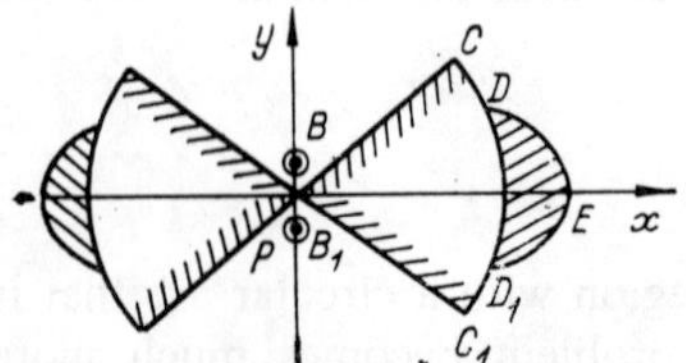

FIG. 2.1

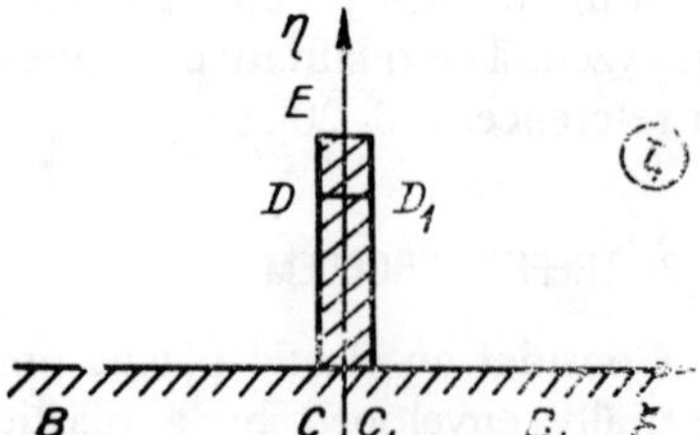

FIG. 2.2.

the determination of the analytical function $\omega(\xi)$ in this region, a boundary problem such as (II.14) is obtained (see Appendix II):

$$\begin{aligned} &\operatorname{Re}\omega = 0 \quad \text{when } \eta = 0,\ |\xi| > \frac{2P}{G(\pi - 2\theta_0)}, \\ &\operatorname{Re}[(\tan\theta_0 + i)\omega] = 0 \quad \text{when } \eta = 0,\ 0 < |\xi| < \frac{2P}{G(\pi - 2\theta_0)}, \\ &|\omega| = r_0 \quad \text{when } \xi = 0,\ \eta < \left(\frac{k}{G}\right) r_0, \\ &|\omega| = \frac{G}{k}\eta \quad \text{when } \xi = 0,\ \eta > \left(\frac{K}{G}\right) r_0, \\ &\omega(\zeta) = Gi\zeta/\tau_\infty + O(\zeta^{-1}) \quad \text{when } \zeta \to \infty. \end{aligned} \tag{2.2.4}$$

Let us present the final formulas for the simplest case of a circular notch on the boundary of the half plane when $P = 0$, $2\theta_0 = \pi$. The contour of the boundary between the elastic and plastic regions, expressed in polar coordinates $z = re^{i\theta}$, has the form

$$\theta\sqrt{\left(\frac{R}{r_0}\right)^2 - \left(\frac{r}{r_0}\right)^2} = \frac{r}{r_0} + \frac{1}{\pi}\int_{-1}^{1} \frac{\ln|t|\sqrt{(R/r_0)^2 - t^2}}{t - r/r_0}\,dt. \tag{2.2.5}$$

Here, R is the distance from the origin to the point of intersection of the plastic zone with the x axis determined from

$$\begin{aligned} \frac{\pi}{2}\ln\frac{\tau_\infty}{k} = &-\arcsin\frac{r_0}{R}\left(1 + \ln\frac{r_0}{R}\right) + \arcsin\frac{r_0}{R}\ln\arcsin\frac{r_0}{R} \\ &- \frac{1}{18}\left(\arcsin\frac{r_0}{R}\right)^3 - \frac{1}{900}\left(\arcsin\frac{r_0}{R}\right)^5 - \cdots \\ &- \frac{2^{2n-1}(2^{2n}-1)B_n}{n(2n+1)!}\left(\arcsin\frac{r_0}{R}\right)^{2n+1} + \cdots, \end{aligned}$$

where B_n is the Bernoulli number.

§ 3. FINITE WIDTH STRIP WITH AN ANGULAR NOTCH OF ARBITRARY OPENING

Consider the elastic–plastic problem for a strip of finite width weakened by an angular notch 2α. The material of the strip is ideal elastic–plastic with the yield limit k under simple shear. Stresses $\tau_{xz} = 0$ and $\tau_{yz} = \tau_\infty$ are applied at infinity (Fig. 2.3). The boundary of the strip and contour of the angular notch are free from loads.

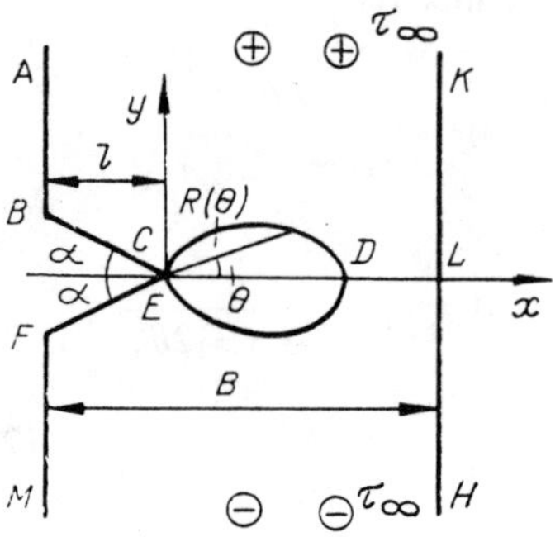

FIG. 2.3

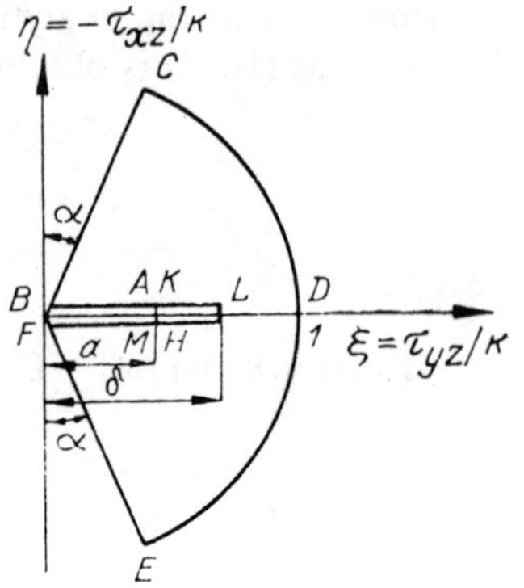

FIG. 2.4.

Using the relationship

$$(x - iy)/l = f'(\zeta), \tag{2.3.1}$$

it is convenient to go from the physical plane to the stress plane. Here,

$$\zeta = \xi + i\eta = -i(\tau_{xz} + i\tau_{yz})/k, \tag{2.3.2}$$

$f(\zeta)$ is the analytical function for all values of ζ corresponding to the stresses in the elastic region, i.e., for $|\zeta| < 1$.

The elastic region on plane ζ is mapped onto the inside of the unit circular sector with a notch (see Fig. 2.4, where $a = \tau_\infty/k$ and $\delta = \tau_L/k$). In the stress plane, the whole boundary is known with the exception of the point that determines the magnitude of stress δ on the free surface opposite the notch.

From the condition of stress continuity with respect to the physical coordinates near point L, it follows that the dimensionless stress δ must be determined from the condition of boundedness $f'(\delta)$.

The boundary conditions for function $\varphi(\xi, \eta) = \mathrm{Re}[f(\zeta)]$ in the stress plane acquire the following form (the Dirichlet problem): $\varphi = 0$ on BC, FE, and CDE,

$$\varphi(\xi, 0) = -\xi, \quad 0 < \xi < a, \quad \varphi = -\beta a + (\beta - 1)\xi, \quad a \le \xi \le \delta, \tag{2.3.3}$$

where $\beta = b/l$.

For the solution of this problem with the function $\Omega(\zeta)$, we map the circular sector of the unit radius with a notch in plane ζ onto the semicircle of a unit radius in plane $\Omega = \omega + i\lambda$ (Fig. 2.5). Evidently,

$$\Omega(\zeta) = i\left(\frac{\zeta^n - \delta^n}{1 - \zeta^n\delta^n}\right)^{1/2}. \tag{2.3.4}$$

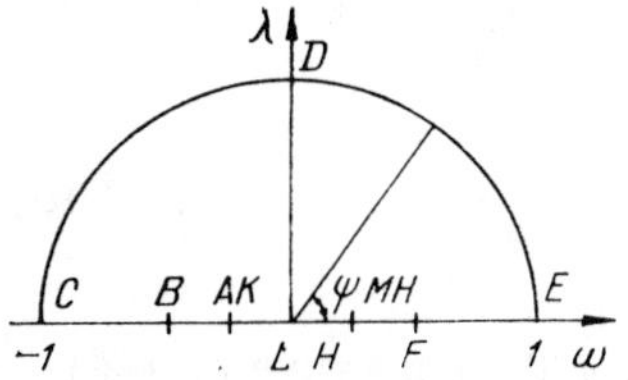

FIG. 2.5.

Here

$$n = 2\pi/(\pi - 2\alpha). \tag{2.3.5}$$

Now we introduce function $\varphi_1(\omega, \lambda) = \varphi(\xi, \eta)$ such that $\varphi_1(\omega, \lambda) = \mathrm{Re}\{f(\zeta)\}$, when ζ is expressed in terms of Ω. Following the conversion of the boundary conditions on plane Ω for function $\varphi_1(\omega, \lambda)$, we obtain the boundary problem

$$\begin{aligned} &\varphi_1(\omega, \lambda) = 0 \quad \text{on } CDE, CB, FE, \\ &\varphi_1(\omega, \lambda) = \varphi_1(\omega, 0) = \varphi(\xi, 0) \quad \text{on } BLF. \end{aligned} \tag{2.3.6}$$

This problem can be solved easily by superposition of two functions $g(\Omega)$ and $g(1/\Omega)$, where

$$g(\Omega) = \frac{1}{\pi i}\int_{BLF} \frac{\varphi_1(\omega, 0)\, d\omega}{\omega - \Omega}. \tag{2.3.7}$$

Thus, $\varphi_1(\omega, \lambda) = \mathrm{Re}[g(\Omega) + g(1/\Omega)]$ is a solution of problem (2.3.3). Therefore,

$$f(\zeta) = g[\Omega(\zeta)] + g[1/\Omega(\zeta)].$$

The last equality can be written in expanded form as

$$\begin{aligned} f(\zeta) &= \frac{1}{\pi i}\int_{BLF} \varphi_1(\omega, 0)\left[\frac{1}{\omega - \Omega(\zeta)} + \frac{1}{\omega - 1/\Omega(\zeta)}\right] d\omega \\ &= \frac{1}{\pi i}\int_{LF} \varphi_1(\omega, 0)\left[\frac{1}{\omega - \Omega(\zeta)} + \frac{1}{\omega - 1/\Omega(\zeta)}\right. \\ &\qquad \left. - \frac{1}{\omega + \Omega(\zeta)} + \frac{1}{\omega + 1/\Omega(\zeta)}\right] d\omega. \end{aligned} \tag{2.3.8}$$

Following the execution of intermediate operations, we find the expression for function $f(\zeta)$,

$$\begin{aligned} f(\zeta) &= -\frac{2}{\pi i}\Omega(\zeta)[1 + \Omega^2(\zeta)] \cdot J, \\ J &= \int_0^{\delta} \frac{\varphi(\xi, 0)\,\omega'(\xi)[\omega^2(\xi) - 1]\, d\xi}{\Omega^2(\zeta)\omega^4(\xi) - [1 + \Omega^4(\zeta)]\omega^2(\xi) + \Omega^2(\zeta)}. \end{aligned} \tag{2.3.9}$$

Here

$$\omega(\xi) = \sqrt{\frac{\delta^n - \xi^n}{1 - \xi^n \delta^n}}\,. \tag{2.3.10}$$

By denoting $\varphi'(\xi,0) = \partial\varphi(\xi,0)/\partial\xi$, the result for the derivative of $f(\zeta) = (x - iy)/l$ is obtained:

$$f'(\zeta) = \frac{2}{\pi}\Omega'(\zeta)\left[1 - \Omega^2(\zeta)\right] \cdot I,$$
$$I = \int_0^\delta \frac{\varphi'(\xi,0)\,\omega(\xi)\left[1 + \omega^2(\xi)\right] d\xi}{\Omega^2(\zeta)\,\omega^4(\xi) - \left[1 + \Omega^4(\zeta)\right]\omega^2(\xi) + \Omega^2(\zeta)}\,. \tag{2.3.11}$$

Relation (2.3.11) can be obtained first by integrating (2.3.9) by parts and then by differentiating.

To find the boundedness condition $f'(\delta)$, it must be noted that $\Omega'(\delta)$ is infinite; therefore, to satisfy the required condition, the coefficient of $\Omega'(\zeta)$ in (2.3.11) must be equal to zero, when $\Omega(\zeta) = \Omega(\delta) \equiv 0$. Thus, the equation to determine δ is obtained:

$$\int_0^\delta \varphi'(\xi,0)\frac{1 + \omega^2(\xi)}{\omega(\xi)}\,d\xi = 0. \tag{2.3.12}$$

Taking into account equations (2.3.3) and (2.3.10) and assuming $t = \xi/\delta$, we finally obtain

$$\beta = \frac{\displaystyle\int_0^1 \frac{1 - \delta^n t^n}{\sqrt{(1 - \delta^{2n} t^n)(1 - t^n)}}\,dt}{\displaystyle\int_{a/\delta}^1 \frac{1 - \delta^n t^n}{\sqrt{(1 - \delta^{2n} t^n)(1 - t^n)}}\,dt}\,. \tag{2.3.13}$$

The solution for the half plane with an angular notch is obtained from the previous equations in which $a = \delta$. The boundedness condition (2.3.13) in this case is not required.

For the elastic–plastic boundary, the following equality holds:

$$x - iy = R(\theta)e^{-i\theta}.$$

Then, taking into account relation (2.3.1), the boundary separating elastic and plastic zones can be determined as

$$R(\theta)/l = e^{i\theta} f'(e^{i\theta}). \tag{2.3.14}$$

Relation (2.3.14) is convenient to express in terms of function $\psi(\theta)$, which is determined by equality $\Omega(e^{i\theta}) = e^{i\psi(\theta)}$:

$$\frac{R(\theta)}{l} = -\frac{4}{\pi}\psi'(\theta)\sin[\psi(\theta)]\int_0^{\delta}\frac{\varphi'(\xi,0)\,\omega(\xi)[\omega^2(\xi)+1]\,d\xi}{\omega^4(\xi)-2\omega^2(\xi)\cos[2\psi(\theta)]+1}. \quad (2.3.15)$$

Using the previous expressions (2.3.3) and (2.3.10) and substituting t for ξ/δ, we obtain the following equations for the elastic–plastic boundary

$$\frac{R}{l} = \frac{2n}{\pi}\frac{\delta^{1+n/2}\cos n\theta/2}{\sqrt{1-2\delta^n\cos n\theta+\delta^{2n}}}[I_1-\beta I_2],$$

$$I_1 = \int_0^1 \frac{(1-\delta^n t^n)\sqrt{(1-t^n)(1-\delta^{2n}t^n)}}{1-2\delta^n t^n\cos n\theta+\delta^{2n}t^{2n}}\,dt, \quad (2.3.16)$$

$$I_2 = \int_{a/\delta}^1 \frac{(1-\delta^n t^n)\sqrt{(1-t^n)(1-\delta^{2n}t^n)}}{1-2\delta^n t^n\cos n\theta+\delta^{2n}t^{2n}}\,dt,$$

for $-(\pi/2-\alpha)<\theta<(\pi/2-\alpha)$.

The equation, obtained from (2.3.16) when $\theta=0$, for the distance of the intersection point of the elastic–plastic boundary with axis Ox from the origin is

$$\frac{x_*}{l} = \frac{2n}{\pi}\cdot\frac{\delta^{1+n/2}}{1-\delta^n}\left[\int_0^1\frac{\sqrt{(1-t^n)(1-\delta^{2n}t^n)}}{1-\delta^n t^n}\,dt\right.$$

$$\left.-\beta\int_{a/\delta}^1\frac{\sqrt{(1-t^n)(1-\delta^{2n}t^n)}}{1-\delta^n t^n}\,dt.\right] \quad (2.3.17)$$

Distribution of strains in the plastic region and the notch opening $2w_0$ at the end of the region have the form

$$\gamma_{\theta z} = \frac{1}{r}\frac{\partial w}{\partial\theta} = \frac{k}{G}\frac{R(\theta)}{r}, \qquad \gamma_{rz} = \frac{\partial w}{\partial r} = 0, \quad (2.3.18)$$

$$\frac{Gw_0}{kl} = \frac{2\beta a}{\pi}\ln\left[\frac{\sqrt{1-a^n\delta^n}+\sqrt{\delta^n-a^n}}{\sqrt{1-a^n\delta^n}-\sqrt{\delta^n-a^n}}\right] + \frac{2n}{\pi}\delta^{1+n/2}(1+\delta^n)$$

$$\times\left[\int_0^1\frac{t^n\,dt}{\sqrt{(1-t^n)(1-\delta^{2n}t^n)}\,(1+\delta^n t^n)}\right.$$

$$\left.-\beta\int_{a/\delta}^1\frac{t^n\,dt}{\sqrt{(1-t^n)(1-\delta^{2n}t^n)}\,(1+\delta^n t^n)}\right]. \quad (2.3.19)$$

Let us consider two limiting cases. When the loading parameter $a=\tau_\infty/k$ is low, the dimensions of the plastic region are small compared with other dimensions. The solution for this case is obtained from the previous equations, in which the terms of orders a and δ are so small compared with unity that they can be neglected. The formulas for determination of δ and x_* have the form

$$\beta\int_{a/\delta}^{1}\frac{dt}{\sqrt{1-t^n}}=\frac{\sqrt{\pi}}{n}\,\frac{\Gamma(1/n)}{\Gamma[(2+n)/2n]}, \tag{2.3.20}$$

where $\Gamma(x)=\int_0^\infty r^{x-1}e^{-r}\,dr$ is the gamma function, and

$$\frac{x_*}{l}=\frac{4n\beta}{\pi(2+n)}\sqrt{\left(\frac{\delta}{a}\right)^n-1}\cdot a^{1+n/2}. \tag{2.3.21}$$

In the case of half plane $\delta=a$, equation (2.3.21) acquires the form

$$\frac{x_*}{l}=\frac{2n}{\sqrt{\pi}(2+n)}\cdot\frac{\Gamma(1/n)}{\Gamma[(2+n)/2n]}\cdot a^{1+n/2}. \tag{2.3.22}$$

Another limiting case arises when $\delta=\tau_L/k=1$. The plastic region in this case completely covers the entire strip width. To maintain equilibrium, the following condition must be satisfied:

$$\tau_\infty\cdot b=k(b-l).$$

The solution in this case is obtained in a similar manner:

$$f(\zeta)=\frac{n}{\pi i}\int_0^1 t^{n-1}\varphi(t,0)\left[\frac{1}{t^n-\zeta^n}+\frac{1}{t^n-\zeta^{-n}}\right]dt, \tag{2.3.23}$$

where $\varphi(t,0)=-t$ for $0<t<1-1/\beta$ and $\varphi(t,0)=-(\beta-1)(1-t)$ for $1-1/\beta<t<1$;

$$\frac{R(\theta)}{l}=\frac{2n}{\pi}\sin n\theta\left[(\beta-1)\int_{1-1/\beta}^{1}\frac{t^n\,dt}{1-2t^n\cos n\theta+t^{2n}}\right.$$

$$\left.-\int_0^{1-1/\beta}\frac{t^n\,dt}{1-2t^n\cos n\theta+t^{2n}}\right]^{1-1/\beta}; \tag{2.3.24}$$

$$\frac{Gw_0}{kl}=\frac{4n}{\pi}\left\{\int_0^{1-1/\beta}\frac{t^n\,dt}{(1-t^n)(1+t^n)}+(\beta-1)\int_{1-1/\beta}^{1}\frac{t^{n-1}(1-t)\,dt}{(1-t^n)(1+t^n)}\right\}. \tag{2.3.25}$$

§ 4. CRACK IN THE HALF PLANE

Consider an elastic–plastic problem for the half plane with a crack of length l at the half plane boundary (Fig. 2.6). The surface of the crack and the boundary of the half plane are free from loads, and the shear stress $\tau_{yz} = \tau_\infty$ acts at infinity. As a parametric plane, it is convenient to select the stress plane

$$\zeta = (G/k) f'(z). \tag{2.4.1}$$

The elastic region is mapped onto a unit semicircle with a slit of length $a = \tau_\infty / k$ on plane ζ (Fig. 2.7). Due to the symmetry of the picture in plane z relative to the x axis, only one half of the region will be discussed.

For the determination of the analytical function

$$z = \omega(\zeta), \tag{2.4.2}$$

we have the boundary conditions

$$\begin{aligned} &\operatorname{Re}\omega(\zeta) = -l, \qquad \zeta \in AB, \\ &\operatorname{Im}\omega(\zeta) = 0, \qquad \zeta \in BC, EA, \\ &\arg \omega(\zeta) = \pi/2 - \varphi, \qquad \varphi = \pi/2 - \theta, \ \zeta \in CE, \\ &\omega(0) = 0, \qquad \omega(i\tau_\infty / k) = \infty. \end{aligned} \tag{2.4.3}$$

Here, $\theta(x, y) = \text{const}$ defines the slip lines extending from the tip of the crack.

The boundary condition on CE is convenient to write in another form. On the boundary of elastic and plastic regions, the condition of stress continuity is fulfilled:

$$f'(z) = -(ki/G) e^{-i\theta}. \tag{2.4.4}$$

Equality (2.4.4) is obtained as follows. If we picture the tip of the crack (point O) as the limit of a convex oval that is contracting to this point, it becomes clear that the slip lines must be radial straight lines ("fan"), i.e., in the plastic region stresses are equal to $\tau_{xz} + i\tau_{yz} = kie^{i\theta}$. Taking into account relations (2.4.1) and (2.4.4), the condition for CE can be written as

$$\operatorname{Re}[\zeta \omega(\zeta)] = 0. \tag{2.4.5}$$

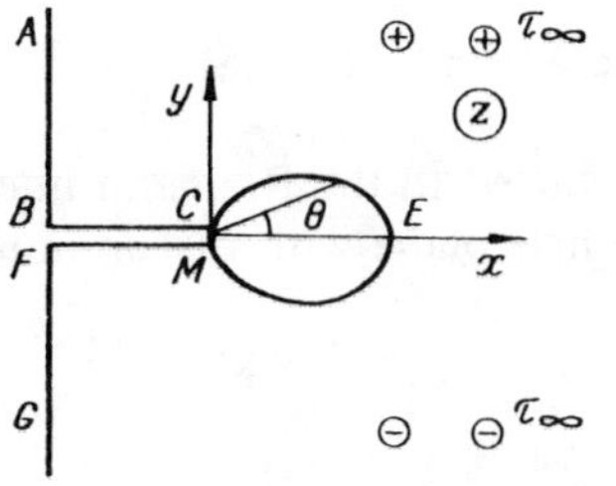

FIG. 2.6

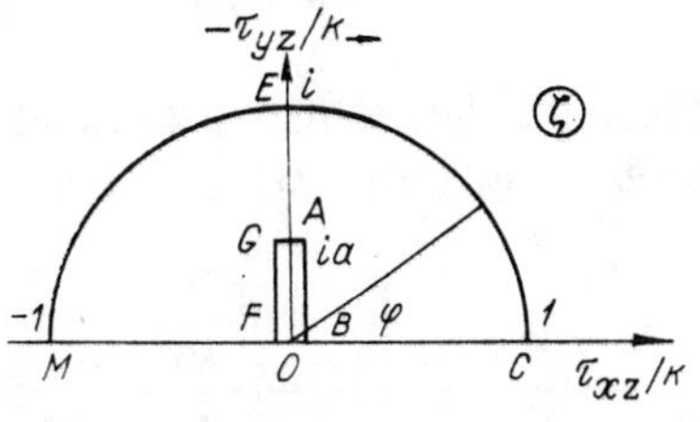

FIG. 2.7.

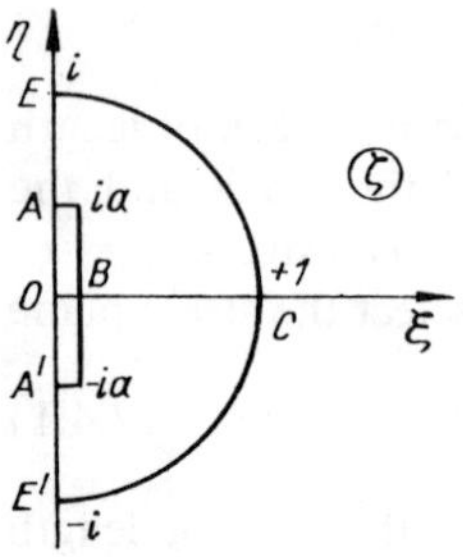

FIG. 2.8.

Let us analytically extend function $\omega(\zeta)$ to the lower semicircle through the diameter of the circle in plane ζ using the relation

$$\omega(\zeta)=\overline{\omega(\zeta)}. \tag{2.4.6}$$

Due to the symmetry of the picture relative to the vertical, only one half is given in Fig. 2.8. The boundary condition (2.4.5) on section $E'CE$ can be written as

$$\omega(\zeta)+(1/\zeta^2)\overline{\omega(\zeta)}=0, \qquad \bar{\zeta}=1/\zeta. \tag{2.4.7}$$

Let us introduce the auxiliary function

$$\varphi(\zeta)=\begin{cases}\varphi^+(\zeta)=\omega(\zeta), & |\zeta|<1,\\ \varphi^-(\zeta)=-(1/\zeta^2)\cdot\overline{\omega(\bar{\xi})}, & |\zeta|>1.\end{cases} \tag{2.4.8}$$

For function $\varphi(\zeta)$, on the circular arc $E'CE$, condition $\varphi^+(\zeta)-\varphi^-(\zeta)=0$ is fulfilled, giving the opportunity to analytically extend function $\omega(\zeta)$ through the circular arc on the half plane (Fig. 2.9).

The boundary problem for the analytical function $\varphi(\zeta)$ on half plane $\operatorname{Re}\zeta>0$ acquires the form

$$\begin{aligned}&\varphi^+(i\eta)+\varphi^-(i\eta)=-2l/\eta^2, \qquad \eta\in A_1B_1,\ B_1'A_1',\\ &\varphi^+(i\eta)-\varphi^-(i\eta)=0, \qquad \eta\in A_1EA,\ A'E'A_1',\\ &\varphi^+(i\eta)+\varphi^-(i\eta)=-2l, \qquad \eta\in ABA',\end{aligned} \tag{2.4.9}$$

$$\varphi(0)=0, \qquad \varphi(\pm ik/\tau_\infty)=\infty, \qquad \varphi(\pm i\tau_\infty/k)=\infty.$$

Thus, the determination of function $\varphi(\zeta)$ was reduced to the Riemann linear problem $\varphi^+(i\eta)=G(i\eta)\varphi^-(i\eta)+g(i\eta)$ for the half plane with discontinuity

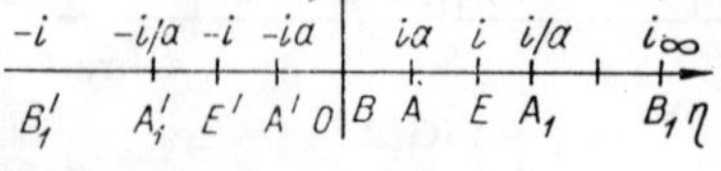

FIG. 2.9.

coefficients $G(i\eta)$ and $g(i\eta)$. Let

$$X(\zeta) = \frac{i}{\sqrt{(\zeta^2 + a^2)(\zeta^2 + a^{-2})}}. \tag{2.4.10}$$

Here, $X(\zeta) = i/\zeta^2 + O(1/\zeta^2)$ when $\zeta \to \infty$. The general solution of the discontinuous problem (2.4.9) for the function $\varphi(\zeta)$ on half plane $\operatorname{Re}\zeta > 0$ is sought as

$$\varphi(\zeta) = \frac{X(\zeta)}{2\pi i}\int_{-i\infty}^{i\infty} \frac{g(t)\,dt}{X^+(t)(t-\zeta)}, \qquad t = i\eta. \tag{2.4.11}$$

By substituting the values of coefficients $g(i\eta)$ on separate sections of an imaginary axis from problem (2.4.9) and the values of $X(i\eta)$ from expression (2.4.10), a solution for all ζ, except $\zeta = 0$, is found as follows:

$$\varphi(\zeta) = \frac{2X(\zeta)}{\pi i}\left[l\zeta(1-\zeta)^2\int_0^a \frac{(t^2-1)\sqrt{(a^2-t^2)(a^{-2}-t^{-2})}}{(t^2-\zeta^2)(1+t^2\zeta^2)}\,dt\right]. \tag{2.4.12}$$

The integral in equation (2.4.12) can be represented in terms of the elliptic integrals of the first, second, and third kind. Omitting the relatively simple but cumbersome intermediate operations, the solution for $\varphi(\zeta)$ in half plane $\operatorname{Re}\zeta > 0$ can be presented as

$$\varphi(\zeta) = \frac{2}{\pi\sqrt{(\zeta^2+a^2)(\zeta^2+a^{-2})}}$$

$$\times\left\{l\zeta(1-\zeta^2)\left[-\frac{1}{a^2\zeta^2}\mathbf{E}\left(\frac{\pi}{2};a^2\right) + \frac{a}{\zeta^2}\left(\frac{1}{a^2} - a^2 - \frac{1}{\zeta^2} - \zeta^2 - 1\right)\right.\right.$$

$$\times\mathbf{F}\left(\frac{\pi}{2};a^2\right) + \frac{a}{1-\zeta^2}\left(\zeta^2 + a^2 + \frac{1}{a^2} + \frac{1}{\zeta^2}\right)$$

$$\left.\left.\times\left\{\Pi\left(\frac{\pi}{2};-\frac{a^2}{\zeta^2};a^2\right) - \frac{1}{\zeta^2}\Pi\left(\frac{\pi}{2};-a^2\zeta^2;a^2\right)\right\}\right]\right\}. \tag{2.4.13}$$

Expression (2.4.13) is cumbersome and ineffective for numerical computations. A more effective form is the asymptotic expansion of functions $\varphi(\zeta)$ in terms of the dimensionless parameter $a = \tau_\infty/k$, which is considered small. Hence, complete elliptic integrals can be computed by expanding the integrand function into series of a and by integrating this series term by term.

Let us present a truncated asymptotic series for function $z = \omega(\zeta)$, which maps the region in the physical plane onto one half of a circle in the stress plane ζ:

$$z = \frac{l(1-\zeta^2)\zeta}{\sqrt{(\zeta^2+a^2)(\zeta^2+a^{-2})}}$$

$$\times\left\{\frac{1}{\zeta(1-\zeta^2)}\left[\sqrt{1+\frac{\zeta^2}{a^2}}\left(1+\frac{1}{2}a^2\zeta^2-\frac{1}{8}a^4\zeta^4+\frac{5}{128}a^6\zeta^6\right)\right.\right.$$

$$\left.-\sqrt{1+\frac{1}{a^2\zeta^2}}\left(1+\frac{a^2}{2\zeta^2}-\frac{a^4}{8\zeta^4}-\frac{5}{128}\frac{a^6}{\zeta^6}\right)\right]$$

$$+\frac{1}{a\zeta^2}\left[1+\frac{a^2}{2}\left(1+\zeta^2+\frac{1}{\zeta^2}\right)+\frac{a^4}{8}\left(1-\zeta^4-\frac{1}{\zeta^4}-\zeta^2-\frac{1}{\zeta^2}\right)\right.$$

$$\left.\left.-\frac{a^6}{16}\left(1+\zeta^2+\frac{1}{\zeta^2}\right)\right]+O(a^7)\right\}. \quad (2.4.14)$$

Here, $\sqrt{1+z^2}=z+O(z^{-1})$ when $z\to\infty$ and expansion (2.4.14) is valid for all ζ, except $\zeta=0$, at small $a=\tau_\infty/k$ (practically up to $a\sim 0.9$ when ζ is neither too small nor too large).

In the limiting case $\tau_\infty=k$, equation (2.4.13) gives

$$z=\frac{2l}{\pi}\left[\frac{\zeta^2-1}{\zeta(\zeta+1)}+\frac{1}{\zeta^2}\arctan\zeta-\arctan\frac{1}{\zeta}\right], \quad (2.4.15)$$

where $\arctan 0 = 0$.

Let us determine the boundary of the plastic region. Since in this case $\zeta=e^{i\varphi}$, it is sufficient to use the terms of equation (2.4.14) up to fifth order. Based on (2.4.14) and (2.4.15), we obtain the equation of the plastic region contour as follows:

for $\tau_\infty\le 0.8k$,

$$\frac{x}{l}=\frac{a(1-\cos 2\varphi)}{2\sqrt{\cos 2\varphi+(1+a^4)/a^2}}\left[1-\frac{a^2}{4}(1+2\cos\varphi)+O(a^4)\right],$$

$$\frac{y}{l}=\frac{a\sin 2\varphi}{2\sqrt{\cos 2\varphi+(1+a^4)/a^2}}\left[1-\frac{a^2}{4}(1+2\cos\varphi)+O(a^4)\right]; \quad (2.4.16)$$

for $\tau_\infty=k$,

$$\frac{x}{l}=\frac{2\sin\varphi}{\pi}\left(\tan\varphi-\frac{\pi}{2}\sin\varphi-\cos\varphi\ln\frac{\cos\varphi}{1+\sin\varphi}\right),$$

$$\frac{y}{l}=\frac{2}{\pi}\left(\sin\varphi-\frac{\pi}{4}\sin 2\varphi-\frac{1}{2}\cos 2\varphi\ln\frac{\cos\varphi}{1+\sin\varphi}\right)+\frac{1}{2}\ln\frac{\cos\varphi}{1-\sin\varphi}. \quad (2.4.17)$$

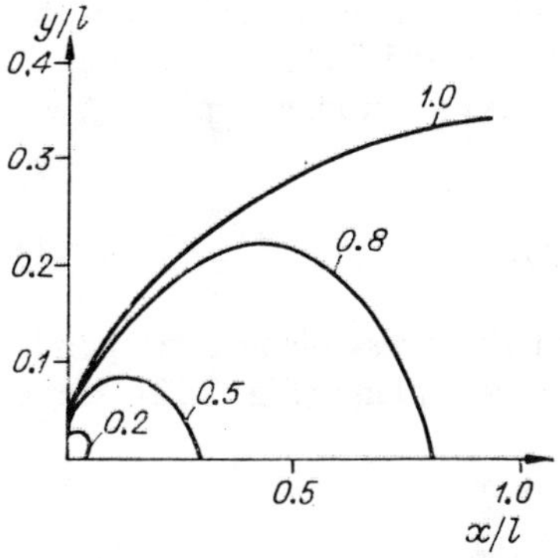

FIG. 2.10.

There is a simple formula for the distance x_* from the origin to the intersection point of the plastic zone contour with the x axis. It is obtained from (2.4.16) for $\varphi = \pi/2$:

$$x_*/l = a^2\left[1 + \tfrac{3}{4}a^2 + O(a^4)\right]. \tag{2.4.18}$$

Figure 2.10 depicts the boundary of plastic and elastic regions, calculated from equations (2.4.16) and (2.4.17), for the values of the loading parameter $a = \tau_\infty/k$ equal to 0.2, 0.5, 0.8, and 1.0.

Let us consider the limiting case. When $a = \tau_\infty/k$ is small, the dimensions of the plastic zone are small compared with the length of the crack. The solution in this case is obtained from the previous equations by equating to zero the terms higher than a^2. In this case the elastic–plastic boundary is a circle with diameter $d = la^2$.

Figure 2.10 shows the tendency for transition from circular plastic zones when plastic deformations are present in a small region to elongated zones for a large plastic region.

§ 5. PERIODIC SYSTEM OF CRACKS

Let us consider the elastic–plastic problem for the half plane with a periodic system of cracks of length l at the boundary of the half plane (Fig. 2.11). The surface of the cracks and boundary of the half plane are free from stresses, and the shear stress $\tau_{yz} = \tau_\infty$ acts at infinity. Due to the periodicity of the problem in the physical plane, it is sufficient to consider the region of one

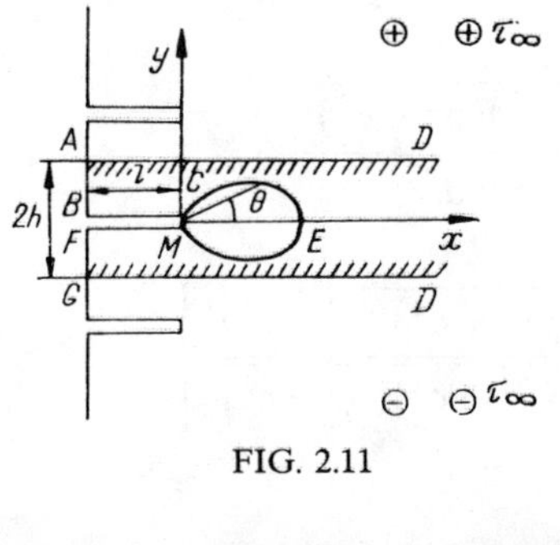

FIG. 2.11

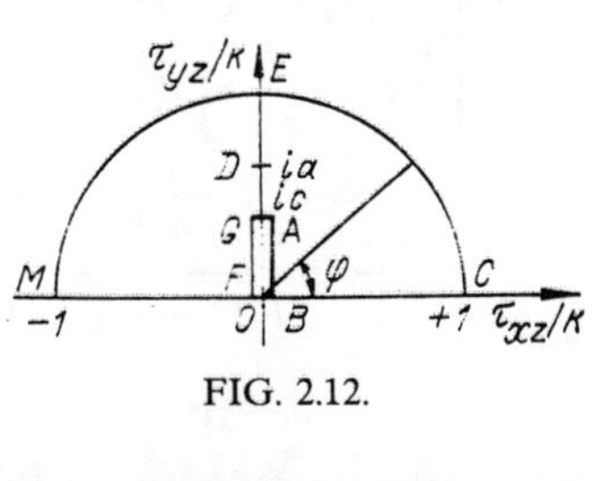

FIG. 2.12.

crack DABCEMFGD. Stress τ_∞ acts at a sufficiently large distance from the crack. This stress remains parallel along sections DA and DG and decreases to the value τ_1 on the boundary of the half plane. As a parametric plane ζ, we can use the locus plane

$$\zeta(z) = (G/k) f'(z). \tag{2.5.1}$$

The known contour of the elastic region in the stress plane corresponds one-to-one to the unknown contour in the physical plane (Fig. 2.12). Here, $a = \tau_\infty/k$, $c = \tau_1/k$ and $k \geq \tau_\infty \geq \tau$.

For the determination of the analytical function

$$z = \omega(\zeta), \tag{2.5.2}$$

we have the boundary problem:

$$\begin{aligned}
&\operatorname{Im}\omega(\zeta) = h, \qquad \zeta \in DA,\\
&\operatorname{Re}\omega(\zeta) = -l, \qquad \zeta \in AB,\\
&\operatorname{Im}\omega(\zeta) = 0, \qquad \zeta \in BC,\ ED,\\
&\arg\omega(\zeta) = \pi/2 - \varphi, \qquad \varphi = \pi/2 - \theta, \quad \zeta \in CE,\\
&\omega(0) = 0, \qquad \omega(i\tau_\infty/k) = \infty.
\end{aligned} \tag{2.5.3}$$

The boundary problem (2.5.3) is solved, as in § 4, with the help of a double analytical extension through the diameter of the semicircle and the arc of the circle. For all ζ, with the exception of $\zeta = 0$, the solution has the form

$$\begin{aligned}
z = {}& \frac{2}{\pi} l\zeta(1-\zeta^2)\frac{\sqrt{(\zeta^2+c^2)(\zeta^2+c^{-2})}}{(\zeta^2+a^2)(\zeta^2+a^{-2})}\\
&\times\left\{-\frac{1}{c\zeta^2}\mathbf{E}\left(\frac{\pi}{2};c^2\right)+\frac{c}{\zeta^2}\left(\frac{1}{c^2}-a^2-\frac{1}{a^2}-1-\zeta^2-\frac{1}{\zeta^2}\right)\mathbf{F}\left(\frac{\pi}{2};c^2\right)\right.\\
&\left.+\frac{c(\zeta^2+a^2)(\zeta^2+a^{-2})}{\zeta^2(1-\zeta^2)}\left[\Pi\left(\frac{\pi}{2};-\frac{c^2}{\zeta^2};c^2\right)-\frac{1}{\zeta^2}\Pi\left(\frac{\pi}{2};-c^2\zeta^2;c^2\right)\right]\right\}\\
&+\frac{h}{\pi}\left[\ln\frac{i\sqrt{(\zeta^2+c^2)(1-a^2c^2)}-\sqrt{(1+\zeta^2c^2)(a^2-c^2)}}{i\sqrt{(\zeta^2+c^2)(1-a^2c^2)}+\sqrt{(1+\zeta^2c^2)(a^2-c^2)}}\right.\\
&\left.-\frac{1}{\zeta^2}\ln\frac{i\sqrt{(1+\zeta^2c^2)(1-a^2c^2)}-\sqrt{(\zeta^2+c^2)(a^2-c^2)}}{i\sqrt{(1+\zeta^2c^2)(1-a^2c^2)}+\sqrt{(\zeta^2+c^2)(a^2-c^2)}}\right].
\end{aligned} \tag{2.5.4}$$

Let us determine the boundary of the plastic region. To do so, we substitute $\zeta = e^{i\varphi}$ in equation (2.5.4). By separating the real and imaginary parts for $z(e^{i\varphi})$, we obtain the equation of the plastic zone contour as follows:

$$\frac{x}{l} = \frac{4}{\pi}\frac{\sqrt{2\cos 2\varphi + c^2 + c^{-2}}}{(2\cos 2\varphi + a^2 + a^{-2})}$$

$$\times \sin^2\varphi\left\{\frac{1}{c}\mathbf{E}\left(\frac{\pi}{2}; c^2\right) + c\left(1 + 2\cos 2\varphi - \frac{1}{c^2} + a^2 + \frac{1}{a^2}\right)\right.$$

$$\times \mathbf{F}\left(\frac{\pi}{2}; c^2\right) + \left(a^2 + \frac{1}{a^2} + 2\cos 2\varphi\right)$$

$$\left.\times\left[\operatorname{Re}\Pi\left(\frac{\pi}{2}; -c^2e^{2i\varphi}; c^2\right) - \cot\varphi \operatorname{Im}\Pi\left(\frac{\pi}{2}; -c^2e^{2i\varphi}; c^2\right)\right]\right\}$$

$$+ \frac{h}{e}\frac{\sin 2\varphi}{\pi}\left[\arctan\frac{\mathbf{A}(\varphi, c, a) + \mathbf{B}(\varphi, c, a)}{\mathbf{C}(\varphi, c, a) + \mathbf{D}(\varphi, c, a)}\right.$$

$$\left. - \arctan\frac{\mathbf{A}(\varphi, c, a) - \mathbf{B}(\varphi, c, a)}{\mathbf{C}(\varphi, c, a) - \mathbf{D}(\varphi, c, a)}\right], \tag{2.5.5}$$

$$\frac{y}{l} = \frac{2}{\pi}\frac{\sqrt{2\cos 2\varphi + c^2 + c^{-2}}}{(2\cos 2\varphi + a^2 + a^{-2})}$$

$$\times \sin^2\varphi\left\{\frac{1}{c}\mathbf{E}\left(\frac{\pi}{2}; c^2\right) + c\left(1 + 2\cos 2\varphi + a^2 + \frac{1}{a^2} - \frac{1}{c^2}\right)\right.$$

$$\times \mathbf{F}\left(\frac{\pi}{2}; c^2\right) + \left(a^2 + \frac{1}{a^2} + 2\cos 2\varphi\right)$$

$$\left.\times\left[\operatorname{Re}\Pi\left(\frac{\pi}{2}; -c^2e^{2i\varphi}; c^2\right) - \cot\varphi \operatorname{Im}\Pi\left(\frac{\pi}{2}; -c^2e^{2i\varphi}; c^2\right)\right]\right\}$$

$$- \frac{2h}{l}\frac{\sin^2\varphi}{\pi}\left[\arctan\frac{\mathbf{A}(\varphi, c, a) + \mathbf{B}(\varphi, c, a)}{\mathbf{C}(\varphi, c, a) + \mathbf{D}(\varphi, c, a)}\right.$$

$$\left. - \arctan\frac{\mathbf{A}(\varphi, c, a) - \mathbf{B}(\varphi, c, a)}{\mathbf{C}(\varphi, c, a) - \mathbf{D}(\varphi, c, a)}\right]. \tag{2.5.6}$$

Here,

$$\operatorname{Re}\Pi\left(\frac{\pi}{2};\,-c^2e^{2i\varphi};\,c^2\right)=\int_0^{\pi/2}\frac{(1+c^2\cos 2\varphi\,\sin^2\alpha)\,d\alpha}{(1+2c^2\cos 2\varphi\,\sin^2\alpha+c^4\sin^2\alpha)\sqrt{1-c^4\sin^2\alpha}},$$

$$\operatorname{Im}\Pi\left(\frac{\pi}{2};\,-c^2e^{2i\varphi};\,c^2\right)=-\int_0^{\pi/2}\frac{c^2\sin 2\varphi\,\sin^2\alpha\,d\alpha}{(1+2c^2\cos 2\varphi\,\sin^2\alpha+c^4\sin^4\alpha)\sqrt{1-c^4\sin^2\alpha}},$$

$$\mathbf{A}(\varphi,c,a)=\sqrt{1-a^2c^2}\sqrt{\sqrt{2\cos 2\varphi+c^2+c^{-2}}-c^{-1}-c\cos 2\varphi}\,,$$

$$\mathbf{B}(\varphi,c,a)=\sqrt{a^2-c^2}\sqrt{\sqrt{2\cos 2\varphi+c^2+c^{-2}}-c-c^{-1}\cos 2\varphi}\,,$$

$$\mathbf{C}(\varphi,c,a)=\sqrt{1-a^2c^2}\sqrt{\sqrt{2\cos 2\varphi+c^2+c^{-2}}+c^{-1}+c\cos 2\varphi}\,,$$

$$\mathbf{D}(\varphi,c,a)=\sqrt{a^2-c^2}\sqrt{\sqrt{2\cos 2\varphi+c^2+c^{-2}}+c+c^{-1}\cos 2\varphi}\,.$$

We present now the formula for the distance x_* from the origin to the intersection point of the contour of the plastic region with the x axis. This formula was obtained from (2.5.5) when $\varphi=\pi/2$:

$$\frac{x_*}{l}=\frac{4}{\pi}\frac{(c^{-1}-c)}{(c^{-1}-a)^2}\left\{\frac{1}{c}\mathbf{E}\left(\frac{\pi}{2};c^2\right)-c\left(\frac{1}{c^2}-a^2-\frac{1}{a^2}-1\right)\mathbf{F}\left(\frac{\pi}{2};c^2\right)\right.$$
$$\left.+\left(\frac{1}{a}-a\right)^2\left[\Pi\left(\frac{\pi}{2};c^2;c^2\right)\right]\right\}+\frac{2h}{\pi}\ln\frac{\sqrt{1-a^2c^2}-\sqrt{a^2-c^2}}{\sqrt{1-a^2c^2}+\sqrt{a^2-c^2}}\,. \tag{2.5.7}$$

If $a=c$, we obtain the solution for one crack.

§ 6. THE HALF PLANE WEAKENED BY A NOTCH

Let us consider an elastic–plastic problem for the half plane weakened by a notch formed by two nonparallel straight-line cuts connected by a circular arc 2α. The material of the half plane is considered ideal elastic–plastic with the yield limit in shear equal to k. The boundary of the half plane and the contour of the notch are free from loads, and stresses $\tau_{xz}=0$, $\tau_{yz}=\tau_\infty$ are applied at infinity (Fig. 2.13).

As shown in Fig. 2.13, we assume that the plastic zone, starting from the straight cuts, includes the entire circular arc.

Using the relationship

$$x-iy=f'(\zeta),\qquad \zeta=\xi+i\eta=-i(\tau_{xz}+i\tau_{yz})/k=\tau e^{i\theta}, \tag{2.6.1}$$

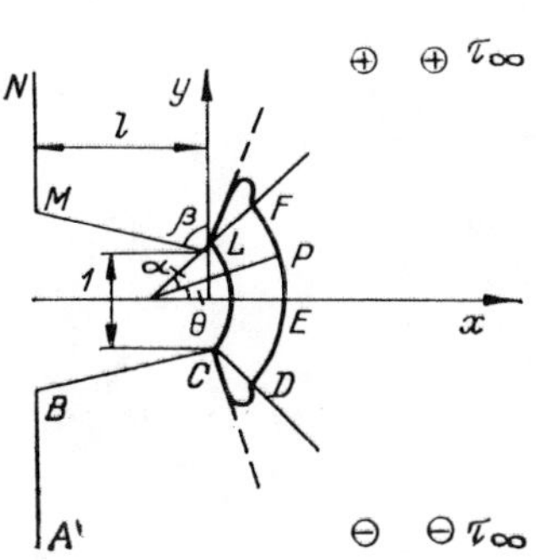

FIG. 2.13

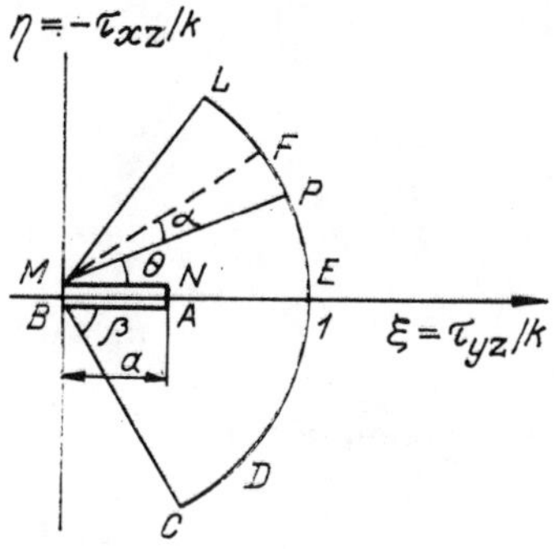

FIG. 2.14.

we go over from the physical plane to the stress plane (Fig. 2.14). Here, $f'(\zeta)$ is an analytical function for all values of ζ that correspond to the stresses in the elastic region, i.e., $|\zeta| < 1$; θ is the angle between the vector of the dimensionless shear stress and the x axis.

On plane ζ, the elastic region is mapped onto the inside of a unit circular sector with a slit of length $a = \tau_\infty/k$. The arc of the unit circle in this case corresponds to the contour separating the elastic and plastic zones (see Fig. 2.14). Thus, by introducing the analytical function $F(\zeta) = \zeta f'(\zeta) + u + iv$, an equation of the elastic–plastic boundary can be derived in the form

$$x - iy = e^{-i\theta} F(e^{i\theta}). \tag{2.6.2}$$

For the analytical function $F(\zeta)$, we obtain the boundary conditions:

$$\begin{aligned}
&\operatorname{Im} F(\zeta) = -\tfrac{1}{2}\cos\theta, \qquad \beta \geq \theta \geq \alpha \text{ on } FL,\\
&\operatorname{Im} F(\zeta) = -\tfrac{1}{2}\cot\alpha \sin\theta, \qquad \alpha \geq \theta \geq -\alpha \text{ on } DEF,\\
&\operatorname{Im} F(\zeta) = \tfrac{1}{2}\cos\theta \qquad -\alpha \geq \theta \geq -\beta \text{ on } CD,\\
&\operatorname{Re} F(\zeta) = \tfrac{1}{2}\tau \sin\beta \quad \text{on } ML, CB,\\
&\operatorname{Re} F(\zeta) = -l\tau \quad \text{on } MN, AB,\\
&F(\tau_\infty/k) = \infty, \qquad F(0) = 0.
\end{aligned} \tag{2.6.3}$$

Equation (2.6.2), which determines the elastic–plastic boundary limits the problem to finding the real part $u(\zeta)$ of function $F(\zeta)$ on the arc of the unit circle, since $v(\zeta)$ is given there by relation (2.6.3).

The required solution can be conveniently presented as a sum of solutions of several problems. We need to determine the real parts of the corresponding functions on the arc of the circle:

For function $F_1(\zeta)$,

$$v_1 = \mp\tfrac{1}{2}\cos\theta, \qquad \beta \geq |\theta| \geq \alpha,$$

$$v_1 = -\tfrac{1}{2}\cot\alpha\sin\theta, \qquad \alpha \geq |\theta|, \tag{2.6.4}$$

$$u_1 \equiv 0 \quad \text{on } ML,\ BC,\ MN,\ AB.$$

For function $F_2(\zeta)$,

$$v_2 \equiv 0 \quad \text{on } CEL,$$

$$u_2 = \tfrac{1}{2}\tau\sin\beta \quad \text{on } ML,\ BC, \tag{2.6.5}$$

$$u_2 \equiv 0 \quad \text{on } MN,\ AB.$$

For function $F_3(\zeta)$,

$$v_3 \equiv 0 \quad \text{on } CEL,$$

$$u_3 \equiv 0 \quad \text{on } ML,\ BC, \tag{2.6.6}$$

$$u_3 = -\tau \quad \text{on } MN,\ AB.$$

On the basis of the superposition principle, the complete solution is written as

$$F(\zeta) = F_1(\zeta) + F_2(\zeta) + lF_3(\zeta).$$

With the help of transformation $\Omega(\zeta)$, we map the circular sector of the unit radius with a slit in the stress plane ζ onto the semicircle of the unit radius in plane Ω. Evidently,

$$\Omega(\zeta) = \left(\frac{\zeta^n - a^n}{1 - \zeta^n a}\right)^{1/2}. \tag{2.6.7}$$

Here, $n = \pi/\beta$. Following the recalculation of the boundary conditions for function $F_1[\zeta(\Omega)] = F_1{}^*(\Omega)$, we obtain the boundary problem

$$v_1^*(\psi) = v_1[\theta(\psi)] \quad \text{on the semicircle},$$

$$u_1 \equiv 0 \quad \text{on the diameter}.$$

Here, ψ is the polar angle in plane Ω $[\Omega(e^{i\theta}) = e^{i\psi(\theta)}]$. The boundary value problems (2.6.5) and (2.6.6) are reduced to the following problem. The real part of the analytical function is given on the diameter, while the imaginary part of this function is equal to zero on the semicircle of the unit circle. This boundary problem for function $F_1(\zeta)$ is easily solved by using an analytical extension through the diameter and the Hilbert formula

$$u_1(\psi) = \frac{4\cos\psi}{\pi}\int_0^{\pi/2}\frac{v_1^*(\sigma)\sin\sigma}{\cos 2\psi - \cos 2\sigma}\,d\sigma. \tag{2.6.8}$$

Taking (2.6.4) into account and expressing ψ and σ in terms of θ and φ in the plane, we find

$$u_1(\theta) = -\frac{2\cos\lambda}{\pi\sqrt{1+b\sin^2\lambda}}\left[\cot\alpha\int_0^{a\pi/2}\sin\frac{2\gamma}{n}\omega(\gamma)\,d\gamma + \int_{a\pi/2}^{\pi/2}\cos\frac{2\gamma}{n}\omega(\gamma)\,d\gamma\right],$$

$$\omega(\gamma) = \frac{\sin\gamma\sqrt{1+b\sin^2\gamma}}{\cos 2\lambda - \cos 2\gamma}, \tag{2.6.9}$$

$$b = \frac{4a^n}{(1-a^n)^2}, \qquad \lambda = \frac{n\theta}{2}, \qquad \gamma = \frac{n\varphi}{2}.$$

The boundary problems obtained for functions $F_2(\zeta)$ and $F_3(\zeta)$ are easily solved with the help of an analytical extension through the semicircle and by solving the Dirichlet problem for the half plane

$$u_2(\theta) = \frac{\sin\beta}{2}\,\frac{2n\cos(n\theta/2)}{c}\cdot I,$$

$$I = \int_0^1\frac{(1+\eta^n)\sqrt{(a^n-\eta^n)(1+a^n\eta^n)}}{1+2\eta^n\cos n\theta + \eta^{2n}}\,d\eta. \tag{2.6.10}$$

$$c = \pi\sqrt{1-2a^n\cos n\theta + a^{2n}},$$

$$u_3(\theta) = \frac{2n\cos n\theta/2}{c}\int_0^a\frac{(1-\xi^2)\sqrt{(a^n-\xi^n)(1-a^n\xi^n)}}{1-2\xi^n\cos n\theta+\xi^{2n}}\,d\xi. \tag{2.6.11}$$

According to reference [9], some qualitative features of the elastic–plastic problem can be established. Let us denote by $R(\theta)$ the length of the characteristic section between the arc and a given point $P(x, y)$. By using $R(\theta)$, it is convenient to characterize the location of the elastic–plastic boundary. From the condition that the entire arc is included in the plastic region, we have $R(\theta) \geq 0$. The plastic region cannot extend beyond the perpendiculars L and C at the ends of the straight sides of the slit [16], so that $R(\pm\beta) = 0$. For the distance $R(\theta)$, we can write

$$R(\theta) = u(\theta) + \tfrac{1}{2}\cot\alpha\cos\alpha - R_0, \qquad \alpha \geq |\theta| \text{ on } DEF,$$

$$R(\theta) = u(\theta) = \tfrac{1}{2}\sin|\theta|, \qquad \beta \geq |\theta| \geq \alpha \text{ on } FL, CD.$$

It can be shown that when $\theta = \pm\alpha$, the elastic–plastic boundary is tangent to the normals to the edges of the arc: $\partial R/\partial\theta = \pm\infty$. Thus, the region of the

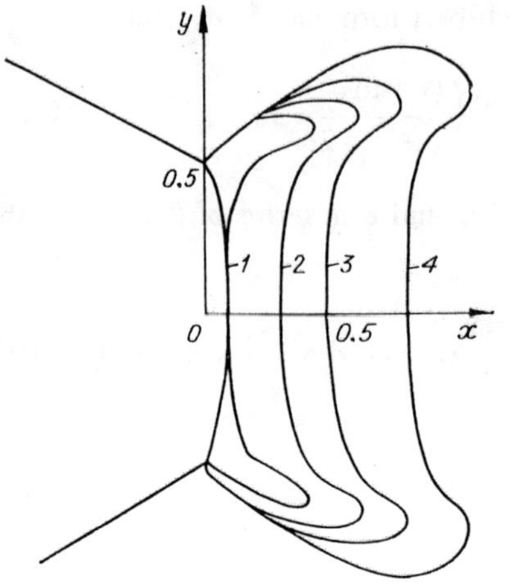

FIG. 2.15

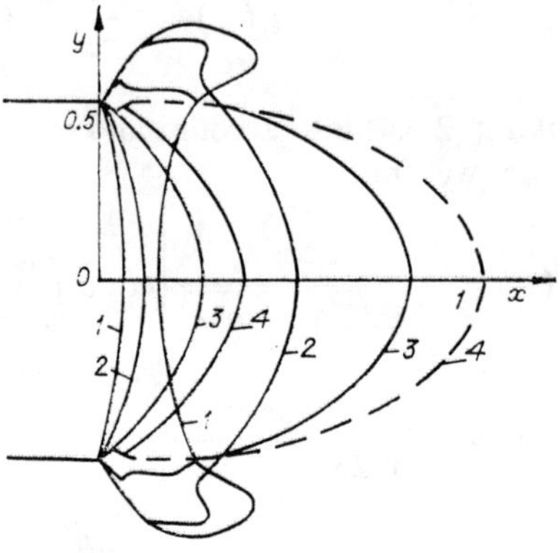

FIG. 2.16.

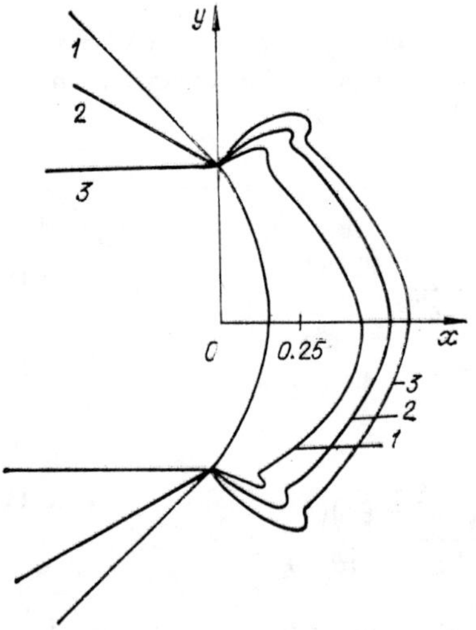

FIG. 2.17.

plastic deformation has characteristic "ears" (see Figs. 2.15–2.18). Therefore, in case of a smooth transition between the circular arc and the straight sides of the slit ($\alpha = \beta$, points L and F, C and D coincide), a plastic region starting from the transition points cannot include the entire arc.

When the inequality $0 < \alpha < \beta$ is satisfied, i.e., in the general case, it is found that the region of plastic deformation for a sufficiently long notch will include the entire arc. It can also be shown that, when $\alpha \to 0$, the following relationship is true:

$$u(0) = 1/\pi \ln \alpha + O(1) + O(l).$$

Therefore, for a notch of any fixed length, the elastic–plastic boundary moves inside the notch when $\alpha \to 0$.

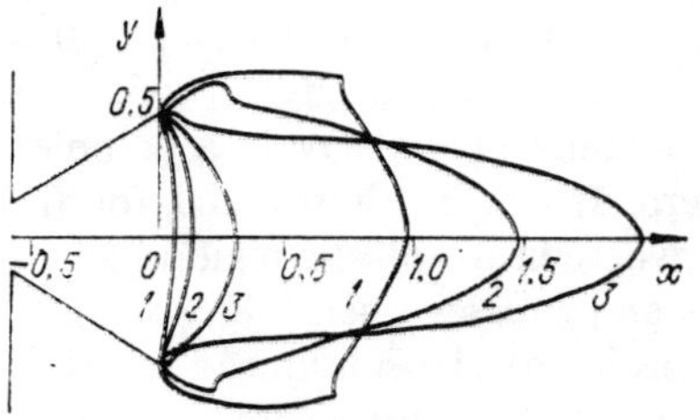

FIG. 2.18.

For a notch of a specified shape, length $l_0(a, \beta, \alpha)$, for which the boundary of the plastic region becomes tangent to the arc of the notch, can be determined. As l increases, the dimensions of the region of plastic strains increase, and at $l > l_0$ the plastic zone will include the entire arc of the notch.

Figure 2.15 shows the elastic–plastic boundary for various lengths of the notch at $\beta = 60°$, $a = 0.3$, $\alpha = 10°$, and $l_0 = 7.66$. Curves 1–4 correspond to the values of $l = 7.66, 10, 12, 15$.

Figure 2.16 shows the elastic–plastic boundaries for a straight notch, i.e., $\beta = 90°$, for $a = 0.5$, $l_0 = 2$. The straight sides of the notch are connected by arcs 1–4 such that $\alpha = 10°$ (1.55), 30° (0.83), 60° (1.14), and 80° (2.48). The values presented in brackets are the characteristic lengths l_0.

For an arc with $\alpha = 80°$ (curve 4), the solution obtained is meaningless, since $l < l_0$ and the contour of the plastic region extends inside the notch. Similarly, it can be shown that with increasing β the size of the plastic region increases. Figure 2.17 shows the elastic–plastic boundaries for various values of β at $a = 0.5$ for a notch with $l = 2$, $\alpha = 30°$. Curves 1–3 correspond to the values of $\beta = 45°$, 60°, and 90°.

Presented in Fig. 2.18 are the contours of the plastic zones that arise near a notch with $\beta = 120°$ at $a = 0.8$, $l = 0.6$. Curves 1–3 correspond to arcs with $\alpha = 10°$ (0.142), 30° (0.165) and 60° (0.465), respectively. The elastic–plastic boundaries corresponding to these arcs are similarly denoted.

§ 7. CONTACT PROBLEM

In the preceding sections of this chapter, we discussed elastic–plastic problems under conditions of complex shear when loads were given on the boundary of the body so that the problems were statically determinate in the plastic region.

Let us consider a statically indeterminate elastic–plastic problem. Let two cylindrical bodies, one of which is absolutely rigid, be in contact. Over the area of contact, a state of adhesion takes place, while the elastic body is under complex shear. The material of the body obeys the Prandtl diagram. Let us assume that a state of adhesion with the rigid punch is given on some section of the smooth contour of the body

$$w = \text{const}, \tag{2.7.1}$$

and the remaining boundary of the body is free from loading.

Let us examine the initial development of plastic zones near the edges of the contact areas. Evidently, the elastic–plastic body in this case can be considered as the upper half plane $\operatorname{Im} z > 0$. Over a part of the surface of this half plane $\operatorname{Re} z < 0$, a constant displacement $w = \text{const}$ is given, while the remaining boundary $\operatorname{Re} z > 0$ is free from loading ($z = x + iy$ is a complex plane). The stress at infinity is equal to zero. It can be shown that for the stated problem no solution continuous everywhere in the plastic region including the boundary, in terms of displacements and stresses exists [18].

Let us construct a solution of the above problem. The solution in terms of displacements and stresses is continuous everywhere in the region except on the boundary there is a singular point $z = 0$ where the displacement and stress are discontinuous. The characteristics in the plastic region are assumed to be radial straight lines; the boundary condition (2.7.1) in the plastic region is then satisfied.

When relation (2.1.3) is used, the required elastic–plastic problem is reduced to the following boundary value problem in the theory of analytical functions:

$$\begin{aligned} &\operatorname{Re} f'(z) = 0, \quad \text{for } \operatorname{Re} z < 0, \\ &\operatorname{Im} f'(z) = 0, \quad \text{for } \operatorname{Re} z > 0, \\ &|f'(z)| = k/G, \qquad \arg[f'(z) + z] = \pi/2, \quad \text{on } L, \\ &f'(z) = P/G\sqrt{lz} + O(\sqrt{z}), \quad \text{for } z \to \infty. \end{aligned} \tag{2.7.2}$$

The condition at infinity indicates that the distribution of displacements and stresses at an infinitely distant point (i.e., at distances that are large compared with the characteristic linear dimension of the plastic region, but small compared with the characteristic linear dimension of the body) is determined from the solution of the elastic problem for a punch of length l to which a force P is applied. (Here, L is an unknown boundary of elastic and plastic zones.)

It is expedient to introduce the plane of locus ζ as a parametric plane

$$\zeta = (G/k) f'(z). \tag{2.7.3}$$

One quarter of the unit circle, which is located in the fourth quadrant of plane ζ, corresponds to the plane of locus ζ of the elastic region. Based on (2.7.2), we obtain the following boundary problem on plane ζ:

$$\begin{aligned} &\operatorname{Im} z(\zeta) = 0, \quad \text{for } \operatorname{Im} \zeta = 0 \text{ and } \operatorname{Re} \zeta = 0, \\ &\zeta^2 z(\zeta) = -\overline{z(\zeta)}, \quad \text{for } |\zeta| = 1, \\ &z(\zeta) = P^2/lk^2\zeta^2, \quad \text{for } \zeta \to 0. \end{aligned} \tag{2.7.4}$$

By utilizing an analytical extension through the contour of the unit circle, we derive the following solution of the problem:

$$z(\zeta) = (P/lk^2)(1/\zeta^2 - 1). \tag{2.7.5}$$

Therefore, $f'(z) = kP/G\sqrt{P^2 + lk^2 z}$.

The elastic and plastic zones are separated by a circle

$$(y_*)^2 + (x_* + 1)^2 = 1, \qquad y_* = (lk^2/P^2)y,\ x_* = (lk^2/P^2)x. \quad (2.7.6)$$

The solution can be generalized for an arbitrary polygonal contour of the body and for any number of punches.

A similar continuous solution with singular points at the edges of the contact area can be constructed for a body with a contour that is concave toward the body. A different picture emerges for a body with a convex contour. Indeed, from equation (2.7.1) and concepts of § 1, it follows that on contour L separating the elastic and plastic regions the tangential component of the stress vector must be equal to zero. In other words, the slip lines must be tangent to contour L. It is not possible to construct a smooth contour, which is based on a convex arc of body boundary, possesses the indicated property, and satisfies the condition $|\tau_{zn}| \le k$ everywhere on the body boundary (τ_{zn} is a boundary load), for any finite number of singular points on the boundary of the body. Apparently, the solution in the plastic zone is always discontinuous in this case. This conclusion agrees with the results of Nikolsky and Taganov [8] for aerodynamics of near-sonic flows, according to which the problem of potential flow around a profile with a local supersonic zone is not correctly formulated.

§ 8. LINEARIZATION UNDER ARBITRARY HARDENING LAW

1

Let us consider the problem of complex shear using the relationships of the deformation theory (1.2.12) under arbitrary hardening law

$$T = T(\Gamma), \qquad dT/d\Gamma > 0,\ dT/d\Gamma < T/\Gamma. \quad (2.8.1)$$

The existence of a unique dependence of Γ on T follows from (2.8.1) and can be written as

$$\Gamma = F(T)/G, \quad (2.8.2)$$

where G is the shear modulus. Accounting for equation (2.2.1) and equality (1.2.12), we have relations

$$\begin{gathered} \tau_{xz}/\tau = \gamma_{xz}/\gamma, \qquad \tau_{yz}/\tau = \gamma_{yz}/\gamma, \\ \tau_{xz} = \partial\psi/\partial y, \qquad \tau_{yz} = -\partial\psi/\partial x, \\ \gamma_{xz} = \partial w/\partial x, \qquad \gamma_{yz} = \partial w/\partial y, \\ \gamma = \left(\gamma_{xz}^2 + \gamma_{yz}^2\right)^{1/2}, \qquad \tau = \left(\tau_{xz}^2 + \tau_{yz}^2\right)^{1/2}. \end{gathered} \quad (2.8.3)$$

Here, $\psi = \psi(x, y)$ is the stress function.

2

Using relation (2.8.2), function $w(x, y)$ is excluded from (2.8.3) by differentiation. As a result, we obtain an equation for the stress function

$$\frac{\partial}{\partial x}\left(\frac{F(\tau)}{\tau}\frac{\partial \psi}{\partial x}\right)+\frac{\partial}{\partial y}\left(\frac{F(\tau)}{\tau}\frac{\partial \psi}{\partial y}\right)=0. \qquad (2.8.4)$$

We linearize this equation applying the Legendre transformation[3] to function $-\psi(x, y)$, i.e., we introduce new independent variables ξ, η and a new independent function $\Phi = \Phi(\xi, \eta)$ in the equations

$$\xi = -\frac{\partial \psi}{\partial x}, \qquad \eta = -\frac{\partial \psi}{\partial y}, \qquad \Phi = x\xi + y\eta + \psi. \qquad (2.8.5)$$

Considering that the Jacobian of transformation

$$\frac{D(\xi, \eta)}{D(x, y)} = \frac{\partial^2 \psi}{\partial x^2}\frac{\partial^2 \psi}{\partial y^2} - \left(\frac{\partial^2 \psi}{\partial x\, \partial y}\right)^2 \equiv \Delta$$

is not equal to zero, and using the expression for the second derivatives of function $\psi(x, y)$,

$$\frac{\partial^2 \psi}{\partial x^2} = -\Delta\frac{\partial^2 \Phi}{\partial \eta^2}, \qquad \frac{\partial^2 \psi}{\partial y^2} = -\Delta\frac{\partial^2 \Phi}{\partial \xi^2}, \qquad \frac{\partial^2 \psi}{\partial x\, \partial y} = \Delta\frac{\partial^2 \Phi}{\partial \xi\, \partial \eta},$$

we obtain a linear equation

$$\frac{\tau F'(\tau) - F(\tau)}{\tau^3}\left[\xi^2\frac{\partial^2 \Phi}{\partial \eta^2} - 2\xi\eta\frac{\partial^2 \Phi}{\partial \xi\, \partial \eta} + \eta^2\frac{\partial^2 \Phi}{\partial \xi^2}\right] + \frac{F(\tau)}{\tau}\left[\frac{\partial^2 \Phi}{\partial \xi^2} + \frac{\partial^2 \Phi}{\partial \eta^2}\right] = 0;$$

$$\tau = (\xi^2 + \eta^2)^{1/2}, \qquad F'(\tau) \equiv dF(\tau)/d\tau. \qquad (2.8.6)$$

such that function $\Phi(\xi, \eta)$ satisfies it. In this case, there exist the inversion formulas

$$x = \frac{\partial \Phi(\xi, \eta)}{\partial \xi}, \qquad y = \frac{\partial \Phi(\xi, \eta)}{\partial \eta}, \qquad \psi = -\xi x - y\eta + \Phi. \qquad (2.8.7)$$

[3] The Legendre transformation is a special case of the contact transformation (see Appendix I).

In polar coordinates r, θ of plane ξ, η, equation (2.8.6) acquires the form

$$g(\tau)\tau^2\frac{\partial^2\Phi}{\partial\tau^2}+\tau\frac{\partial\Phi}{\partial\tau}+\frac{\partial^2\Phi}{\partial\theta^2}=0,$$
$$g(\tau)\equiv F(\tau)/(F'(\tau)\tau). \qquad (2.8.8)$$

Conditions $F>0$ and $F'(\tau)>0$ ensure that this equation is elliptical.

The inversion relations (2.8.7) can be written as

$$x=\frac{\partial\Phi}{\partial\tau}\cos\theta-\frac{1}{\tau}\frac{\partial\Phi}{\partial\theta}\sin\theta, \qquad y=\frac{\partial\Phi}{\partial\tau}\sin\theta+\frac{1}{\tau}\frac{\partial\Phi}{\partial\theta}\cos\theta,$$
$$\psi=-\tau\frac{\partial\Phi}{\partial\tau}+\Phi. \qquad (2.8.9)$$

Function $w=w(\tau,\theta)$ is determined from the simultaneous system based on (2.8.8):

$$\frac{\partial w}{\partial\tau}=\frac{F(\tau)}{G}\left[\frac{1}{\tau}\frac{\partial^2\Phi}{\partial\tau\partial\theta}-\frac{1}{\tau^2}\frac{\partial\Phi}{\partial\theta}\right]=-\frac{F(\tau)}{G\tau^2}\frac{\partial\psi}{\partial\theta},$$
$$\frac{\partial w}{\partial\theta}=\frac{F(\tau)}{G}\left[\frac{1}{\tau}\frac{\partial^2\Phi}{\partial\theta^2}+\frac{\partial\Phi}{\partial\tau}\right]=\frac{F^2(\tau)}{G\tau F'(\tau)}\frac{\partial\psi}{\partial\tau}. \qquad (2.8.10)$$

From equalities $\tau_{xz}=-\tau\sin\theta$ and $\tau_{yz}=\tau\cos\theta$, it follows that $\theta=\omega-\pi/2$, where ω is the angle formed by vector (τ_{xz},τ_{yz}) and the x axis (Fig. 2.19).

Assuming in equation (2.8.6),

$$\Phi_m(\tau,\theta)=Z_m(\tau)\exp(\pm im\lambda\theta), \qquad i^2=-1, m=0,\pm1,\pm2,\ldots \qquad (2.8.11)$$

we find that function $Z_m(\tau)$ satisfies equation

$$g(\tau)\tau^2\frac{d^2Z_m(\tau)}{d\tau^2}+\tau\frac{dZ_m(\tau)}{d\tau}-\lambda^2m^2Z_m(\tau)=0, \qquad m=0,\pm1,\pm2,\ldots. \qquad (2.8.12)$$

Knowing the solution of this equation, special solutions of equation (2.8.6) can

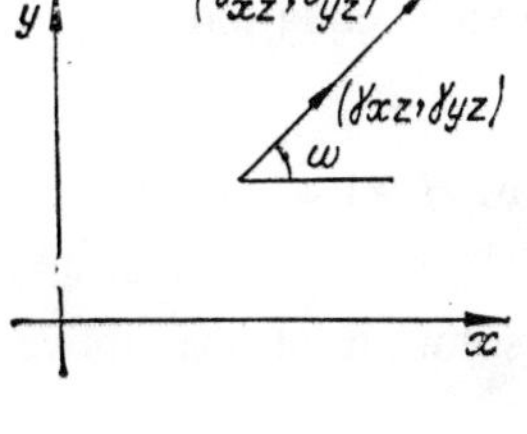

FIG. 2.19.

be constructed in the form

$$\sum_m Z_m(\tau)(a_m \cos m\lambda\theta + b_m \sin m\lambda\theta), \tag{2.8.13}$$

where a_m, b_m are arbitrary constants.

Let us examine some examples.

a. Let

$$F(\tau) = \tau(1 - \tau^2/\tau_*^2)^{(\nu^2-1)/2}, \qquad \tau < \min(\tau_*, \tau_*/\nu), \tag{2.8.14}$$

where τ_*, ν are constants. Function

$$g(\tau) = (\tau_*^2 - \tau^2)/(\tau_*^2 - \nu^2\tau^2). \tag{2.8.15}$$

Equation (2.8.8) in this case coincides with the equation of gas dynamics for a polytropic gas [6]. In this case

$$\nu^2 = (\varkappa + 1)/(\varkappa - 1), \qquad \tau_* = 2c_\infty^2\left[1 + M_\infty^2(\varkappa - 1)/2\right]/(\varkappa - 1),$$

where $\varkappa$ is a polytropic index, c_∞ is the velocity of sound for the gas stream at an infinitely distant point, and M_∞ is the Mach number at the same point.

Equation (2.8.12) has the form

$$\tau^2(\tau_*^2 - \tau^2)\frac{d^2Z_m(\tau)}{d\tau^2} + \tau(\tau_*^2 - \tau^2\nu^2)\frac{dZ_m(\tau)}{d\tau} - \lambda^2m^2(\tau_*^2 - \nu^2\tau^2)Z_m(\tau) = 0. \tag{2.8.16}$$

Solution of this equation can be written as

$$Z_m(\tau) = \tau^{\lambda m\varepsilon}H(\alpha, \beta, \mu, \tau^2/\tau_*^2), \qquad \mu = \lambda m\varepsilon + 1,$$

$$\alpha = \frac{\lambda m\varepsilon}{2} + \frac{\nu^2 - 1}{4} + \frac{\delta}{2}\sqrt{\frac{(\nu^2-1)^2}{4} + \lambda^2m^2\nu^2}, \tag{2.8.17}$$

$$\beta = \frac{\lambda m\varepsilon}{2} + \frac{\nu^2 - 1}{4} - \frac{\delta}{2}\sqrt{\frac{(\nu^2-1)^2}{4} + \lambda^2m^2\nu^2}, \qquad \varepsilon = \pm 1,\ \delta = \pm 1.$$

Here, function $H(\alpha, \beta, \mu, t)$ satisfies Gauss's hypergeometric equation

$$t(t-1)\frac{d^2H}{dt^2} + [(\alpha + \beta - 1)t - \mu]\frac{dH}{dt} + \alpha\beta H = 0. \tag{2.8.18}$$

The following hypergeometric function is a special solution of the above equation:

$$\chi(\alpha, \beta, \mu, t) = 1 + \sum_{p=1}^{\infty} \frac{C_{\alpha+p-1}^{p} C_{\beta+p-1}^{p}}{C_{\mu+p-1}^{p}} t^{p}.$$

b. By executing the limit transition $\nu \to 0$ in equality (2.8.14), we obtain

$$F(\tau) = \frac{\tau}{\sqrt{1 - (\tau/\tau_*)^2}}, \qquad \tau < \tau_*. \tag{2.8.19}$$

This relation, proposed by Sokolovsky [11], possesses a number of remarkable properties that allow construction of a solution for problems of complex shear when corresponding solutions for the linear case $F(\tau) = \tau$ are known.

Special solutions for dependence (2.8.19) of type (2.8.11) are constructed with the help of function $Z_m(\tau)$, determined by (2.8.17) with $\nu = 0$.

c. For a general power dependence

$$T = A\Gamma^n, \qquad F(\tau) = GA^{-1/n}\tau^{1/n}, \qquad g(\tau) = n, \qquad 0 < n \leq 1, \tag{2.8.20}$$

equation (2.8.12) acquires form

$$\tau^2 \frac{d^2 Z_m(\tau)}{d\tau^2} + \frac{\tau}{n} \frac{dZ_m(\tau)}{d\tau} - \frac{\lambda^2 m^2}{n} Z_m(\tau) = 0.$$

This is the Euler equation. Its general solution is

$$Z_m(\tau) = c_1 \tau^p + c_2 \tau^q,$$

$$p = (1 - n^{-1})/2 + \sqrt{(1 + n^{-1})^2/4 + \lambda^2 m^2/n}, \tag{2.8.21}$$

$$q = (1 - n^{-1})/2 - \sqrt{(1 + n^{-1})^2/4 + \lambda^2 m^2/n}.$$

d. Combination of linear and general power equations

$$F(\tau) = \tau + d\tau^s, \tag{2.8.22}$$

where d, s are positive constants. In this case, equation (2.8.12) has the form

$$\tau^2[d\tau^{s-1} + 1] \frac{d^2 Z_m(\tau)}{d\tau^2} + \tau[sd\tau^{s-1} + 1] \frac{dZ_m(\tau)}{d\tau}$$

$$-\lambda^2 m^2 [sd\tau^{s-1} + 1] Z_m(\tau) = 0,$$

and its solution is

$$Z_m(\tau) = \tau^{\varepsilon m\lambda} H(\alpha, \beta, \mu, -d\tau^{s-1}),$$

$$\mu = 1 + \frac{2m\lambda\varepsilon}{s-1}, \qquad \alpha = \frac{m\lambda}{s-1} + \frac{1}{2} + \delta\sqrt{\frac{1}{4} + \frac{\lambda^2 m^2 s}{(s-1)^2}}, \tag{2.8.23}$$

$$\beta = \frac{m\lambda}{s-1} + \frac{1}{2} - \delta\sqrt{\frac{1}{4} + \frac{\lambda^2 m^2 s}{(s-1)^2}}, \qquad \varepsilon = \pm 1, \delta = \pm 1,$$

where function $H(\alpha, \beta, \mu, t)$ satisfies equation (2.8.18).

3

Using relation (2.8.1), we exclude function $\psi(x, y)$ from (2.8.4) by differentiation. As a result, we obtain an equation for function $w(x, y)$:

$$\frac{\partial}{\partial x}\left[\frac{T(\gamma)}{\gamma}\frac{\partial w}{\partial x}\right] + \frac{\partial}{\partial y}\left[\frac{T(\gamma)}{\gamma}\frac{\partial w}{\partial y}\right] = 0. \tag{2.8.24}$$

Now we linearize this equation by applying the Legendre transformation to function $w(x, y)$, i.e., we introduce new independent variables $\gamma_{xz} = \partial w/\partial x$ and $\gamma_{yz} = \partial w/\partial y$ and a new unknown function $\psi = \psi(\gamma_{xz}, \gamma_{yz})$,

$$\Psi = x\frac{\partial w}{\partial x} + y\frac{\partial w}{\partial y} - w. \tag{2.8.25}$$

Let us assume that the Jacobian of transformation (2.8.25),

$$\frac{D(\gamma_{xz}, \gamma_{yz})}{D(x, y)} = \frac{\partial^2 w}{\partial x^2}\frac{\partial^2 w}{\partial y^2} - \left(\frac{\partial^2 w}{\partial x\, \partial y}\right)^2,$$

is different from zero. Denoting polar coordinates in plane γ_{xz}, γ_{yz} by γ, ω and following the procedure adopted in subsection 2, we obtain a linear equation that is satisfied by function $\Psi = \Psi(\gamma, \omega)$:

$$h(\gamma)\gamma^2\frac{\partial^2\Psi}{\partial\gamma^2} + \gamma\frac{\partial\Psi}{\partial\gamma} + \frac{\partial^2\Psi}{\partial\omega^2} = 0,$$

$$h(\gamma) = T(\gamma)/(T'(\gamma)\gamma). \tag{2.8.26}$$

This equation was obtained by Rice [27] using a different method.

There exist inversion formulas

$$x = \frac{\partial \Psi}{\partial \gamma}\cos\omega - \frac{1}{\gamma}\frac{\partial \Psi}{\partial \omega}\sin\omega,$$

$$y = \frac{\partial \Psi}{\partial \gamma}\sin\omega + \frac{1}{\gamma}\frac{\partial \Psi}{\partial \omega}\cos\omega, \qquad (2.8.27)$$

$$w = \gamma\frac{\partial \Psi}{\partial \gamma} - \Psi.$$

Function $\psi = \psi(\gamma, \omega)$ is determined from the simultaneous system based on equation (2.8.26):

$$\frac{\partial \psi}{\partial \gamma} = T(\gamma)\left[\frac{1}{\gamma}\frac{\partial^2 \Psi}{\partial \gamma \partial \omega} - \frac{1}{\gamma^2}\frac{\partial \Psi}{\partial \omega}\right] = \frac{T(\gamma)}{\gamma^2}\frac{\partial w}{\partial \omega},$$

$$(2.8.28)$$

$$\frac{\partial \psi}{\partial \omega} = T(\gamma)\left[\frac{1}{\gamma}\frac{\partial^2 \Psi}{\partial \gamma \partial \omega} + \frac{\partial \Psi}{\partial \gamma}\right] = -\frac{T^2(\gamma)}{T'(\gamma)\gamma}\frac{\partial w}{\partial \gamma}.$$

From relations $\gamma_{xz} = \gamma\cos\omega$ and $\gamma_{yz} = \gamma\sin\omega$, it follows that ω is the angle which the vector $(\gamma_{xz}, \gamma_{yz})$ forms with the x axis. Accounting for the collinearity of vectors (τ_{xz}, τ_{yz}) and $(\gamma_{xz}, \gamma_{yz})$, we conclude that $\theta = \omega - \pi/2$, where θ is the polar angle in plane ξ, η introduced in subsection 2.

Assuming $\Psi = \Omega_m(\gamma)\exp(\pm im\lambda\omega)$ in equation (2.8.26), we find that $\Omega_m(\gamma)$ satisfies equation

$$h(\gamma)\gamma^2\frac{d^2\Omega_m(\gamma)}{d\gamma^2} + \gamma\frac{d\Omega_m(\gamma)}{d\gamma} - \lambda^2 m^2\Omega_m(\gamma) = 0. \qquad (2.8.29)$$

In the following section we present applications of the preceding results.

§ 9. NEUBER PROBLEM

We shall now consider the problem of the stress distribution under complex shear in a body with a notch of various shapes for an arbitrary hardening law.

1. Solution for sharp notches

Let us take the solution of equation (2.8.8) of type

$$\Phi(\tau, \theta) = Z_m(\tau)\cos(\lambda m\theta).$$

Function $Z_m(\tau)$ satisfies equation (2.8.12). From (2.8.9) and (2.8.10) we obtain

$$x = Z'_m(\tau)\cos\theta\cos(\lambda m\theta) + Z_m(\tau)\tau^{-1}\sin\theta\sin(\lambda m\theta)\lambda m,$$

$$y = Z'_m(\tau)\sin\theta\cos(\lambda m\theta) - Z_m(\tau)\tau^{-1}\cos\theta\sin(\lambda m\theta)\lambda m,$$

$$\psi = (-\tau Z'_m(\tau) + Z_m(\tau))\cos\lambda m\theta, \qquad (2.9.1)$$

$$w = F(\tau)G\tau^{-1}\left[Z'_\tau - Z_m(\tau)(\lambda m)^2\right]\sin(\lambda m)/(\lambda m),$$

$$Z'_m(\tau) = dZ_m(\tau)/d\tau.$$

Let us assume $\lambda m = \pi/(\pi-\beta)$, then $\psi = 0$ on the straight lines $y = \tan(\beta/2)$ and $y = -x\cot(\beta/2)$. Formulas (2.9.1) describe the stress state under complex shear in a prismatic body containing a notch with an angle β. In this case, we have $\theta = 0$ and $x = dZ_m(\tau)/d\tau$ on the positive half-axis x. The latter relationship yields the dependence of τ on x (stress diagram).

We shall present some special solutions that are obtained in a closed form:

a) For Hooke's law: $Z_m(\tau) = A\tau^{-\lambda m}$, A is a constant; the stress diagram is shown in Fig. 2.20 for $\lambda m = 3/2$ ($\beta = \pi/3$).

b) For equation (2.8.19):

$$Z_m(\tau) = \frac{A}{1-\alpha^2}\left[h^\alpha - n_*^\alpha h^{-\alpha} - \alpha\tau(h^\alpha + n_*^\alpha h^{-\alpha})/F(\tau)\right]$$

$$\alpha = \lambda m, \qquad n_* = 2^{-2}\tau_*^{-2}, \qquad h = \left(1 + \sqrt{1-(\tau/\tau_*)^2}\right)\Big/(2\tau);$$

the stress lines $\psi = \text{const}$ and stress diagram for $\lambda m = 3/2$ are presented in Fig. 2.21.

c) For the general case $F = F(\tau)$ and $\beta = 0$ ($\lambda m = 1$):

$$Z_m(\tau) = A\tau\int_{\tau_0}^{\tau}\frac{d\xi}{\xi^2 F(\xi)};$$

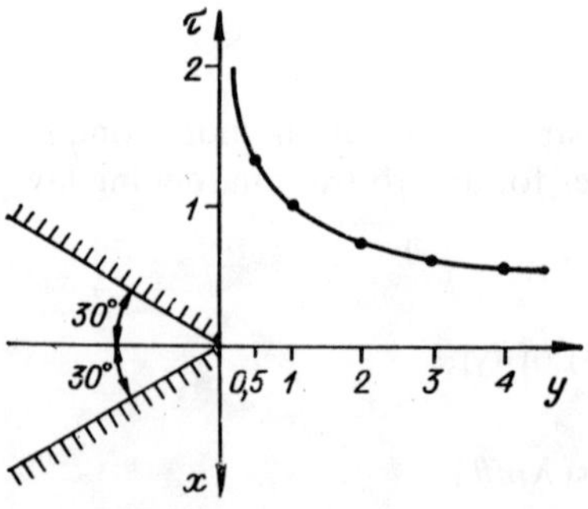

FIG. 2.20.

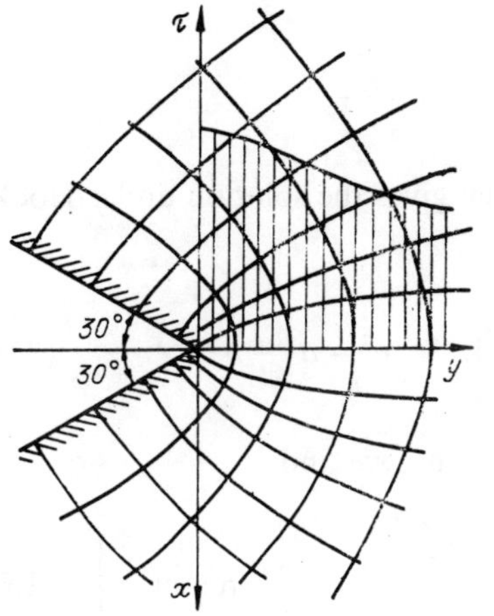

FIG. 2.21

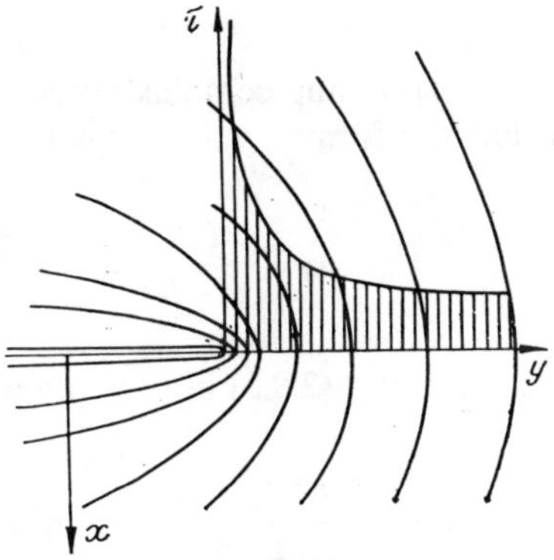

FIG. 2.22.

the stress lines and stress diagram for $F = \tau + d\tau^s$, $d = 4.4 \times 10^4$; $s = 10$ are presented in Fig. 2.22.

2. Sokolovsky's method

Equation (2.8.19) is written in the form

$$\tau = k\gamma \Big/ \sqrt{1 + (2n\gamma)^2}\,. \tag{2.9.2}$$

Here, n and k are material constants. Let us assume $z = x + iy$, $\omega = \varphi + i\psi$, $\zeta = \xi + i\eta$, and $\varphi = kw$, and reduce the problem to equation [11]

$$dz = d\zeta - \left(\frac{n}{k}\right)^2 \left(\overline{\frac{d\omega}{d\zeta}}\right)^2 d\bar{\zeta}, \qquad t\exp(-i\theta) = \frac{d\omega}{d\zeta}, \tag{2.9.3}$$

where t is determined from

$$\tau = \frac{t}{1 + n^2t^2/k^2}, \qquad k\gamma = \frac{t}{1 - n^2t^2/k^2}, \tag{2.9.4}$$

and $nt \leq k$.

Using equations (2.9.3) and (2.9.4) we can obtain a solution of the problem of stress concentration under complex shear for the nonlinear law (2.9.2), when the solution of the same problem for a linear law is known.

Let us present solutions for some shapes of notches in a half plane and strip [2, 11].

a. Half plane with a notch having the shape of half ellipse. Let us consider a half plane weakened by a cut in the shape of half ellipse with semiaxes a and *b*. The stresses applied at infinity are

$$x = 0, \qquad \tau_{xz} = 0, \qquad \tau_{yz} = \tau_\infty.$$

The corresponding complex potential, which gives the solution under Hooke's law, has the form

$$\omega = \frac{it_\infty}{a-b}\left(b\zeta - a\sqrt{\zeta^2 - c^2}\right), \qquad c^2 = a^2 - b^2. \tag{2.9.5}$$

By substituting (2.9.5) in (2.9.3) and integrating, we find

$$z = \zeta + \frac{n^2 t_\infty^2}{k^2(a-b)^2}\left[(a^2+b^2)\bar{\zeta} - 2ab\sqrt{\bar{\zeta}^2 - c^2} - a^2 c \operatorname{arctanh}\frac{\bar{\zeta}}{c}\right]. \tag{2.9.6}$$

The arbitrary integration constant is omitted here.

Equation (2.9.3), subject to (2.9.5), is written as

$$t\exp(-i\theta) = \frac{it_\infty}{a-b}\left(b - \frac{a\zeta}{\sqrt{\zeta^2 - c^2}}\right). \tag{2.9.7}$$

A new complex value $\lambda = \alpha + i\beta$ is introduced for convenience.

By separating the real and imaginary parts in relations (2.9.5)–(2.9.7), we obtain

$$\varphi = kw = \frac{t_\infty}{a-b}\left(a\frac{\cosh\alpha}{\cos 2\beta} - b\right)\eta,$$

$$\psi = \frac{t_\infty}{a-b}\left(b - a\frac{\cos\beta}{\cosh\alpha}\right)\xi, \tag{2.9.8}$$

$$x = \xi + \frac{n^2 t_\infty^2}{k^2(a-b)^2}\left[\left(a^2 + b^2 - 2ab\frac{\cos\beta}{\cosh\alpha}\right)\xi - a^2 c\alpha\right],$$

$$y = \eta - \frac{n^2 t_\infty^2}{k^2(a-b)^2}\left[\left(a^2 + b^2 - 2ab\frac{\cosh\alpha}{\cos\beta}\right)\eta + a^2 c\beta\right], \tag{2.9.9}$$

$$t^2 = \frac{t_\infty}{(a-e)^2}\left[(b - a\cosh\alpha\cos\beta)^2 + a^2\sinh^2\alpha\sin^2\beta\right],$$

$$\tan\theta = \frac{b}{a\sin\alpha\sin\beta} - \cot\alpha\cot\beta. \tag{2.9.10}$$

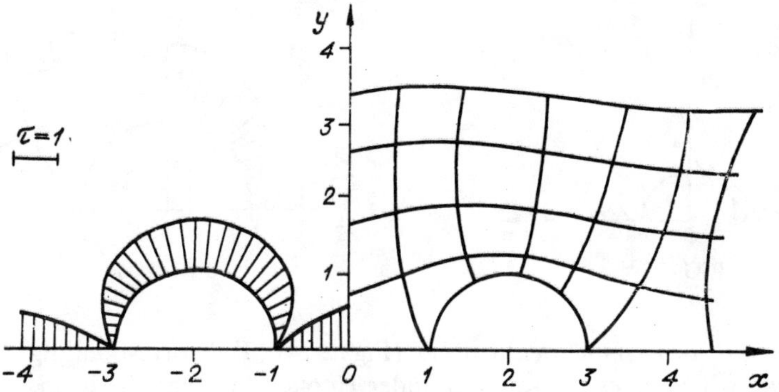

FIG. 2.23.

b. Half plane with two oval notches. Stresses $y=0$, $\tau_{xz}=0$, and $\tau_{yz}=\tau_\infty$ are given at infinity. The origin is chosen in the middle of the lines joining the centers of the notches (Fig. 2.23). The solution has the form

$$z=\zeta-\frac{n^2t_\infty}{k^2}\left[\bar{\zeta}+2\beta(\bar{\lambda}_1+\bar{\lambda}_2)\right.$$

$$\left.-\beta^2\left(\frac{\bar{\zeta}}{\alpha_0^2}\bar{\lambda}_1\bar{\lambda}_2+\frac{1}{3}\bar{\lambda}_1^3+\frac{1}{3}\bar{\lambda}_2^3+\frac{1}{2\alpha_0^3}\ln\frac{\bar{\lambda}_2}{\bar{\lambda}_1}\right)\right], \tag{2.9.11}$$

$$\lambda_1=\alpha_1+i\beta_1,\qquad \lambda_2=\alpha_2+i\beta_2,$$

$$t^2=t_\infty^2\left\{\left[1+\beta_0\left(\beta_1^2+\beta_2^2-\alpha_1^2-\alpha_2^2\right)\right]^2+4\beta_0^2\left(\alpha_1\beta_1+\alpha_2\beta_2\right)^2\right\},$$

$$\tan\theta=\frac{2\beta(\alpha_1\beta_1+\alpha_2\beta_2)}{1+\beta\left(\beta_1^2+\beta_2^2-\alpha_1^2-\alpha_2^2\right)}, \tag{2.9.12}$$

$$w=\frac{t_\infty}{G}\left[\xi+\beta(\alpha_1+\alpha_2)\right],\qquad \psi=t_\infty\left[\eta+\beta(\beta_1+\beta_2)\right],$$

where α_0 and β_0 are geometrical parameters of the notches.

From relation (2.9.11), it is easy to obtain a solution for a half plane weakened by a hole. For this purpose one must substitute $i\alpha$ for α in (2.9.11). Let us present a numerical example from reference [2] for $n/k=0.2$, $t_\infty=1.0$, $\alpha_0=2.12$, and $\beta_0=1.12$. In this case, the depth of the notch is equal to 1.02 and the shortest distance between the notches amounts to two diameters. Figure 2.23 shows the distribution of stresses along the contour as well as the lines of constant warping and the stress lines.

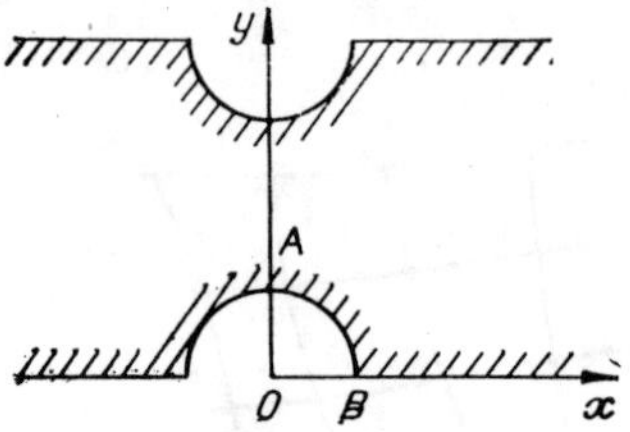

FIG. 2.24

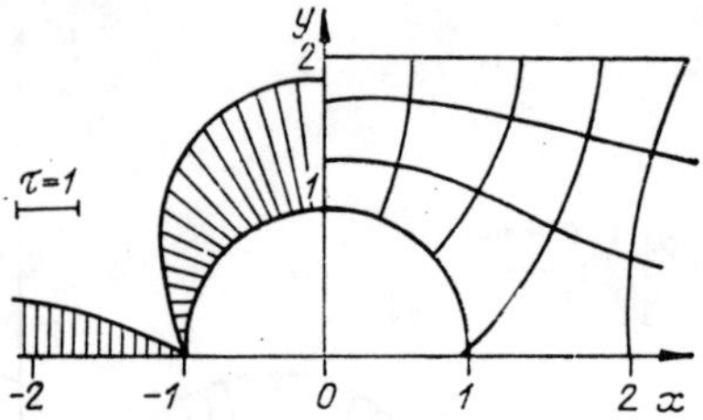

FIG. 2.25.

c. Strip with two symmetrical notches (Fig. 2.24). The corresponding complex potential, which gives the solution under Hooke's law, has the form

$$\omega(\zeta) = t_\infty(\zeta + \beta_0 \coth \alpha_0 \zeta). \tag{2.9.13}$$

For the nonlinear law (2.9.2), the solution is written as

$$z = \zeta - \frac{n^2 t_\infty^2}{k^2}\left[\bar{\zeta} + \beta_0(2 + \alpha_0\beta_0)\bar{\lambda} - \frac{\alpha_0\beta_0^2}{3}\bar{\lambda}^3\right], \tag{2.9.14}$$

$$\lambda = \alpha + i\beta = \coth \alpha_0 \zeta,$$

$$\alpha = \frac{\sinh 2\alpha_0\xi}{2(\cosh^2\alpha_0\xi - \cos^2\alpha_0\eta)}, \qquad \beta = -\frac{\sin 2\alpha_0\eta}{2(\cosh^2\alpha_0\xi - \cos^2\alpha_0\eta)},$$

$$t = t_\infty^2\left\{\left[1 + \alpha_0\beta_0(1 - \alpha^2 + \beta^2)\right]^2 + 4\alpha_0^2\beta_0^2\alpha^2\beta^2\right\},$$

$$\tan\theta = \frac{2\alpha_0\beta_0\alpha\beta}{1 + \alpha_0\beta_0(1 + \beta^2 - \alpha^2)},$$

$$w = \frac{t_\infty}{k}(\xi + \beta_0\alpha), \qquad \psi = t_\infty(\eta + \beta_0\beta).$$

To obtain the solution of the problem for the half plane with periodically spaced notches αi must be substituted for α in (2.9.13) and (2.9.14). Figure 2.25 shows the stress distribution along the contour as well as the lines of constant warping and stress lines. Numerical calculations have been conducted for a strip with two notches at $n/k = 0.2$, $t_\infty = 1.0$, $\alpha_0 = 0.82$, and $\beta_0 = 1.40$ [2]. The width of the strip was taken as 4 and the notch depth as 1.

3. Rice's solution

Let us consider the problem of complex shear for the half plane with an angle 2α and depth l. At infinity, stresses $\tau_{xz} = 0$ and $\tau_{yz} = \tau_\infty$ are given (Fig. 2.26). We assume that the material at the initial stage of deformation is linear and that, in the hardening region, the dependence $\tau = \tau(\gamma)$ is general.

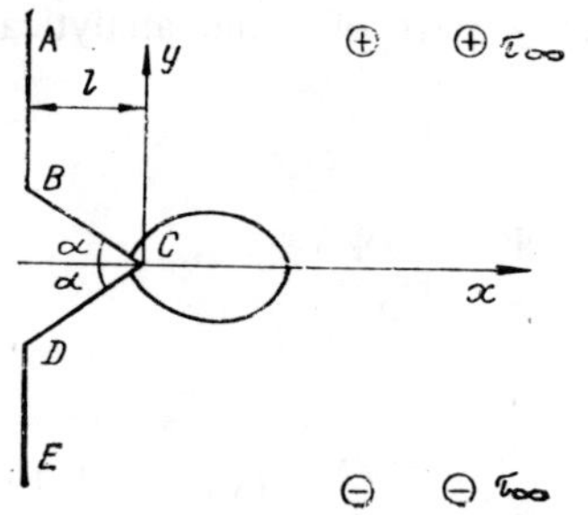

FIG. 2.26

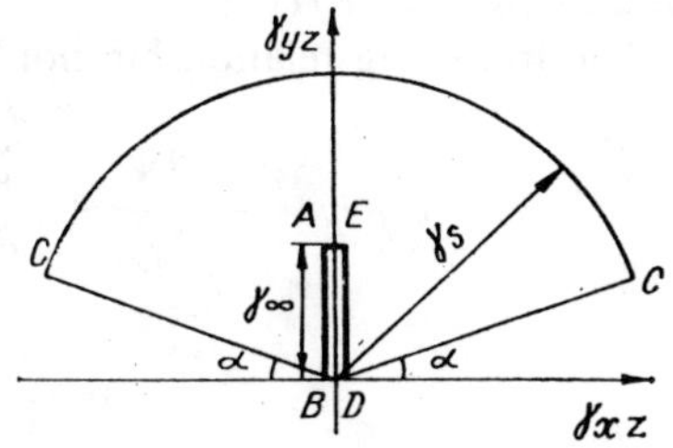

FIG. 2.27.

The elastic region in the strain plane is mapped onto the inside of a circular sector with a slit (Fig. 2.27). The boundary conditions in this case are

$$\left.\frac{\partial \Psi}{\partial \gamma_{xz}}\right|_{\gamma_{xz}=0} = -\frac{1}{\gamma}\left.\frac{\partial \Psi}{\partial \theta}\right|_{\theta=0} = -l \quad \text{on } AB \text{ and } DE, \tag{2.9.15}$$

$$\begin{aligned} &\times\left[\left.\frac{\partial \Psi}{\partial \theta}\right|_{\theta=\pi/2-\alpha} = 0 \quad \text{on } BC,\right. \\ &\left.\frac{\partial \Psi}{\partial \theta}\right|_{\theta=\pi/2+\alpha} = 0 \quad \text{on } CD. \end{aligned} \tag{2.9.16}$$

Function ψ satisfies Laplace's equation $\Delta\Psi = 0$ in the elastic region and equation (2.8.26) in the plastic region.

a. Solution in the elastic region ($\gamma < \gamma_s$). With the help of transformation

$$\zeta = \xi + i\eta = \left(\frac{\gamma_{yz} - i\gamma_{xz}}{\gamma_s}\right)^{\lambda} = \left(\frac{\gamma}{\gamma_s}\right)^{\lambda} e^{i\lambda\theta}, \tag{2.9.17}$$

we map the sector with a slit (Fig. 2.27) onto a semicircle with a slit in plane ζ (Fig. 2.28). The slit occupies section $0 \le \xi \le s$ of axis ξ. Here, λ and s are determined from

$$\lambda = \pi/(\pi - 2\alpha), \qquad s = (\gamma_\infty/\gamma_s)^{\lambda} = (\tau_\infty/\tau_s)^{\lambda}. \tag{2.9.18}$$

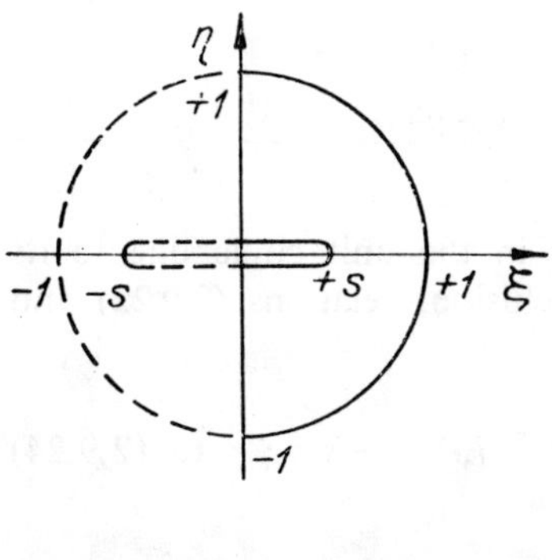

FIG. 2.28.

Function Ψ can be considered as the imaginary part of some analytical function $\Psi = \mathrm{Im}\{f(\zeta)\}$.

The following relationship holds:

$$-x+iy=-\frac{\partial\Psi}{\partial\gamma_{xz}}+i\frac{\partial\Psi}{\partial\gamma_{yz}}=e^{-i\theta}\left(\frac{1}{\gamma}\frac{\partial\Psi}{\partial\theta}+i\frac{\partial\Psi}{\partial\gamma}\right)$$

$$=\frac{\lambda}{\gamma_s}\zeta^{(\lambda-1)/\lambda}\left(\frac{\partial\Psi}{\partial\eta}+i\frac{\partial\Psi}{\partial\xi}\right)=\frac{\lambda}{\gamma_s}\zeta^{(\lambda-1)/\lambda}\cdot f'(\zeta). \quad (2.9.19)$$

Accounting for (2.9.19), we find that the boundary conditions (2.9.15) and (2.9.16) transform to condition

$$\left.\frac{\partial\Psi}{\partial\xi}\right|_{\xi=0}=0 \quad \text{on axis } \eta, \quad (2.9.20)$$

and the boundary condition in the slit (2.9.15) to condition

$$\left.\frac{\partial\Psi}{\partial\eta}\right|_{\eta=0}=\frac{\gamma_s}{\lambda}\xi^{-(\lambda-1)/\lambda}\cdot l, \qquad 0<\xi<s. \quad (2.9.21)$$

Let us consider an analytical extension for function $f(\zeta)$ into region $\xi<0$. Based on equality $2\,\partial\psi/\partial\eta=f'(\zeta)+\bar{f}'(\bar{\zeta})$ and the boundary conditions on the upper face of the slit and its analytical extension, we obtain

$$[f'(\xi)]^{+}+[\bar{f}'(\xi)]^{-}=(2\gamma_s/\lambda)\,|\xi|^{-(\lambda-1)/\lambda}\cdot l, \qquad -s<\xi<s. \quad (2.9.22)$$

Here, $[F(\xi)]^{+}$ and $[F(\xi)]^{-}$ are the values of $F(\zeta)$ when ζ tends to axis ξ from above and below, respectively.

For the lower face of the slit we have

$$[f'(\xi)]^{-}+[f'(\xi)]^{+}=(2\gamma_s/\lambda)\,|\xi|^{-(\lambda-1)/\lambda}\cdot l, \qquad -s<\xi<s. \quad (2.9.23)$$

From relations (2.9.22) and (2.9.23) we find that $f'(\zeta)-\bar{f}'(\zeta)$ is analytic inside the unit circle $|\zeta|\le 1$ including the points along the slit.

Since

$$\frac{\partial\Psi}{\partial\xi}(\xi,0)=f'(\xi)-\bar{f}'(\xi), \qquad 1>|\xi|>s,$$

function $f'(\zeta)-\bar{f}'(\zeta)$ equals zero everywhere in the unit circle due to its analyticity. Taking into account the last conclusion, relations (2.9.22) and (2.9.23) can be reduced to

$$[f'(\zeta)]^{+}+[f'(\zeta)]^{-}=(2\gamma_s/\lambda)\,|\xi|^{-(\lambda-1)/\lambda}\cdot l, \qquad -s<\xi<s. \quad (2.9.24)$$

Thus, determination of function $f'(\zeta)$ has been reduced to the Riemann problem. It can be shown that the solution of problem (2.9.24) has the form

$$f'(\zeta)=\frac{\gamma_s l}{\pi\lambda\sqrt{\zeta^2-s^2}}\int_{-s}^{s}\frac{\sqrt{s^2-\mu^2}\,|\mu|^{-(\lambda-1)/\lambda}}{\mu-\zeta}d\mu+\frac{\zeta g(\zeta)}{\sqrt{\zeta^2-s^2}}. \quad (2.9.25)$$

Here, $g(\zeta)$ is analytic everywhere inside circle $|\zeta|<1$ and its expansion into a Taylor series contains only even powers.

Assuming $t=(\mu/s)^{-\lambda}$, expression (2.9.19) can be rewritten in this fashion:

$$-x+iy=\frac{\lambda}{\gamma_s}\zeta^{(2\lambda-1)/\lambda}(\zeta^2-s^2)^{-1/2}\left[g(\zeta)+\frac{2}{\pi}\gamma_s l s^{(1+\lambda)/\lambda}\int_0^1\frac{(1-t^{2\lambda})^{1/2}\,dt}{s^2t^{2\lambda}-\zeta^2}\right]. \quad (2.9.26)$$

b. Solution in the plastic region ($\gamma>\gamma_s$). The solution of equation (2.8.26) is sought by separation of variables:

$$\Psi=\sum_{k=1}^{\infty}D_k f_k(\gamma)\sin(2k-1)\lambda\theta. \quad (2.9.27)$$

Substituting ψ into (2.8.26), we obtain the following equation for $f_k(\gamma)$:

$$\frac{\tau(\gamma)}{\gamma\tau'(\gamma)}f_k''(\gamma)+\frac{1}{\gamma}f_k'(\gamma)-\frac{(2k-1)^2\lambda^2}{\gamma^2}f_k(\gamma)=0 \quad (2.9.28)$$

under conditions

$$f_k(\gamma_s)=1,\qquad f_k'(\infty)=0, \quad (2.9.29)$$

For the physical coordinates on the elastic–plastic boundary we can write

$$(-x+iy)|_{\gamma=\gamma_s}=\frac{e^{-i\theta}}{2\gamma_s}\sum_{k=1}^{\infty}D_k\{[(2k-1)\lambda+\gamma_s f_k'(\gamma_s)]\cdot e^{i(2k-1)\lambda\theta} \quad (2.9.30)$$
$$+[(2k-1)\lambda-\gamma_s f_k'(\gamma_s)]e^{-i(2k-1)\lambda\theta}\}.$$

c. Combination of solutions. Both solutions previously obtained must coincide on the elastic–plastic boundary $\gamma=\gamma_s$. The unknown function $g(\zeta)$ and constants D_k can be determined from this coincidence condition.

Let us denote the value of ζ at $\gamma=\gamma_s$ by $\sigma=e^{i\lambda\theta}$. Substitute σ for ζ in equation (2.9.26) and equate the latter to (2.9.30). Thus, the required condition of solution continuity will have the form

$$g(\sigma) = \frac{2}{\pi}\gamma_s l s^{(1+\lambda)/\lambda}\sigma^{-2}\int_0^1 \frac{(1-t^{2\lambda})^{1/2}\,dt}{1-(s^2\sigma^{-2})t^{2\lambda}} + \frac{1}{2\lambda}(1-s^2\sigma^{-2})^{1/2}$$

$$\times \sum_{k=1}^{\infty} D_k\{[(2k-1)\lambda + \gamma_s f_k'(\gamma_s)]\sigma^{2(k-1)} \tag{2.9.31}$$

$$+ [(2k-1)\lambda - \gamma_s f_k'(\gamma_s)]\sigma^{-2k}\}.$$

The right-hand side of (2.9.31) can be expanded into a series in terms of even powers of σ including negative and positive exponents. The left-hand side, on the other hand, contains even powers of σ including only positive exponents. Consequently, assuming the coefficients of terms $\sigma^{-2}, \sigma^{-4}, \ldots, \sigma^{-2k}$ are equal to zero, we obtain an infinite system of linear equations for determination of D_k. After finding D_k, substituting their values in (2.9.31), and substituting ζ for σ, we determine $g(\zeta)$. Knowing D_k and function $g(\zeta)$, it is easy to determine the physical coordinates of the points that are deformed in the elastic and plastic regions.

Since it is hard to obtain a closed-form solution for the system of equations for D_k, it is expedient to utilize the perturbation methods.

Let us present a solution of equation (2.9.31) with an accuracy of the order s^{12} [27]:

$$D_1 = -\frac{\lambda\gamma_s l s^{(1+\lambda)/\lambda}}{\lambda - \gamma_s f_1'(\gamma_s)}\left\{B_0 + \frac{1}{2}B_0C_1s^2 + \frac{1}{4}B_0C_1^2s^4\right.$$

$$+\frac{1}{8}\left[\left(B_1 + \frac{1}{2}B_0\right)C_2 + B_0C_1^3\right]s^6$$

$$+\frac{1}{16}\left[\left(B_1 + \frac{5}{4}B_0\right)C_2C_1 + B_0C_1^4\right]s^8$$

$$+\frac{1}{16}\left[\left(B_2 + \frac{1}{2}B_1 + \frac{3}{8}B_0\right)C_3\right.$$

$$\left.\left. + \frac{1}{2}(B_1 + 2B_0)C_1 + \frac{1}{2}B_0C_1^5\right]s^{10}\right\},$$

$$D_2 = -\frac{\lambda\gamma_s l s^{(1+\lambda)/\lambda}}{3\lambda - \gamma_s f_2'(\gamma_s)}\left\{\left(B_1 + \frac{1}{2}B_0\right)s^2 + \frac{3}{8}B_0C_1s^4\right.$$

$$+\frac{3}{16}B_0C_1^2s^6 + \frac{1}{8}\left[\left(B_1 + \frac{1}{2}B_0\right)C_2 + \frac{3}{4}B_0C_1^3\right]s^8$$

$$\left.+\frac{3}{64}\left[\left(B_1 + \frac{3}{2}B_0\right)C_1C_2 + B_0C_1^4\right]s^{10}\right\},$$

$$D_3 = -\frac{\lambda\gamma_s l s^{(\lambda+1)/\lambda}}{5\lambda - \gamma_s f_3'(\gamma_s)}\left\{\left(B_2 + \frac{1}{2}B_1 + \frac{3}{8}B_0\right)s^4 + \frac{5}{16}B_0C_1s^6\right. \tag{2.9.32}$$

$$\left. + \frac{5}{32}B_0C_1^2s^8 + \frac{5}{128}\left[\left(B_1 + \frac{1}{2}B_0\right)C_2 + \frac{2}{3}B_0C_1^3\right]s^{10}\right\},$$

$$D_4 = -\frac{\lambda\gamma_s l s^{(1+\lambda)/\lambda}}{7\lambda - \lambda_s f_4'(\gamma_s)}\left\{\left(B_3 + \frac{1}{2}B_2 + \frac{3}{8}B_1 + \frac{5}{16}B_0\right)s^6\right.$$

$$\left. + \frac{35}{128}B_0C_1s^8 + \frac{35}{256}B_0C_1^2s^{10}\right\},$$

$$D_5 = -\frac{\lambda\gamma_s l s^{(1+\lambda)/\lambda}}{9\lambda - \gamma_s f_5'(\gamma_s)}\left\{\left(B_4 + \frac{1}{2}B_3 + \frac{3}{8}B_2 + \frac{5}{16}B_1\right.\right.$$

$$\left.\left. + \frac{35}{128}B_0\right)s^8 + \frac{63}{256}B_0C_1s^{10}\right\},$$

$$D_6 = -\frac{\lambda\gamma_s l s^{(1+\lambda)/\lambda}}{11\lambda - \gamma_s f_6'(\gamma_s)}\left(B_5 + \frac{1}{2}B_4 + \frac{3}{8}B_3 + \frac{5}{16}B_2 + \frac{35}{128}B_1 + \frac{63}{256}B_0\right)s^{10},$$

$$D_k = 0, \qquad k = 7, 8, 9, \ldots;$$

$$C_k = \frac{(2k-1)^k + \gamma_s f_k'(\gamma_s)}{(2k-1)\lambda - \gamma_s f_k'(\gamma_s)}; \tag{2.9.33}$$

$$B_j = \frac{\Gamma(j + 1/2\lambda)}{\lambda\sqrt{\pi}\,\Gamma(j + 1/2\lambda + 3/2)}, \qquad \Gamma(x) \text{ is the gamma function} \tag{2.9.34}$$

$$g(\zeta) = -\frac{1}{2}\gamma_s l s^{(1+\lambda)/\lambda}\sum_j^{\infty} G_j(\zeta)s^{2j}, \tag{2.9.35}$$

$$G_j(\zeta) = \sum_{q=0}^{j} d_{q+1,j}C_{q+1}\zeta^{2q} + \sum_{p=0}^{j-1}\left(\sum_{p=0}^{j-1}\beta_{j-i}d_{j+p+1-i,i}C_{j+p+1-i}\right)\zeta^{2p},$$

$$D_k = -\frac{\lambda\gamma_s l s^{(1+\lambda)/\lambda}}{(2k-1)\lambda - \gamma_s f_k'(\gamma_s)}\sum_{j=k-1}^{m} d_{k,j}s^{2j}, \tag{2.9.36}$$

$k = 1, 2, \ldots, m+1,$

$D_k = 0,\ k = m+2, m+3, \ldots,$

$$\beta_1 = \frac{1}{2}, \qquad \beta_2 = \frac{1}{8}, \ldots, \qquad \beta_p = \frac{1}{4^{p-1}}\frac{(2p-3)!}{(p-2)!\,p!}, \qquad p \geq 2.$$

By comparing formulas (2.9.32) with equations (2.9.36), we can find $d_{k,j}$ and hence determine function $g(\zeta)$ from relations (2.9.35).

Let us consider a limiting case. When the level of the loading parameter $s=(\tau_\infty/\tau_s)^\lambda$ is low, the size of the plastic region is negligibly small compared with depth l of the notch. In this case, by neglecting all terms of order s^2 which are small compared with unity, we find

$$D_1 = -\frac{\lambda\gamma_s l B_0 s^{(1+\lambda)/\lambda}}{1-\gamma_s f_1'(\gamma_s)}, \tag{2.9.37}$$

$$D_k = 0, \qquad k = 2,3,4,\ldots,$$

$$\Psi = -\frac{\lambda\gamma_s l B_0 s^{(1+\lambda)/\lambda}}{\lambda-\gamma_s f_1'(\gamma_s)} f_1(\gamma)\sin\lambda\theta. \tag{2.9.38}$$

The physical coordinates in the plastic region can be determined from equations (2.9.13)

$$x = \frac{\lambda\gamma_s l B_0 s^{(1+\lambda)/\lambda}}{\lambda-\gamma_s f_1'(\gamma_s)}\left[\frac{\lambda f_1(\gamma)}{\gamma}\cos\theta\cos\lambda\theta + f_1'(\gamma)\sin\theta\sin\lambda\theta\right],$$

$$y = \frac{\lambda\gamma_s l B_0 s^{(1+\lambda)/\lambda}}{\lambda-\gamma_s f_1'(\gamma_s)}\left[\frac{\lambda f_1(\gamma)}{\gamma}\sin\theta\cos\lambda\theta - f_1'(\gamma)\cos\theta\sin\lambda\theta\right]. \tag{2.9.39}$$

When $\alpha=0$, we obtain a crack of length l. In this case $\lambda=1$. The function determined from equations (2.9.28) and (2.9.29) using one quadrature has the form

$$f_1(\gamma) = \frac{\gamma}{\gamma_s}\left[\int_{\gamma_s}^{\infty}\frac{du}{u^2\tau(u)}\right]^{-1}\int_{\gamma}^{\infty}\frac{du}{u^2\tau(u)}. \tag{2.9.40}$$

This result can be readily checked by substitution. Substituting (2.9.40) in (2.9.39) and noting that, according to (2.9.34), $B_0=1$, we find

$$\Psi = -\gamma_s\tau_s s^2 l\left[\gamma\int_{\gamma}^{\infty}\frac{du}{u^2\tau(u)}\right]\sin\theta. \tag{2.9.41}$$

In fracture mechanics, $K_{\mathrm{III}}=\tau_\infty\sqrt{\pi l}=\tau_s s\sqrt{\pi l}$ is called the stress intensity factor [27, 28]. Expression (2.9.41) can be rewritten as

$$\Psi = -\frac{\gamma_s}{\pi\tau_s}K_{\mathrm{III}}^2\left[\gamma\int_{\gamma}^{\infty}\frac{du}{u^2\tau(u)}\right]\sin\theta. \tag{2.9.42}$$

Taking into account expression (2.9.42), the physical coordinates corresponding to the strains in the plastic zone can be determined from formulas (2.9.13):

$$x=\frac{K_{\mathrm{III}}^2}{2\pi\tau_s^2}\left\{\frac{\gamma_s\tau_s}{\gamma\tau(\gamma)}\cos 2\theta+\left[2\gamma_s\tau_s\int_\gamma^\infty\frac{du}{u^2\tau(u)}-\frac{\gamma_s\tau_s}{\gamma\tau(\gamma)}\right]\right\},$$

$$y=\frac{K_{\mathrm{III}}^2}{2\pi\tau_s^2}\frac{\gamma_s\tau_s}{\gamma\tau(\gamma)}\sin 2\theta. \qquad (2.9.43)$$

In case of a crack, the elastic–plastic boundary is a circle with center

$$X(\gamma_s)=\frac{K_{\mathrm{III}}^2}{2\pi\tau_s^2}\left[2\gamma_s\tau_s\int_{\gamma_s}^\infty\frac{du}{u^2\tau(u)}-1\right]$$

and radius $R(\gamma_s)=K_{\mathrm{III}}^2/2\pi\tau_s^2$.

To conclude the present section, we shall present numerical results [27]. For a material with a power law strain hardening, $\tau(\gamma)=\tau_s(\gamma/\gamma_s)^N$, the solution of equations (2.9.28) and (2.9.29) for $f_k(\gamma)$ has the following form:

$$f_k(\gamma)=\left(\frac{\gamma}{\gamma_s}\right)^{-\mu_k},$$

TABLE II.1

τ_∞/τ_s	$-D_1/\gamma_s l$	$-D_2/\gamma_s l$	$-D_3/\gamma_s l$	$-D_4/\gamma_s l$	$-D_5/\gamma_s l$	$-D_6/\gamma_s l$
			$N=0$			
0.2	0.0408	0.0004	0	0	0	0
0.4	1.1740	0.0070	0.0006	0.0001	0	0
0.6	0.4414	0.0398	0.0071	0.0016	0.0004	0.0001
0.8	0.9735	0.1553	0.0489	0.0186	0.0078	0.0028
			$N=0.1$			
0.2	0.0370	0.0004	0	0	0	0.2
0.4	0.1557	0.0057	0.0004	0	0	0
0.6	0.3851	0.0318	0.0055	0.0012	0.0003	0.0001
0.8	0.8059	0.1182	0.0363	0.0139	0.0058	0.0022
			$N=0.3$			
0.2	0.0311	0.0003	0	0	0	0
0.4	0.1286	0.0046	0.0004	0	0	0
0.6	0.3072	0.0249	0.0044	0.0010	0.0002	0.0001
0.8	0.6022	0.0869	0.0270	0.0105	0.0046	0.0019
			$N=0.5$			
0.2	0.0268	0.0002	0	0	0	0
0.4	0.1096	0.0040	0.0003	0	0	0
0.6	0.2556	0.0212	0.0037	0.0008	0.0002	0.0001
0.8	0.4810	0.0709	0.0223	0.0088	0.0039	0.0017

TABLE II.2

τ_∞/τ_s	$-D_1/\gamma_s l$	$-D_2/\gamma_s l$	$-D_3/\gamma_s l$	$-D_4/\gamma_s l$	$-D_5/\gamma_s l$	$-D_6/\gamma_s l$
			$N=0$			
0.2	0.0248	0.0001	0	0	0	0
0.4	0.1297	0.0026	0.0001	0	0	0
0.6	0.3670	0.0221	0.0028	0.0004	0.0001	0
0.8	0.8804	0.1156	0.0311	0.0103	0.0037	0.0012
			$N=0.1$			
0.2	0.0220	0.0001	0	0	0	0
0.4	0.1141	0.0021	0.0001	0	0	0
0.6	0.3157	0.0175	0.0021	0.0003	0.0001	0
0.8	0.7160	0.0862	0.0227	0.0075	0.0027	0.0009
			$N=0.3$			
0.2	0.0184	0.0001	0	0	0	0
0.4	0.0942	0.0018	0.0001	0	0	0
0.6	0.2535	0.0139	0.0017	0.0003	0	0
0.8	0.5405	0.0645	0.0171	0.0057	0.0021	0.0008
			$N=0.5$			
0.2	0.0160	0	0	0	0	0
0.4	0.0814	0.0015	0.0001	0	0	0
0.6	0.2151	0.0121	0.0015	0.0002	0	0
0.8	0.4423	0.0537	0.0143	0.0048	0.0018	0.0007

TABLE II.3

τ_∞/τ_s	$-D_1/\gamma_s l$	$-D_2/\gamma_s l$	$-D_3/\gamma_s l$	$-D_4/\gamma_s l$	$-D_5/\gamma_s l$	$-D_6/\gamma_s l$
			$N=0$			
0.2	0.0089	0	0	0	0	0
0.4	0.0721	0.0004	0	0	0	0
0.6	0.2571	0.0072	0.0004	0	0	0
0.8	0.7216	0.0656	0.0129	0.0032	0.0009	0.0002
			$N=0.1$			
0.2	0.0077	0	0	0	0	0
0.4	0.0618	0.0003	0	0	0	0
0.6	0.2168	0.0056	0.0003	0	0	0
0.8	0.5769	0.0483	0.0093	0.0023	0.0006	0.0002
			$N=0.3$			
0.2	0.0064	0	0	0	0	0
0.4	0.0511	0.0003	0	0	0	0
0.6	0.1766	0.0046	0.0003	0	0	0
0.8	0.4471	0.0374	0.0072	0.0018	0.0005	0.0001
			$N=0.5$			
0.2	0.0056	0	0	0	0	0
0.4	0.0449	0.0002	0	0	0	0
0.6	0.1534	0.0041	0.0003	0	0	0
0.8	0.3378	0.0320	0.0062	0.0015	0.0004	0.0001

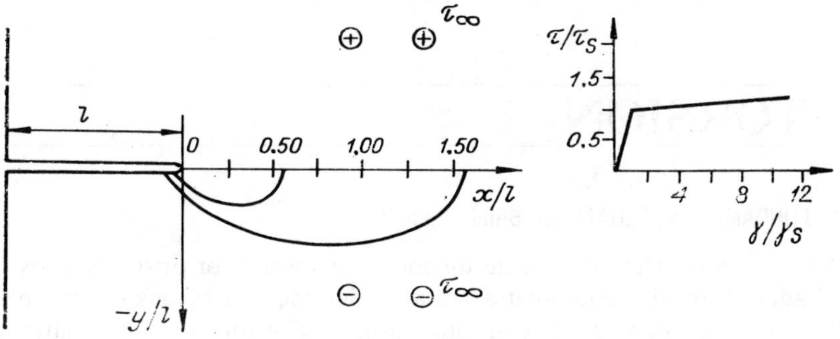

FIG. 2.29.

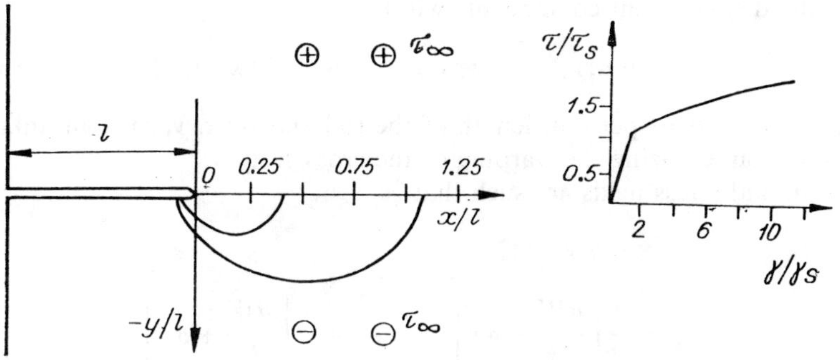

FIG. 2.30.

where

$$\mu_k = \left((1-N^2)/4 + (2k-1)^2\lambda^2 N\right)^{1/2} - (1-N)/2. \qquad (2.9.44)$$

In this case, constants D_k can be numerically determined using equations (2.9.32)–(2.9.35) for various values of the notch angle and hardening exponent.

Tables II.1–II.3 show the results for (a) crack $\alpha = 0$, (b) 45° V-shaped notch ($\alpha = 22.5°$), and (c) 90° V-shaped notch ($\alpha = 45°$), respectively. The values of D_k for a wide range of problems can be determined by interpolating the values presented in the tables.

Figures 2.29 and 2.30 depict the plastic zone boundaries for low strain hardening ($N = 0.1$) and high strain hardening ($N = 0.3$) materials and two loading values, i.e., $\tau_\infty = 0.6\tau_s$ and $\tau_\infty = 0.8\tau_s$, respectively. Figures 2.29 and 2.30 show the tendency from almost circular shape for small plastic zones to elongated shapes for large plastic zones.

3. TORSION

§ 1. FUNDAMENTAL EQUATIONS. BRIEF REVIEW

Let us consider elastic–plastic torsion of cylindrical or prismatic rods. We shall adopt Cartesian coordinates xyz and align the z axis along the axis of the rod. In accordance with the conventional theory of torsion of prismatic rods [9], we assume that all cross sections are experiencing rigid twist in their planes and are distorted in the direction of the z axis. Within the adopted assumptions, the displacement components will be

$$u = -\alpha z y, \qquad v = \alpha z x, \qquad w = W(x, y, z), \tag{3.1.1}$$

where α is the twist per unit length of the rod and $W(x, y, z)$ is an unknown function characterizing the warping of the cross section.

Strain and stress fields are such that

$$\begin{gathered}
\varepsilon_x = \varepsilon_y = \varepsilon_z = \varepsilon_{xy} = 0, \\
\varepsilon_{xz} = \frac{1}{2}\left[\frac{\partial W}{\partial x} - \alpha y\right], \qquad \varepsilon_{yz} = \frac{1}{2}\left[\frac{\partial W}{\partial y} + \alpha x\right], \\
\sigma_x = \sigma_y = \sigma_z = \tau_{xy} = 0, \\
\tau_{xz} = \tau_{xz}(x, y), \qquad \tau_{yz} = \tau_{yz}(x, y).
\end{gathered} \tag{3.1.2}$$

The equilibrium equations in this case are reduced to a single differential equation

$$\frac{\partial \tau_{xz}}{\partial x} + \frac{\partial \tau_{yz}}{\partial y} = 0. \tag{3.1.3}$$

In the elastic region, we have the compatibility condition, which leads to the expression

$$\frac{\partial \tau_{xz}}{\partial y} - \frac{\partial \tau_{yz}}{\partial x} = -2G\alpha. \tag{3.1.4}$$

In the plastic region, we have the condition of plasticity [32]

$$\tau_{xz}^2 + \tau_{yz}^2 = \tau_s^2 \tag{3.1.5}$$

and the associated law of plastic flow

$$\tau_{yz}\left(\frac{\partial W}{\partial x} - \alpha y\right) - \tau_{xz}\left(\frac{\partial W}{\partial y} + \alpha x\right) = 0. \tag{3.1.6}$$

Here, $\tau_s = \sigma_s/\sqrt{3}$ according to the Mises condition or $\tau_s = \sigma_s/2$ according to the Tresca–Saint Venant condition, and σ_s is the yield point under tension.

Let us introduce the stress function ψ such that

$$\tau_{xz} = \tau_s \frac{\partial \Psi}{\partial y}, \qquad \tau_{yz} = -\tau_s \frac{\partial \Psi}{\partial x}.$$

To determine Ψ in the elastic region, we have the differential equation of torsion

$$\frac{\partial^2 \Psi}{\partial x^2} + \frac{\partial^2 \Psi}{\partial y^2} = -\frac{2G\alpha}{\tau_s}. \tag{3.1.7}$$

From the plasticity condition for the stress function in the plastic region, we obtain the differential equation

$$\left(\frac{\partial \Psi}{\partial x}\right)^2 + \left(\frac{\partial \Psi}{\partial y}\right)^2 = 1. \tag{3.1.8}$$

We assume that all the components of stresses and displacements remain continuous during transition through the boundary between the elastic and plastic zones.

Since the lateral surface of the twisted rod is free from stresses, the contour can be considered as one of the stress lines and the vector of the shear stress is tangent to the stress line

$$\frac{\tau_{yz}}{\tau_{xz}} = \frac{dy}{dx}. \tag{3.1.9}$$

The torque M_0 can be expressed as

$$M_0 = \iint_F (x\tau_{yz} - y\tau_{xz})\, dx\, dy = 2\tau_s \iint_F \Psi\, dx\, dy. \tag{3.1.10}$$

The integration here is performed along the entire surface of the cross section F.

For the elastic–plastic torsion of prismatic rods, there exists an analogy established by Nadai (see [17] as referenced in Chapter 1). The essence of this analogy is as follows. A surface, corresponding to the stress function ψ in the plastic region, is constructed. A membrane, loaded by a uniformly distributed pressure, is pressed against this surface. The function that corresponds to the shape acquired by the membrane satisfies the same differential equation as does the stress function ψ in the elastic region. The sections of contact between the membrane and the surface will correspond to the plastic region. The remaining part will correspond to the elastic region.

The problem of elastic–plastic torsion has a number of variational formulations.

According to the Haar–Karman principle, the stress function is determined from the following variational problem [30]. Let us minimize

$$I(\Psi) = \iint_F f\,dx\,dy, \qquad f \equiv \left(\frac{\partial \Psi}{\partial x}\right)^2 + \left(\frac{\partial \Psi}{\partial y}\right)^2 - 4 \cdot \frac{G\alpha}{\tau_s}\Psi,$$

on the set of functions

$$K: \left\{ \Psi \in H_0^1(F), \left(\frac{\partial \Psi}{\partial x}\right)^2 + \left(\frac{\partial \Psi}{\partial y}\right)^2 \le 1 \right\}.$$

Here, $H_0^1(F)$ is the Sobolev space, i.e., the set of functions that together with their first derivatives are quadratically integrable and vanish on the boundary Γ of region F.

An equivalent formulation of this problem is the following. Minimize $I(\Psi)$ within the set

$$K^* = \left\{ \Psi \in H_0^1(F), \Psi \le u_\Gamma(x, y) \right\},$$

where $u = u_\Gamma(x, y)$ is the distance of point (x, y) to curve Γ. The solution of these variational problems is equivalent to the solution of a corresponding variational inequality (see [3], Chapter I).

Let us examine another variational formulation of the problem of elastic–plastic torsion [4].

It is required to find the minimum of an integral with a variable region of integration $D \subset F$,

$$I_D(\Psi) = \iint_D f\,dx\,dy,$$

where $\psi = u_\Gamma(x, y)$ on the boundary L of region D. The required curve L is assumed piecewise smooth, and function ψ, which is determined in $D + L$, as well as its first derivatives, are quadratically integrable.

In torsion problems, semiinverse methods have been proposed: Some characteristics of the sought solution are given beforehand; the solution itself and the corresponding boundary shape of the body are then accordingly reconstructed (Sokolovsky [24], Galin [13]).

As the preceding chapter shows, the elastic–plastic problem for complex shear can be adequately investigated by analytical means. In a more complicated torsion problem, when the plastic zone becomes comparable with the size of the rod cross section, results are rather scarce. First of all, mention should be made here of the exact solution by Sokolovsky for a rod of near-elliptical oval shape [24]. The solution was obtained in 1942 by the semiinverse method. Galin [13], using another semiinverse method, solved several elastic–plastic problems for rods with a near-polygonal section (particularly, near-rectangular section). Galin also reduced the torsion problem for a rod with a polygonal section to the solution of a Fuchsian differential equation [12], which allowed him to find an efficient solution of some problems (for example, for a square section).

Kachanov obtained a solution for a rod with square cross section using a variational method [16]. Southwell and Takaci, using relaxation methods [36], solved the elastic–plastic problem for a L-shaped, square, and triangular profiles as well as for a round rod with six semicircular grooves [38]. Christopherson [25], Shaw [35], and Okudo [33] discussed other types of profiles using the same methods.

Davis and Tuba investigated, using the method of elastic solutions, torsion of solid and hollow shafts with internal and external notches for a rather arbitrary type of unique stress–strain diagram [26]. The indicated method was also discussed by Amosov [8].

Banichuk, Petrov, and Chernousko, using the method of local variations, solved the elastic–plastic problem for a square section and also for a polygon of a particular shape [11]. The indicated method was also used in references [10] and [15] for the solution of elastic–plastic problems.

Perlin presented a solution for an oval-shaped rod under conditions of partial enclosure of the elastic core by the plastic zone [21].

Annin and Sadovsky, using the method of straight lines, obtained an approximate solution for a rectangular section with the plastic zones developing near one pair of sides only [7].

Annin [1] reduced the problem of an oval rod in torsion to a boundary value problem for the Monge–Ampere equation, which allowed construction of an efficient numerical algorithm for its solution. Rods of a multiply connected section in torsion were analysed by Lanchon [30].

The existence of a solution for elastic–plastic torsion of prismatic rods was discussed by Galin [14], Annin [1,2], and other researchers [17,20,27,30,39,40].

§ 2. OVAL ROD UNDER TORSION

1

Let the cross section F of a rod in torsion be an oval Γ (Fig. 3.1), which is presented in the parametric form

$$x_\Gamma = x_\Gamma(\beta) = \frac{dM(\beta)}{d\beta}\cos\beta + M(\beta)\sin\beta,$$
$$y_\Gamma = y_\Gamma(\beta) = \frac{dM(\beta)}{d\beta}\sin\beta - M(\beta)\cos\beta, \tag{3.2.1}$$

where β is an angle between the tangent to Γ and the x axis, $\pi/2 \le \beta < 5/2\pi$,

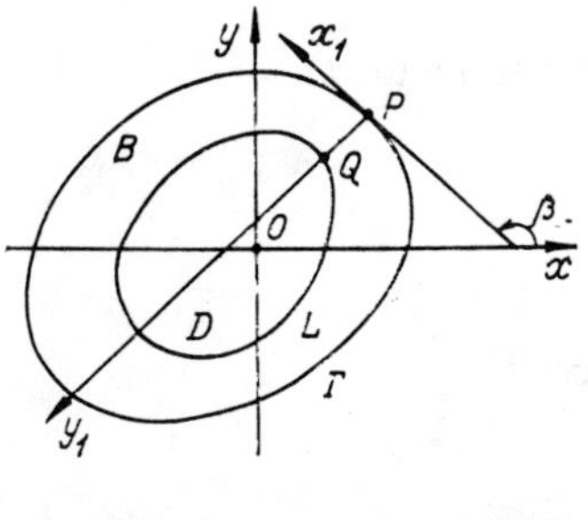

FIG. 3.1.

$M(\beta)$ is the support function of Γ (distance from the origin $O \in F$ to the tangent to Γ as a function of β); $M(\beta)$ is a periodic function with period 2π, which is sufficiently smooth (at least continuously differentiable twice), and its radius of curvature

$$R(\beta) = \frac{d^2 M(\beta)}{d\beta^2} + M(\beta)$$

satisfies the condition

$$0 < R_{\min} \le R \le R_{\max} < \infty. \tag{3.2.2}$$

Here, $R_{\min}$ and $R_{\max}$ are minimum and maximum values of the radius of curvature of Γ. Without loss of generality, we shall assume that angle $\beta = \pi/2$ at the point of intersection of the positive semiaxis x with Γ. We shall now present expression $M(\beta)$ for a number of contours.
a. Lame's curves:

$$(x/p)^{2m} + (y/q)^{2m} = 1,$$

where m is a positive integer,

$$M(\beta) = \left[(p \sin \beta)^{2m/(2m-1)} + (q \cos \beta)^{2m/(2m-1)}\right]^{(2m-1)/(2m)}.$$

For an ellipse ($m = 1$),

$$M(\beta) = \sqrt{p^2 \sin^2\beta + q^2 \cos^2\beta}\,. \tag{3.2.3}$$

b. Sokolovsky's oval:

$$x = [b + c(\cos 2\beta + 2)] \sin \beta,$$

$$y = -[b + c(\cos 2\beta - 2)] \cos \beta, \tag{3.2.4}$$

$$M(\beta) = b - c \cos 2\beta.$$

Here, constants b and c satisfy the condition $R_{\min} = b - 3c > 0$.
c. Contour Γ_1 with the support function $M_1(\beta)$ is parallel to contour Γ with the support function $M(\beta)$. Then

$$M_1(\beta) = M(\beta) + d,$$

where d is the distance between Γ_1 and Γ.
d. Contour Γ is given parametrically (in terms of parameter t):

$$x = f(t),\ y = g(t).$$

In this case

$$M(\beta) = f(t) \sin \beta - g(t) \cos \beta,$$

where t is determined in terms of β from equation

$$\tan\beta = \frac{dg(t)}{dt} \Big/ \frac{df(t)}{dt}.$$

e. Contour Γ is given in polar coordinates r, φ of plane x, y:

$$r = r(\varphi).$$

The support function is determined in the form $(0 \le \theta \le 2\pi)$

$$M\left(\frac{\pi}{2} + \theta\right) = \max_{\varphi}\left[r(\varphi)\cos(\varphi - \theta)\right]. \tag{3.2.5}$$

The maximum in terms of φ is taken in the range $\theta - \pi/2 < \varphi < \theta + \pi/2$.

2

Let us assume that when a rod with a cross section $F + \Gamma$ is in torsion, we have an elastic core where the simply connected region D with boundary L lies entirely inside F (Fig. 3.1). Let us denote the doubly connected region bounded by Γ and L by B, where $B = F - (D + L)$. In this region, the material is in the plastic state.

The problem of elastic–plastic torsion designated here as problem A is formulated below as in [2].

Problem A. Given a simply connected region F bounded by Γ, find a simply connected region D that belongs, together with its boundary L, to region F. Also find, with an accuracy up to a constant, function $\Psi(x, y)$, which is determinate and continuous in $F + \Gamma$ and has continuous first-order derivatives $\partial\Psi/\partial x$, $\partial\Psi/\partial y$ in $F + \Gamma$ and finite second-order derivatives in D subject to the conditions

$$\frac{\partial^2\Psi}{\partial x^2} + \frac{\partial^2\Psi}{\partial y^2} = -\frac{1}{a}, \quad \left(\frac{\partial\Psi}{\partial x}\right)^2 + \left(\frac{\partial\Psi}{\partial y}\right)^2 < 1 \quad \text{in region } D, \tag{3.2.6}$$

$$\left(\frac{\partial\Psi}{\partial x}\right)^2 + \left(\frac{\partial\Psi}{\partial y}\right)^2 = 1 \quad \text{in } L + B + \Gamma, \tag{3.2.7}$$

$$\overrightarrow{\operatorname{grad}}\,\Psi = \overrightarrow{n_\Gamma} \quad \text{on contour } \Gamma. \tag{3.2.8}$$

Here, $\overrightarrow{n_\Gamma}$ is a unit vector of the inner normal to contour Γ; $a = \tau_s/2G\alpha$. Condition (3.2.8), which follows from (3.1.9), can be replaced by the condition of function Ψ vanishing on Γ:

$$\Psi(x, y)|_\Gamma = 0. \tag{3.2.9}$$

Here α is a parameter such that if $\alpha < \tau_s/GR_{\max}$, the following condition is observed when deformation at all points of the rod is assumed elastic:

$$(\tau_{xz})^2 + (\tau_{yz})^2 < \tau_s^2.$$

If, on the other hand

$$\alpha > \tau_s / GR_{\min}, \tag{3.2.10}$$

there must exist some points in the rod for which the following condition holds when the entire rod is deformed elastically:

$$(\tau_{xz})^2 + (\tau_{yz})^2 \geq \tau_s^2.$$

Annin [1,2] showed that when condition (3.2.10) is satisfied, there exists a unique solution of Problem A. The proof of this theorem provides a constructive way to obtain a solution of the problem of elastic–plastic torsion for oval Γ, which we shall discuss below.

3

Solving equation (3.2.7) subject to condition (3.2.9) by Cauchy's method, we obtain

$$x = \frac{dM(\beta)}{d\beta}\cos\beta + [M(\beta) - u]\sin\beta,$$

$$y = \frac{dM(\beta)}{d\beta}\sin\beta - [M(\beta) - u]\cos\beta, \tag{3.2.11}$$

$$\Psi = u; \qquad \frac{\partial \Psi}{\partial x} = -\sin\beta, \qquad \frac{\partial \Psi}{\partial y} = \cos\beta.$$

Parameter β changes within the range $\pi/2 \leq \beta < 5/2\pi$. Let us establish the geometrical meaning of parameter u and its variability range. When β is fixed in the first two formulas of (3.2.11) and parameter u is varied from 0 to ∞, we obtain the equation of the internal normal to Γ at point P (Fig. 3.1) with unit vector $\overline{n}_\Gamma$. Therefore, parameter u in (3.2.11) for a fixed β varies from 0, which corresponds to point P of curve Γ, to some value $N(\beta)$ corresponding to point Q of curve L. Point Q is the intersection of curve L and the normal previously mentioned. We adopt as point Q the intersection point of this normal with curve L nearest to P.

Dependence $u = N(\beta)$, $\pi/2 \leq \beta < 5/2\pi$, determines the required curve L in the plane of parameters β and u. Thus, we assume that curve L allows a parametric representation of the form

$$x_L = x_L(\beta) = \frac{dM(\beta)}{d\beta}\cos\beta + [M(\beta) - N(\beta)]\sin\beta,$$

$$y_L = y_L(\beta) = \frac{dM(\beta)}{d\beta}\sin\beta - [M(\beta) - N(\beta)]\cos\beta, \tag{3.2.12}$$

$$\pi/2 \leq \beta < 5/2\pi,$$

with the required periodical function $N(\beta)$ of period 2π.

From the Jacobian

$$\frac{D(x, y)}{D(\beta, u)} = R(\beta) - u,$$

we can assume hereafter that $N(\beta) < R(\beta)$.

We should note one more geometrical interpretation of parameter u. Let point $q = (x, y) \in QP$ (Fig. 3.1) and let $u = |qP| < R_{\min}$. The circle of radius $R_{\min}$, which is internally tangential to curve Γ at point P, belongs entirely to region F bounded by Γ; therefore for $0 < u < R_{\min}$,

$$|qP| < |qP^*|,$$

where point P^* runs along path Γ. This means that at point q,

$$\Psi(x, y) = u_\Gamma(x, y) = \min_{P^*} \sqrt{(x - x^*)^2 + (y - y^*)^2}. \tag{3.2.13}$$

The minimum here is taken along the entire $P^* \in \Gamma$.

4

From the parametric representation (3.2.11) we have the following relations in region $B + L + \Gamma$, i.e., for $\pi/2 \le \beta < 5/2\pi$, $0 \le u \le N(\beta)$:

$$\frac{\partial \Psi}{\partial x} = -\sin\beta, \qquad \frac{\partial \Psi}{\partial y} = \cos\beta, \tag{3.2.14}$$

$$\Psi - x\frac{\partial \Psi}{\partial x} - y\frac{\partial \Psi}{\partial y} = M(\beta), \tag{3.2.15}$$

$$-\frac{\partial \Psi}{\partial x}y + \frac{\partial \Psi}{\partial y}x = \frac{dM(\beta)}{d\beta}. \tag{3.2.16}$$

Based on these relations, we can construct exact solutions for the problem of elastic–plastic torsion. By choosing an arbitrary solution of Poisson's equation (3.2.6),

$$\Psi = \hat{\Psi}(x, y), \tag{3.2.17}$$

we determine curve $\hat{L}$ in plane xy from the equation

$$\left(\frac{\partial \hat{\Psi}}{\partial x}\right)^2 + \left(\frac{\partial \hat{\Psi}}{\partial y}\right)^2 = 1.$$

Let this curve bound the singly connected region $\hat{D}$. We introduce the following parametrization on curve $\hat{L}$:

$$x_L = \hat{x}(\gamma), \qquad y_L = \hat{y}(\gamma), \qquad \gamma_0 \le \gamma \le \gamma_1,$$

and determine functions $\beta = \hat{\beta}(\gamma)$ and $M = \hat{M}(\gamma)$ from (3.2.14) and (3.2.15). By eliminating γ from these relations, we find $M = \hat{M}(\beta)$.

If this function is determined for $\pi/2 \le \beta \le 5/2\pi$, periodical with period 2π, independent of a, and

$$\frac{d^2\hat{M}(\beta)}{d\beta^2} + \hat{M}(\beta) > 0,$$

then it can be used as the support function for some oval $\hat{\Gamma}$ determined by (3.2.1). The stress function in region $\hat{D}$ is determined by (3.2.17), and in region $F - \hat{D}$, where $\hat{F}$ is a convex region bounded by $\hat{\Gamma}$, by (3.2.11). We thus obtain an exact solution for the problem of elastic–plastic torsion for a rod with cross section $\hat{F} + \hat{\Gamma}$.

This inverse method was proposed, in somewhat different form, by Sokolovsky [24], who also discussed the example given below.

Example. Let

$$\hat{\Psi} = -x^2/2A - y^2/2B - 1/4(A+B) + b, \tag{3.2.18}$$

where constants A and B are determined from equations $1/A + 1/B = 1/a$; $A - B = 4c$. Here, b and c are arbitrary positive constants and $b - 3c > 0$. Curve $\hat{L}$ is an ellipse $x^2/A^2 + y^2/B^2 = 1$, parametrized in the usual way: $x = A\cos\gamma$ and $y = B\sin\gamma$. Further, we find $\hat{\beta}(\gamma) = \gamma + \pi/2$ and $\hat{M}(\gamma) = b + c\cos 2\gamma$. Therefore, $\hat{M}(\beta) = b - c\cos 2\beta$. This support function determines the Sokolovsky's oval (3.2.4).

Some modification of the inverse method is possible by first determining $d\hat{\beta}(\gamma)/d\gamma$ from relation (3.2.16) and from relation (3.2.15), differentiated with respect to γ,

$$\left[-\frac{\partial\hat{\Psi}}{\partial x}\hat{y}(\gamma) + \frac{\partial\hat{\Psi}}{\partial y}\hat{x}(\gamma)\right]\frac{d\hat{\beta}}{d\gamma} = \frac{d}{d\gamma}\left[\hat{\Psi} - \hat{x}(\gamma)\frac{\partial\hat{\Psi}}{\partial x} - \hat{y}(\gamma)\frac{\partial\hat{\Psi}}{\partial y}\right],$$

and then obtaining $\hat{\beta}$ as a function of γ by a quadrature.

5

Since relations (3.2.14) and (3.2.15) are valid on L for function $\psi(x, y)$ in region $D + L$, we have a problem with an unknown boundary, where two conditions are given. Using the Legendre transformation, we reduce this problem to a boundary value problem in a known region.

Assuming $(x, y) \in D + L$, we introduce new independent variables ξ, η and a new unknown function $\phi = \phi(\xi, \eta)$:

$$\xi = -\frac{\partial\Psi(x, y)}{\partial x}, \qquad \eta = -\frac{\partial\Psi(x, y)}{\partial y}, \tag{3.2.19}$$

$$\Phi = -x\frac{\partial\Psi}{\partial x} - y\frac{\partial\Psi}{\partial y} + \Psi. \tag{3.2.20}$$

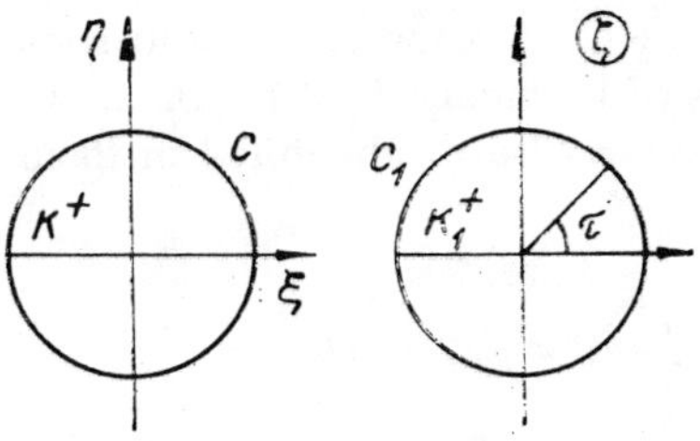

FIG. 3.2.

In this case

$$\frac{\partial^2\Psi}{\partial x^2} = -\Delta\frac{\partial^2\Phi}{\partial \eta^2}, \qquad \frac{\partial^2\Psi}{\partial y^2} = -\Delta\frac{\partial^2\Phi}{\partial \xi^2},$$

$$\frac{\partial^2\Psi}{\partial x\,\partial y} = \Delta\frac{\partial^2\Phi}{\partial \xi\,\partial \eta};$$

$$\Delta \equiv \left[\frac{\partial^2\Phi}{\partial \xi^2}\cdot\frac{\partial^2\Phi}{\partial \eta^2} - \left(\frac{\partial^2\Phi}{\partial \xi\,\partial \eta}\right)^2\right]^{-1}.$$

A one-to-one mapping of region $D+L$ of plane (x, y) onto circle $K^+ + C$: $\xi^2+\eta^2 \le 1$ of plane (ξ, η) can be implemented by using formulas (3.2.19) (Fig. 3.2). Function $\Phi(\xi, \eta)$, which is continuous with respect to $K^+ + C$, satisfies the Monge–Ampere equation in circle K^+,

$$\frac{\partial^2\Phi}{\partial \xi^2}\cdot\frac{\partial^2\Phi}{\partial \eta^2} - \left(\frac{\partial^2\Phi}{\partial \xi\,\partial \eta}\right)^2 = a\left(\frac{\partial^2\Phi}{\partial \xi^2} + \frac{\partial^2\Phi}{\partial \eta^2}\right), \tag{3.2.21}$$

and inequalities

$$\frac{\partial^2\Phi}{\partial \xi^2} > 0, \qquad \frac{\partial^2\Phi}{\partial \eta^2} > 0. \tag{3.2.22}$$

On circle C: $\rho = 1$ the following condition is satisfied:

$$\Phi|_C = M(\theta + \pi/2). \tag{3.2.23}$$

Here, ρ, θ are polar coordinates in plane (ξ, η). If $M(\beta)$ is a function, continuously differentiable four times over $[\pi/2, 5/2\pi]$, we shall suppose, hereafter, that the boundary value problem (3.2.21)–(3.2.23) for function Φ is uniquely solvable. Function $\Phi(\xi, \eta)$ is continuously differentiable twice in $K^+ + C$.

If function Φ is known, equations

$$x = -\frac{\partial\Phi(\xi,\eta)}{\partial\xi}, \qquad y = -\frac{\partial\Phi(\xi,\eta)}{\partial\eta} \tag{3.2.24}$$

accomplish one-to-one mapping of circle $K^+ + C$ of plane (ξ, η) onto some region $D+L$ of plane (x, y). The equations of boundary L of region D are obtained from (3.2.22) when $\rho = 1$ and $\theta = \beta - \pi/2$ are substituted in them, i.e., we assume

$$\xi = \cos\theta = \sin\beta, \qquad \eta = \sin\theta = -\cos\beta, \qquad \pi/2 \le \beta < 5/2\pi.$$

Thus, curve L is determined parametrically in the form

$$x_L(\beta) = \frac{\partial\Phi(1, \beta - \pi/2)}{\partial\rho}\sin\beta + \frac{dM(\beta)}{d\beta}\cos\beta,$$

$$y_L(\beta) = -\frac{\partial\Phi(1, \beta - \pi/2)}{\partial\rho}\cos\beta + \frac{dM(\beta)}{d\beta}\sin\beta, \tag{3.2.25}$$

$$\frac{\pi}{2} \le \beta < \frac{5}{2}\pi.$$

Here $[\partial\Phi(1, \theta)]/\partial\rho$ is a normal derivative on circle C. Equation (3.2.25) can also be presented in form (3.2.12) with the following value of $N(\beta)$:

$$N(\beta) = M(\beta) - \frac{\partial\Phi(1, \beta - \pi/2)}{\partial\rho}. \tag{3.2.26}$$

In this case, for any $\pi/2 \le \beta < 5/2\pi$, the following estimation holds:

$$R_{\min} - 2a \le N(\beta) \le R(\beta) - a. \tag{3.2.27}$$

From (3.2.27) and from inequality (3.2.10), it follows that $N(\beta) \ge 0$ for any $\pi/2 \le \beta < 5/2\pi$. The stress function $\Psi(x, y)$ in region $D+L$ and its first derivatives are equal to

$$\Psi = -\xi\frac{\partial\Phi(\xi, \eta)}{\partial\xi} - \eta\frac{\partial\Phi(\xi, \eta)}{\partial\eta} + \Phi(\xi, \eta),$$

$$\frac{\partial\Psi}{\partial x} = -\xi, \qquad \frac{\partial\Psi}{\partial y} = -\eta. \tag{3.2.28}$$

Hence, the shear stresses in the elastic region, i.e., in $D+L$, are equal to

$$\tau_{xz} = -\tau_s\eta, \qquad \tau_{yz} = \tau_s\xi. \tag{3.2.29}$$

In these formulas, one must substitute expressions obtained from equations (3.2.24) for ξ and η.

Now we shall determine warping $W = W(x, y)$, $(x, y) \in F + \Gamma$. Let us denote warping in the elastic region $D+L$ by W^e and in the plastic region

$L+B+\Gamma$ by W^p. On the elastic–plastic boundary L,

$$W^e = W^p.$$

By integrating equation (3.1.2), we obtain for $(x, y) \in D + L$,

$$W^e(x, y) = \int_{(x_0, y_0)}^{(x,y)} \left(\frac{\tau_{xz}}{G} + \alpha y \right) dx + \left(\frac{\tau_{yz}}{G} - \alpha x \right) dy + W_0. \quad (3.2.30)$$

Here, τ_{xz}, τ_{yz} should be considered as functions of x, y, determined from (3.2.24) and (3.2.29); constant w_0 is the value of warping at some point $P_0 = (x_0, y_0) \in D$, which belongs to region D for all values of α. Such points do exist. If we take, for example, P_0 at the center of curvature of curve Γ, where $R_{\min}$ is realized, then, according to (3.2.27), point P_0 for all values of α will belong to region D. Henceforth, we shall assume that P_0 coincides with the origin of the coordinates. Now we shall determine $W^p(x, y)$ for a known W^e. In region B, formulas (3.2.11) are true and

$$0 \le u \le N(\beta), \qquad \pi/2 \le \beta < 5/2\pi, \quad (3.2.31)$$

where $N(\beta)$ is determined from (3.2.26) and the shear stresses are

$$\tau_{xz} = \tau_s \cos\beta, \qquad \tau_{yz} = \tau_s \sin\beta. \quad (3.2.32)$$

By taking β and u as new independent variables [see (3.2.11)] and assuming $W^p = W^p(\beta, u)$, we obtain from (3.2.6) and (3.2.32):

$$\frac{\partial W^p(\beta, u)}{\partial u} + \alpha \frac{dM(\beta)}{d\beta} = 0,$$
$$0 \le u \le N(\beta), \qquad \pi/2 \le \beta < 5/2\pi. \quad (3.2.33)$$

By integrating (3.2.33) subject to condition

$$W^p(\beta, u)|_{u=N(\beta)} = W^e(x_L(\beta), y_L(\beta)),$$

where $W^e(x_L(\beta), y_L(\beta))$ is the value of warping (3.2.20) at point L corresponding to a given β, we obtain

$$W^p(u, \beta) = W^e(x_L(\beta), y_L(\beta)) - \alpha \frac{dM(\beta)}{d\beta}(u - N(\beta)). \quad (3.2.34)$$

Taking into account equalities (3.2.29) and (3.2.32), we determine the torque relative to the origin from equation (3.1.10)

$$\frac{M_0}{\tau_s} = \int_{\pi/2}^{5\pi/2} N(\beta) \left[M(\beta)R(\beta) - (M(\beta) + R(\beta)) \frac{N(\beta)}{2} + \frac{N^2(\beta)}{3} \right] d\beta$$
$$+ a \int_0^{2\pi} \int_0^1 \frac{\partial \Phi}{\partial \rho} \left(\frac{\partial^2 \Phi}{\partial \rho^2} + \frac{1}{\rho} \frac{\partial \Phi}{\partial \rho} + \frac{1}{\rho^2} \frac{\partial^2 \Phi}{\partial \theta^2} \right) \rho^2 \, d\rho \, d\theta. \quad (3.2.35)$$

Thus, the stresses, warping, and torque in a rod of oval section under torsion with support function $M(\beta)$ for twist angles satisfying inequality (3.2.10) are completely determined through function $\Phi(\xi, \eta)$, which is the solution in circle $K^+ + C$ of equation (3.2.21) with conditions (3.2.22) and (3.2.23).

In conclusion, we should mention that if we put

$$\Phi(\xi, \eta) = \omega(\xi, \eta) + \tfrac{1}{2}a(\xi^2 + \eta^2) - \tfrac{1}{2}a, \tag{3.2.36}$$

function $\omega = \omega(\xi, \eta)$ satisfies, in circle K^+: $\xi^2 + \eta^2 < 1$, the Monge–Ampere equation in its simplest form,

$$\frac{\partial^2\omega}{\partial\xi^2}\frac{\partial^2\omega}{\partial\eta^2} - \left(\frac{\partial^2\omega}{\partial\xi\,\partial\eta}\right)^2 = a^2 \tag{3.2.37}$$

and inequalities

$$\frac{\partial^2\omega}{\partial\xi^2} > 0, \qquad \frac{\partial^2\omega}{d\eta^2} > 0. \tag{3.2.38}$$

On circle C: $\xi^2 + \eta^2 = 1$ the following condition is met:

$$\omega|_C = M(\theta + \pi/2). \tag{3.2.39}$$

§ 3. SOKOLOVSKY'S SOLUTION

1

Let the cross section be an oval whose equation has the form (3.2.4). The support function for this oval is equal to

$$M(\beta) = b - c\cos 2\beta, \qquad \pi/2 \le \beta < 5/2\pi, \tag{3.3.1}$$

where constants b and c satisfy condition

$$c > 0, \qquad R_{\min} = b - 3c > 0.$$

By solving equation (3.2.21) with conditions (3.2.22) and (3.2.23), a quadratic polynominal is obtained:

$$\Phi(\xi, \eta) = A\xi^2/2 + B\eta^2/2 - (A + B)/4 + b, \tag{3.3.2}$$

where

$$A = a + 2c + (a^2 + 4c^2)^{1/2}, \qquad B = a - 2c + (a^2 + 4c^2)^{1/2}, \qquad a = \tau_s/(2G\alpha).$$

From formula (3.2.25), we obtain an equation of the elastic–plastic boundary

$$x_L(\beta) = A\sin\beta, \qquad y_L(\beta) = -B\cos\beta, \qquad \pi/2 \le \beta < 5/2\pi. \tag{3.3.3}$$

Curve L is an ellipse with semiaxes A and B, such that when $\alpha \to \infty$, i.e., $a \to 0$, it degenerates into section $-4c \leq x \leq 4c$ of the x axis. The value of $N(\beta)$, determined from formulas (3.2.26), is equal to

$$N(\beta) = b - a - \sqrt{a^2 + 4c^2} + c\cos 2\beta. \tag{3.3.4}$$

The link between ξ, η and x, y, expressed by formulas (3.2.24), has the form

$$x = A\xi, \qquad y = B\eta. \tag{3.3.5}$$

Function $\Psi = \Psi(x, y)$, $(x, y) \in D + L$, found from (3.2.28), agrees with function $\hat{\Psi}(x, y)$, which is determined by formula (3.2.18).

The shear stresses in the elastic region in accordance with (3.2.29) and (3.3.5) are

$$\tau_{xz} = -\tau_s y/B, \qquad \tau_{yz} = \tau_s x/A. \tag{3.3.6}$$

Warping in the elastic region, computed from (3.2.30) and (3.3.6), is

$$W^e(x, y) = -\alpha(1 - 2aA^{-1})xy + W_0.$$

Accounting for (3.3.3), on L for a given β we have

$$W^e(x_L(\beta), y_L(\beta)) = (c\tau_s/G)\sin 2\beta + W_0. \tag{3.3.7}$$

Taking into consideration (3.3.4) and (3.3.7), we determine warping in the plastic region from (3.2.34).

Finally, the torque is found from (3.2.35):

$$M_0 = \tfrac{2}{3}\pi\tau_s\left(b^3 - \tfrac{9}{2}bc^2 + 4c^3\right) - \tfrac{8}{3}\pi\tau_s c^3$$
$$\times\left[1 + a^3/4c^3 - (1 - a^2/2c^2)(1 + a^2/4c^2)^{1/2}\right]. \tag{3.3.8}$$

Condition (3.2.10), which ensures the existence of an elastic region totally enclosed by the plastic zone, gives the following limit on the twist α:

$$\alpha > \tau_s/G(b - 3c).$$

This is a sufficient condition; it can be defined more accurately in the discussed example. Inequality $A < b + c$, ensuring that ellipse L is inside the oval, gives the following limit on α:

$$\alpha \geq \frac{\tau_s}{G}\,\frac{b - c}{(b + c)(b - 3c)}.$$

Figure 3.3 depicts the relationship between M_0 and α computed from (3.3.8) for $c/b = 0.05$, 0.10, 0.15 [24].

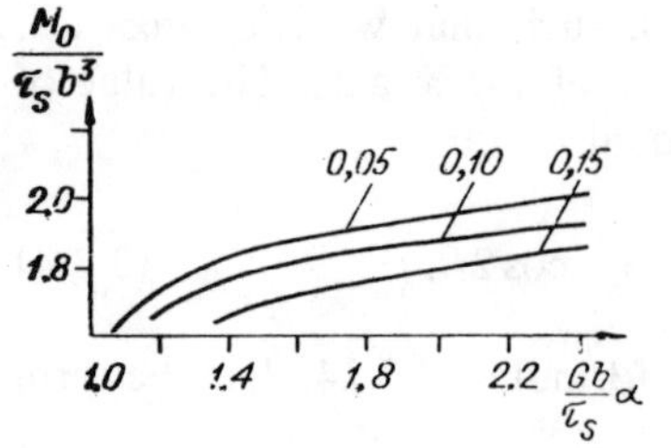

FIG. 3.3.

2

This solution can also be used as an approximate solution of the elastic–plastic problem for contours resembling Sokolovsky's oval.

Let oval Γ have only four extreme points of curvature, be elongated along the x axis, and symmetrical relative to the axes x, y, i.e., its support function $M(\beta)$ satisfies condition

$$M(3\pi/2-\theta)=M(\pi/2-\theta)=M(\pi/2+\theta), \qquad 0\le\theta<2\pi.$$

Let us assume further that $M(\beta)$ is continuously differentiable four times. By expanding it into a Fourier series over interval $[\pi/2, 5/2\pi]$, we obtain

$$M(\beta)=b-c\cos 2\beta+f(\beta),$$

$$b=\frac{1}{2\pi}\int_0^{2\pi}M\left(\theta+\frac{\pi}{2}\right)d\theta, \tag{3.3.9}$$

$$c=\frac{1}{\pi}\int_0^{2\pi}M\left(\theta+\frac{\pi}{2}\right)\cos 2\theta\, d\theta,$$

and $f(\beta)$ is a remainder. We denote the minimum and maximum values of function $f(\beta)$ in the interval $[\pi/2, 5/2\pi]$ by $f_{\min}$ and $f_{\max}$ and assume that

$$c>0, \qquad b-3c>0. \tag{3.3.10}$$

Let, again, $\Phi(\xi,\eta)$ denote the solution of equation (3.2.21) subject to conditions (3.2.22) and (3.2.23). Let us set up the difference

$$\delta(\xi,\eta)=\Phi(\xi,\eta)-\Phi_*(\xi,\eta).$$

Here, $\Phi_*(\xi,\eta)$ is determined by (3.3.2) and (3.3.9). Function $\delta(\xi,\eta)$ satisfies in K^+ a second-order homogeneous linear equation of elliptical type; therefore, based on the properties of the maximum and minimum, we obtain the following inequality for all $(\xi,\eta)\in K^+ + C$:

$$\Phi_*(\xi,\eta)+f_{\min}\le\Phi(\xi,\eta)\le\Phi_*(\xi,\eta)+f_{\max}.$$

If the minimum value of the modulus of function $\Phi_*(\xi,\eta)$ in $K^+ + C$, which is equal to $h=|b-(A+B)/4|$, is much higher than d, which is equal to the largest of the numbers $|f_{\min}|$, $|f_{\max}|$, a similarity of solution of the elastic–plastic problem for oval Γ and for Sokolovsky's oval is to be expected. For a

TABLE III.1

t	1	0.8	0.6	0.5	0.4	0.2
b/p	1	0.90	0.81	0.77	0.73	0.67
c/p	0	0.11	0.20	0.25	0.29	0.38
d/h	0	0.00	0.02	0.04	0.09	0.39

rigorous proof of this statement, one must show that the difference

$$\left| \frac{\partial \Phi(1,\theta)}{\partial \rho} - \frac{\partial \Phi_*(1,\theta)}{\partial \rho} \right|,$$

is small; this difference determines the proximity of the elastic–plastic boundaries.

Example. Oval Γ is an ellipse with semiaxes p and q, for which the support function is equal to

$$M(\beta) = \sqrt{p^2 \sin^2 \beta + q^2 \cos^2 \beta}\,.$$

Table III.1 shows the effect of $t = q/p$ on b/p and c/p, where b and c are computed from equations (3.3.9), and on d/h at $\alpha = \tau_s/(GR_{\min})$, i.e., at $a/p = t^2/2$. The table indicates that even at $t = 0.5$ condition (3.3.10) is met and the difference $|\Phi - \Phi_*|/\Phi_*$ does not exceed 4 percent. It should also be noted that for $t < 0.5$ we can consider that

$$b = (p+q)/2, \qquad c = (p-q)/2.$$

§ 4. ALGORITHM OF THE NUMERICAL SOLUTION FOR OVAL SECTION. EXAMPLES

1

Now we shall present an algorithm of the numerical solution [5, 6] of the boundary value problem (3.2.21)–(3.2.23) which allows to find function $\Phi(\xi,\eta)$ in circle $K^+ + C$. This function completely determines the stress, warping, and torque of an oval with support function $M(\beta)$.

Equation (3.2.21) can be written as

$$\Delta\Phi = 2a + \sqrt{4a^2 + (\Delta\Phi)^2 - 4I\Phi}\,, \qquad (3.4.1)$$

where Δ is the Laplacian and I is the Jacobian,

$$I\Phi = \frac{\partial^2\Phi}{\partial\xi^2} \cdot \frac{\partial^2\Phi}{\partial\eta^2} - \left(\frac{\partial^2\Phi}{\partial\xi\,\partial\eta}\right)^2.$$

When equation (3.2.21) is written in this form, condition (3.2.22) is satisfied automatically. Equation (3.4.1) subject to condition (3.2.23), is solved by the following iterative process. An arbitrary initial approximation $\Phi^0 = \Phi^0(\rho, \theta)$ is assumed. Considering that the nth approximation $\Phi^n = \Phi^n(\rho, \theta)$ is known, we construct the $(n+1)$th approximation $\Phi^{n+1} = \Phi^{n+1}(\rho, \theta)$, $(\rho, 0) \in K^+ + C$, as a solution of the problem

$$\tilde{\Delta}\Phi^{n+1} = f^n, \tag{3.4.2}$$

$$\begin{aligned} f^n &\equiv 2a + \left(4a^2 + (\tilde{\Delta}\Phi^n)^2 - 4\tilde{I}\Phi^n\right)^{1/2}, \\ \Phi^{n+1}(1, \theta) &= M(\theta + \pi/2). \end{aligned} \tag{3.4.3}$$

Here, $\tilde{\Delta}$ and $\tilde{I}$ are difference approximations of the Laplacian and the Jacobian. The process is terminated when

$$\max_{K^+ + C} |\Phi^{N+1} - \Phi^n| < \varepsilon,$$

where ε is a fixed and sufficiently small number.

2

For a numerical implementation of this algorithm, an efficient method of solving the Dirichlet problem for the Poisson equation in a circle is needed. Annin, et al. [5] solved the Poisson equation on a rectangular mesh alternately by the triangular method and by the Arakawa approximation of the Jacobian. The case of an ellipse with semiaxes equal to 4 and 2, for which the support function is equal to

$$M(\beta) = \left(16 \sin^2\beta + 4 \cos^2\beta\right)^{1/2}$$

and $R_{\min} = 1$, was discussed. Condition (3.2.10) defined a limit: $a < 0.5$. A circle of radius 1 having its center at point (1.5; 1.5) was placed in a square (0, 0), (0, 3), (3, 0), (3, 3). The square was covered with a 40×40 mesh, i.e., with a pitch equal to 0.075. The value of ε was equal to 10^{-2}. The iterative process of solving the Dirichlet problem for the Poisson equation was terminated when the maximum difference in the mesh region, approximating a circle, between two successive approximations was less than 10^{-4}. A number of parameter a values were used, particularly 0.45, 0.10, and 0.025. Computer time for a run amounted to 6 min on BESM-6. Figure 3.4 shows a quarter of the ellipse and locations of the elastic–plastic boundary for $a = 0.45$ (curve 1) and $a = 0.10$

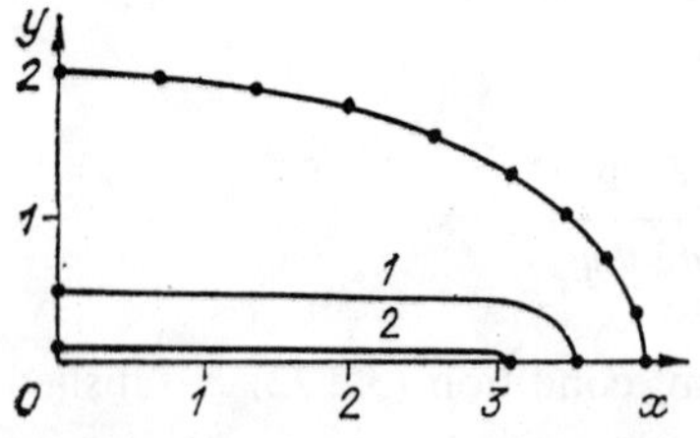

FIG. 3.4.

(curve 2). For $a = 0.025$, the elastic region practically coincides with a section of the x axis, connecting the centers of curvature of those points on the ellipse which are located on the x axis. For comparison, we shall discuss the Sokolovsky's oval with $b = 0.38$, $c = 1$ (see Table III.1). The values of A and B for the elastic–plastic boundaries defined by equations (3.3.3) are equal to

$$A = 4.5, \qquad B = 0.40 \quad \text{for } a = 0.45,$$

$$A = 4.1, \qquad B = 0.10 \quad \text{for } a = 0.10.$$

It is apparent that the Sokolovsky solution gives an unsatisfactory approximation near the x axis and a satisfactory approximation near the y axis.

3

Annin and Sadovsky [6] solved the Dirichlet problem for Poisson's equation on a polar mesh:

$$\rho_k = (k + 0.5)h, \qquad k = 0, 1, \ldots, k^*,$$

$$\theta_l = l \cdot \delta, \qquad l = 0, 1, \ldots, l^*,$$

$$h = (k^* + 0.5)^{-1}, \quad \delta = 2\pi/l^*.$$

Here, k^*, l^* are integers and l^* is even. The following mesh functions were considered for a fixed n:

$$\Phi_{k,l} \equiv \Phi^{n+1}(\rho_k, \theta_l), \qquad f_{k,l} = f^n(\rho_k, \theta_l), \qquad m_l = M(\theta_l + \pi/2).$$

The difference equation of the second-order accuracy for determining $\Phi_{k,l}$, when $f_{k,l}$ and m_l are known, has the form

$$\begin{aligned} &(k+1)(\Phi_{k+1,l} - \Phi_{k,l}) + k(\Phi_{k-1,l} - \Phi_{k,l}) \\ &\qquad + (k+0.5)^{-1}\delta^{-2}(\Phi_{k,l+1} + 2\Phi_{k,l} + \Phi_{k,l-1}) = (k+0.5)h^2 f_{k,l}, \\ &\qquad \Phi_{1,l} - \Phi_{0,l} + 2\delta^{-2}(\Phi_{0,l+1} - 2\Phi_{0,l} + \Phi_{0,l-1}) = 2^{-1}h^2 f_{0,l}, \\ &\qquad \Phi_{k^*l} = m_l, \qquad k = 1, 2, \ldots, k^* - 1, \qquad l = 0, 1, \ldots, l^* - 1. \end{aligned} \tag{3.4.4}$$

We now introduce the following vectors in the l^*-dimensional Euclidean space:

$$\vec{\Phi}_k = (\Phi_{k,0}, \Phi_{k,1}, \ldots, \Phi_{k,l^*-1}),$$

$$\vec{f}_k = (f_{k,0}, f_{k,1}, \ldots, f_{k,l^*-1}),$$

$$\vec{m} = (m_0, m_1, \ldots, m_{l^*-1}), \qquad k = 0, 1, \ldots, k^*,$$

$$\vec{e}_j = \begin{cases} (1, \cos j\delta, \cos 2j\delta, \ldots, \cos j(l^*-1)\delta) & \text{for } j = 0, 1, \ldots, l^*/2, \\ (0, \sin j\delta, \sin 2j\delta, \ldots, \sin j(l^*-1)\delta) & \text{for } j = l^*/2 + 1, \ldots, l^* - 1. \end{cases}$$

Vectors $\vec{\Phi}_k$, $\vec{f}_k$, $\vec{m}$ were expanded in terms of the orthogonal basis $\vec{e}_0, \vec{e}_1, \ldots, \vec{e}_{l^*-1}$:

$$\vec{\Phi}_k = \sum_{j=0}^{l^*-1} \alpha_{j,k}\vec{e}_j; \qquad \vec{f}_k = \sum_{j=0}^{l^*-1} \beta_{j,k}\vec{e}_j, \qquad \vec{m} = \sum_{j=0}^{l^*-1} \gamma_j\vec{e}_j.$$

The expansion coefficients $\alpha_{j,k}$ of vectors Φ_k were determined from the following system obtained from (3.4.4):

$$(k+1)(\alpha_{j,k+1} - \alpha_{j,k}) + k(\alpha_{j,k-1} - \alpha_{j,k})$$

$$+2(k+0.5)^{-1}\delta^{-2}(\cos j\delta - 1)\alpha_{j,k} = (k+0.5)h^2\beta_{j,k},$$

$$(\alpha_{j,1} - \alpha_{j,0}) + 4\delta^2(\cos j\delta - 1)\alpha_{j,0} = 2^{-1}h^2\beta_{j,0},$$

$$\alpha_{j,k^*} = \gamma_j, \qquad k = 1, 2, \ldots, k^*-1, \qquad j = 0, 1, \ldots, l^*-1. \quad (3.4.5)$$

For a fixed j, system (3.4.5) was solved by the elimination method. Discussed was the following example: contour Γ, in polar coordinates r, φ of plane

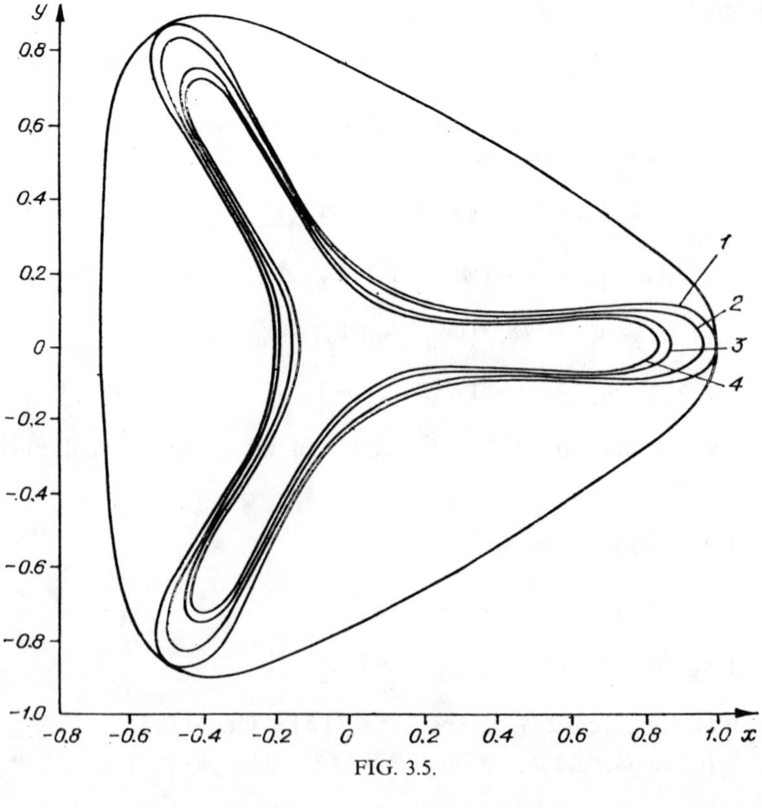

FIG. 3.5.

TABLE III.2

$\alpha G/\tau_s$	3.5	4	5	6
M_0/τ_s	1.066	1.072	1.079	1.082

(x, y) is described by

$$r^2 - \lambda r^s \cos s\varphi = 1 - \lambda.$$

Here, $s \geq 3$ is an integer and λ a parameter. When an appropriate parameter λ is selected, contour Γ is convex and is considered to be a good approximation of a regular s-sided polygon.

The stress function for this rod in pure elastic torsion is equal to [see (3.2.6)]

$$\Psi = \frac{G\alpha}{\tau_s}\left[-2^{-1}r^2 + 2^{-1}\lambda r^s \cos s\varphi + 2^{-1}(1-\lambda)\right].$$

Calculations were conducted for two cases: $s = 3$, $\lambda = 0.36$ and $s = 4$, $\lambda = 0.185$. The support function was computed from formula (3.2.5). It was assumed that $k^* = 20$ and $l^* = 120$. The computer time on BESM-6 mainframe for each case was approximately 8 min.

Figure 3.5 shows contour Γ with parameters $s = 3$ and $\lambda = 0.36$, which resembles a regular triangle. This figure also shows the location of the elastic–plastic boundaries for $a = 0.250$ (curve 1), for $a = 0.167$ (curve 2), for $a = 0.125$ (curve 3), and for $a = 0.100$ (curve 4). Given in Table III.2 are the values of M_0/τ_s calculated from equation (3.2.35). The dependence of M_0/τ_s on $\alpha G/\tau_s$ is illustrated in Fig. 3.6. The values of M_0/τ_s and a, at which the yield point is first reached in the elastic rod, are equal to 0.683 and 0.494,

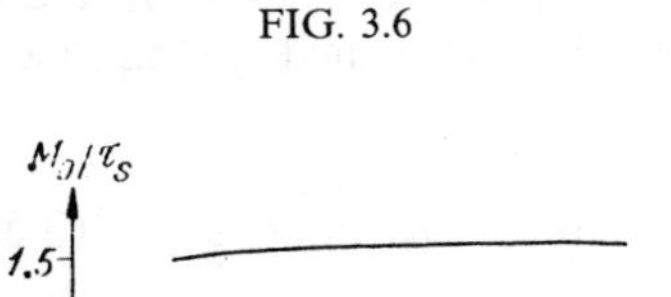

FIG. 3.6

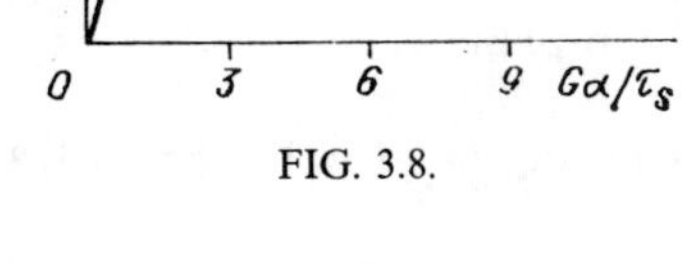

FIG. 3.8.

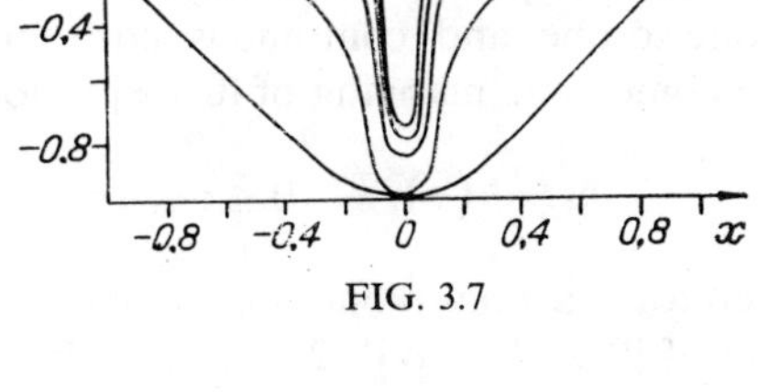

FIG. 3.7

TABLE III.3

$\alpha G/\tau_s$	2	3	4	5
M_0/τ_s	1.500	1.533	1.553	1.562

respectively. Figure 3.7 depicts contour Γ with parameters $s = 4$ and $\lambda = 0.185$, which resembles a square, along with the elastic–plastic boundary for $a = 0.143$ (curve 1), for $a = 0.125$ (curve 2), for $a = 0.100$ (curve 3), and for $a = 0.083$ (curve 4). Given in Table III.3 are the values of M_0/τ_s. The dependence of M_0/τ_s on $\alpha G/\tau_s$ is illustrated in Fig. 3.8. The values of the torque and parameter a at which yield is first reached are equal to 1.102 and 0.537, respectively.

§ 5. GALIN'S INTEGRAL REPRESENTATION

In this section we shall present another approach to the solution of Problem A for a rod with an oval section. This approach is based on the methods of the theory of complex-variable functions.

Formulas (3.2.19), as already mentioned, realize a one-to-one and continuous mapping of an unknown region $D + L$ in plane (x, y) onto a unit circle $K^+ + C$ in plane (ξ, η) (Fig. 3.2). In this mapping, a one-to-one and continuous correspondence between L and C is established. Let us denote the length of arc L by s, and the entire length of curve L by S. The point of reference for the arc length is chosen at $B \in L$, which corresponds to the point of curve C with coordinates $\rho = 1$, $\theta = 0$ for mapping (3.2.19). The correspondence between L and C gives rise to homeomorphic mapping of $[0, S]$ onto $[0, 2\pi]$:

$$\theta = f(s), \qquad 0 \le s \le S, \qquad f(0) = 0, \qquad f(S) = 2\pi. \qquad (3.5.1)$$

Let the function

$$x + iy = w_1(\zeta), \qquad i^2 = -1, \qquad (3.5.2)$$

conformally map a unit circle $K_1^+ + C_1$: $|\zeta| \le 1$ in the complex variable plane ζ onto an unknown region $D + L$, and let τ be the polar angle in plane ζ (Fig. 3.2). The parameters of the conformal mapping are selected such that point $\zeta = 0$ corresponds to some point $A = (x_A, y_A) \in D$, and the point of circle C_1 with coordinates $|\zeta| = 1$, $\tau = 0$ corresponds to point $B \in L$, which is selected as the reference point for arc length s. In this mapping between curves L and C_1, a one-to-one and continuous correspondence is established giving rise to homeomorphic mapping of $[0, 2\pi]$ onto $[0, S]$:

$$s = h(\tau), \qquad 0 \le \tau \le 2\pi, \qquad h(0) = 0, \qquad h(2\pi) = S. \qquad (3.5.3)$$

Therefore, a one-to-one correspondence, giving rise to homeomorphic mapping of $[0, 2\pi]$ onto $[0, 2\pi]$, is established between the points of unit circle C in plane (ξ, η) and the points of unit circle C_1 in plane ζ:

$$\theta = f(h(\tau)) \equiv g(\tau), \qquad 0 \le \tau \le 2\pi, \qquad g(0) = 0, \qquad g(2\pi) = 2\pi. \qquad (3.5.4)$$

Let us introduce in $K_1 + C_1$ an analytical complex-variable function ζ:

$$\left(2a\frac{\partial \Psi}{\partial y} + y\right) + i\left(2a\frac{\partial \Psi}{\partial x} + x\right) = w_2(\zeta).$$

The Cauchy–Riemann conditions are satisfied by virtue of (3.2.6). As a consequence of equalities (3.2.14), (3.5.2), and (3.5.4), the following condition is met on circle C_1: $\zeta = e^{i\tau}$:

$$-\frac{\partial \Psi}{\partial x} = \mathrm{Re}\left[\frac{1}{2a}w_1(\zeta) + \frac{i}{2a}w_2(\zeta)\right]'\Bigg|_{\zeta=e^{i\tau}} = \cos g(\tau),$$
$$-\frac{\partial \Psi}{\partial y} = \mathrm{Re}\left[-\frac{i}{2a}w_1(\zeta) - \frac{1}{2a}w_2(\zeta)\right]\Bigg|_{\zeta=e^{i\tau}} = \sin g(\tau). \tag{3.5.5}$$

Using the Schwarz formula, we find

$$w_1(\zeta) = \frac{a}{2\pi}\int_0^{2\pi} e^{ig(\tau)}\frac{e^{i\tau}+\zeta}{e^{i\tau}-\zeta}d\tau + ca,$$
$$w_2(\zeta) = -\frac{ai}{2\pi}\int_0^{2\pi} e^{-ig(\tau)}\frac{e^{i\tau}+\zeta}{e^{i\tau}-\zeta}d\tau + i\bar{c}a. \tag{3.5.6}$$

Constant c is equal to

$$c \equiv c_1 + ic_2 = \frac{\partial \Psi(x_A, y_A)}{\partial x} + \frac{x_A}{a} + i\left(\frac{\partial \Psi(x_A, y_A)}{\partial y} + \frac{y_A}{a}\right), \tag{3.5.7}$$

and the bar denotes the conjugate value. Assuming that $\zeta = e^{i\sigma}$ in the first of formulas (3.5.6), we obtain a parametric representation of contour L:

$$x_L(\tau) = a\left[\cos g(\tau) + \frac{1}{2\pi}\int_0^{2\pi}\sin g(\sigma)\cot\frac{\sigma-\tau}{2}d\sigma\right] + c_1 a,$$
$$y_L(\tau) = a\left[\sin g(\tau) - \frac{1}{2\pi}\int_0^{2\pi}\cos g(\sigma)\cot\frac{\sigma-\tau}{2}d\sigma\right] + c_2 a, \qquad 0 \le \tau \le 2\pi. \tag{3.5.8}$$

The real constants c_1, c_2 are determined from coordinates of an arbitrary point $A \in D$ and the value of function Ψ at this point. During the transformation, determined by formulas (3.2.19), point A becomes point $E = (\xi_E, \eta_E) \in K^+$. Evidently, it can be assumed that E is an arbitrary point of circle K^+. From equations (3.2.19), (3.2.24), (3.2.36), and (3.5.7) it follows that

$$a(c_1 + ic_2) = \frac{\partial \omega(\xi_E, \eta_E)}{\partial \xi} + i\frac{\partial \omega(\xi_E, \eta_E)}{\partial \eta},$$

where function $\omega(\xi, \eta)$ is determined from (3.2.37)–(3.2.39).

If contour Γ is symmetrical with respect to axes x and y, $M(\pi/2-\theta)=M(\pi/2+\theta)=M(3\pi/2-\theta)$, $0\le\theta\le 2\pi$. Therefore, for all points $(\xi,\eta)\in K^{+}+C$, there exists the relationship $\omega(-\xi,\eta)=\omega(\xi,\eta)=\omega(\xi,-\eta)$. Therefore, at point $\xi=\eta=0$,

$$\frac{\partial\omega(0,0)}{\partial\xi}=\frac{\partial\omega(0,0)}{\partial\eta}=0.$$

Assuming $\xi_E=0$, $\eta_E=0$, we shall have

$$c_1=c_2=0. \tag{3.5.9}$$

In this instance, point A is an inverse image of point $E=(0,0)$ for transformation (3.2.19). Formulas (3.5.6) and (3.5.8) give a parametric representation of curve L and function $\Psi(x,y)$, $(x,y)\in D+L$, in terms of one real variable function $g(\tau)$, $0\le\tau<2\pi$. This representation for $c_1=c_2=0$ was first obtained by Galin [14]. The geometrical meaning of function $g(\tau)$ was presented in [3].

Sometimes it is convenient to represent functions $w_1(\zeta)$ and $w_2(\zeta)$ in terms of a Cauchy type integral. Such a representation, obtained from (3.5.6) by substituting Cauchy kernel for Schwarz kernel, has the form ($t=e^{i\tau}$)

$$w_1(\zeta)=\frac{1}{2\pi i}\int_{c_1}\frac{2ae^{ig(\tau)}}{t-\zeta}dt+d_1,$$

$$w_2(\zeta)=\frac{1}{2\pi i}\int_{c_1}\frac{(-2ai)e^{-ig(\tau)}}{t-\zeta}dt+d_2, \tag{3.5.10}$$

$$d_1=ca-\frac{a}{2\pi}\int_0^{2\pi}e^{ig(\tau)}\,d\tau,\qquad d_2=i\bar{c}a+\frac{a}{2\pi}\int_0^{2\pi}e^{-ig(\tau)}\,d\tau.$$

Let, for example,

$$e^{ig(\tau)}=\gamma_0 t^{n}\prod_{m=1}^{n_1}\frac{1-\bar{\gamma}_m t}{t-\gamma_m}\prod_{m=1}^{n_2}\frac{1-\bar{\delta}_m t}{t-\delta_m},$$

$$\gamma_0=\prod_{m=1}^{n_1}\frac{1-\gamma_m}{1-\bar{\gamma}_m}\prod_{m=1}^{n_2}\frac{1-\delta_m}{1-\bar{\delta}_m},\qquad t=e^{i\tau}. \tag{3.5.11}$$

Here, γ_m, δ_m are constants, which can be equal and $0<|\gamma_m|<1$, $|\delta_m|>1$. Integers n, n_1, n_2 are linked by relation $n+n_2-n_1=1$, which ensures that condition

$$\underset{0\le\tau\le 2\pi}{\Delta}\arg e^{ig(\tau)}=2\pi$$

is satisfied. Thus, when τ varies from 0 to 2π, function $g(\tau)$ also varies from 0 to 2π. By expanding Blaschke's product (3.5.11) into simple fractions in terms of t, we shall have

$$e^{ig(\tau)} = P(t) + \sum_{m=1}^{n_1^*} Q_m\left(\frac{1}{t-\gamma_m}\right) + \sum_{m=1}^{n_2^*} R_m\left(\frac{1}{t-\delta_m}\right). \qquad (3.5.12)$$

$P(t)$, $Q_m(t)$, $R_m(t)$ are polynomials with respect to t. Substituting (3.5.12) in the first of formulas (3.5.10) gives

$$w_1(\zeta) = 2aP(\zeta) + 2a\sum_{m=1}^{n_2^*} R_m\left(\frac{1}{\zeta-\delta_m}\right) + d_1.$$

By substituting the conjugate value (3.5.12) in the second of formulas (3.5.10), we obtain

$$w_2(\zeta) = (-2ai)\sum_{m=1}^{n_1^*} \overline{Q}_m\left(\frac{\zeta}{1-\overline{\gamma}_m\zeta}\right) + d_2 + \overline{P(0)}.$$

§ 6. SINGULAR INTEGRAL EQUATION

By substituting (3.5.5) and (3.5.8) in (3.2.16), we obtain a singular nonlinear integral equation for function $g(\tau)$:

$$\frac{a}{2\pi}\int_0^{2\pi}\cos[g(\sigma)-g(\tau)]\cot\frac{\sigma-\tau}{2}\,d\sigma$$

$$= -\frac{dM(g(\tau)+\pi/2)}{dg} + a[-c_1 + \sin g(\tau) + c_2\cos g(\tau)],$$

$$0 \le \tau \le 2\pi. \qquad (3.6.1)$$

This equation was obtained by Annin [1] for $c_1 = c_2 = 0$ (see also [3]). If, on the other hand, we substitute (3.5.5) and (3.5.8) in (3.2.15) and take into account equality $\Psi = N(\beta)$ valid on L, we obtain

$$N\left(g(\tau)+\frac{\pi}{2}\right) = M\left(g(\tau)+\frac{\pi}{2}\right) - a - ac_1\cos g(\tau) - ac_2\sin g(\tau)$$

$$-\frac{a}{2\pi}\int_0^{2\pi}\sin[g(\sigma)-g(\tau)]\cot\frac{\sigma-\tau}{2}\,d\sigma, \qquad 0 \le \tau \le 2\pi.$$

$$(3.6.2)$$

Let us consider cases for which equation (3.6.1) can be solved.

1. Γ is a circle of radius R_0. By choosing the origin of Cartesian coordinates xy at its center, we shall have the following expression for the support function: $M(\beta) = R_0$. Assuming $c_1 = c_2 = 0$, from (3.6.1) we find $g(\tau) = \tau$, $0 \le \tau \le 2\pi$.

2. Oval Γ is symmetrical with respect to axes x and y and is similar to a circle of radius R_0. Let its support function be equal to

$$M(\beta) = R_0(1 + \varepsilon m(\beta)),$$
$$m(\beta) \equiv \sum_{j=1}^{n} p_j \cos 2j\beta, \qquad \pi/2 \le \beta \le \tfrac{5}{2}\pi. \tag{3.6.3}$$

Here, p_j are constants and ε is perturbation, such that

$$\varepsilon \max_{\pi/2 \le \beta < 5\pi/2} m(\beta) \ll 1,$$

$$\varepsilon \max_{\pi/2 \le \beta < 5\pi/2} \left[m(\beta) + \frac{d^2 m(\beta)}{d\beta^2} \right] \ll 1.$$

We shall consider those values of ε and range of parameter a that allow us to assume $\varepsilon \ll a/R_0$ in this range. Assuming $c_1 = c_2 = 0$, we seek a solution of equation (3.6.1) in the form

$$g(\tau) = \tau + \varepsilon h(\tau), \qquad 0 \le \tau \le 2\pi,$$
$$h(\tau) = \frac{\beta_0}{2} + \sum_{j=1}^{\infty} (\beta_j \cos j\tau + \gamma_j \sin j\tau), \qquad h(0) = h(2\pi) = 0. \tag{3.6.4}$$

Function $h(\tau)$ is assumed to satisfy the Hölder condition. By retaining the first order terms with respect to ε we shall have

$$\cos g(\tau) = \cos\tau - \varepsilon h(\tau)\sin\tau,$$
$$\sin g(\tau) = \sin\tau + \varepsilon h(\tau)\cos\tau,$$
$$\frac{1}{R_0} \cdot \frac{dM(g + \pi/2)}{dg} = \varepsilon \frac{dm(\tau + \pi/2)}{d\tau} \tag{3.6.5}$$
$$= \varepsilon(2p_1 \sin 2\tau - 4p_2 \sin 4\tau - \cdots - 2np_n \sin 2n\tau).$$

By substituting (3.6.4) in (3.6.1) and accounting for

$$\frac{1}{2\pi} \int_0^{2\pi} \cos n\sigma \cot\frac{\sigma - \tau}{2} d\sigma = -\sin n\tau, \qquad n = 0, 1, \ldots,$$

$$\frac{1}{2\pi} \int_0^{2\pi} \sin n\sigma \cot\frac{\sigma - \tau}{2} d\sigma = \cos n\tau, \qquad n = 1, 2, \ldots,$$

we obtain

$$h(\tau) = -\frac{1}{a}\frac{dm(\tau+\pi/2)}{d\tau}$$

$$= \frac{1}{a}(-2p_1 \sin 2\tau + 4p_2 \sin 4\tau + \cdots + 2np_n \sin 2n\tau). \quad (3.6.6)$$

By substituting (3.6.6) and (3.6.5) in (3.5.8), we find an equation for L. Therefore, for a contour Γ resembling a circle and having a parametric equation of the form [see (3.2.1) and (3.6.3)]

$$x_\Gamma(\beta) = R_0\left[\sin\beta + \varepsilon\left(m(\beta)\sin\beta + \frac{dm(\beta)}{d\beta}\cos\beta\right)\right],$$

$$y_\Gamma(\beta) = R_0\left[\cos\beta + \varepsilon\left(-m(\beta)\cos\beta + \frac{dm(\beta)}{d\beta}\sin\beta\right)\right],$$

the elastic–plastic boundary is determined as

$$x_L(\tau) = 2a\cos\tau + \varepsilon[-2p_1\cos 3\tau + 4p_2\cos 5\tau + \cdots + 2np_n\cos(n+1)\tau]R_0,$$

$$y_L(\tau) = 2a\sin\tau + \varepsilon[-2p_1\sin 3\tau + 4p_2\sin 5\tau + \cdots + 2np_n\sin(n+1)\tau]R_0,$$

$$0 \le \tau \le 2\pi.$$

Ignoring the second-order terms ε^2, we find

$$(x_L^2 = y_L^2)^{1/2} = 2a + \varepsilon R_0(-p_1\cos 2\tau + 2p_2\cos 4\tau + \cdots + np_n\cos 2n\tau).$$

By substituting (3.6.5) and (3.6.6) in (3.5.5), we determine functions $w_1(\zeta)$ and $w_2(\zeta)$ in $K_1 + C_1$:

$$w_1(\zeta) = 2a\zeta + \varepsilon[-2p_1\zeta^3 + 4p_2\zeta^5 + \cdots + 2np_n\zeta^{2n+1}]R_0,$$

$$w_2(\zeta) = i\varepsilon[-2p_1\zeta + 4p_2\zeta^3 + \cdots + 2np_n\zeta^{2n-1}]R_0.$$

Therefore, the stress function in the elastic region is completely determined. The stress function in the plastic region is determined by formulas (3.2.11) if function $N = N(\beta)$, which determines the variation limit of parameter u, is known. This function is parametrically found from (3.6.2) and (3.6.4) in the form

$$N = N(\tau) = R_0 + \varepsilon m(\tau + \pi/2)R_0 - x_L(\tau)[\cos\tau - \varepsilon h(\tau)\sin\tau]$$

$$-y_L(\tau)[\sin\tau + \varepsilon h(\tau)\cos\tau],$$

$$\beta = \beta(\tau) = \pi/2 + \tau + \varepsilon h(\tau), \qquad 0 \le \tau \le 2\pi.$$

3. Oval Γ_ε is symmetrical with respect to axes x and y and is similar to oval Γ, which is symmetrical with respect to the same axes and for which the solution of equation (3.6.1) is known. Let the support function of oval Γ_ε be equal to

$$M_\varepsilon(\beta) = M(\beta) + \varepsilon m(\beta),$$

where $M(\beta)$ is the support function of oval Γ, ε is perturbation, and $m(\beta)$ is a periodic function with period 2π, which is continuously differentiable twice. Let us denote the solution of equation (3.6.1) for Γ by $g(\tau)$. We seek a solution of the same equation for Γ_ε in the form

$$g_\varepsilon(\tau) = g(\tau) + \varepsilon h(\tau), \qquad g(0) = 0,\ g(2\pi) = 2\pi,$$

$$h(0) = h(2\pi) = 0, \qquad 0 \le \tau \le 2\pi.$$

Function $h(\tau)$ is supposed to satisfy the Hölder condition. By retaining the first order terms with respect to ε, we have

$$\cos g_\varepsilon(\tau) = \cos g(\tau) - \varepsilon h(\tau)\sin g(\tau),$$

$$\sin g_\varepsilon(\tau) = \sin g(\tau) + \varepsilon h(\tau)\cos g(\tau),$$

$$\frac{dM(g_\varepsilon(\tau) + \pi/2)}{dg_\varepsilon} = \frac{dM(g(\tau) + \pi/2)}{dg} + \varepsilon\left[\frac{dm(g(\tau) + \pi/2)}{dg} + \frac{d^2M}{dg^2}h\right]. \tag{3.6.7}$$

By substituting (3.6.7) in (3.6.1), we obtain the following second order Fredholm linear integral equation with respect to function $h(\tau)$:

$$-h(\tau)P(\tau) + \frac{1}{2\pi}\int_0^{2\pi} Q(\sigma, \tau)h(\sigma)\, d\sigma = R(\tau), \tag{3.6.8}$$

$$P(\tau) \equiv \frac{1}{2\pi}\int_0^{2\pi} \sin[g(\sigma) - g(\tau)]\, d\sigma + \frac{1}{a}\frac{d^2M(g(\tau) + \pi/2)}{dg^2},$$

$$Q(\sigma, \tau) \equiv \frac{\sin[g(\sigma) - g(\tau)]}{\sin(\sigma - \tau)/2}\cos\frac{\sigma - \tau}{2},$$

$$R(\tau) \equiv \frac{1}{a}\frac{dm(g(\tau) + \pi/2)}{dg}.$$

If we assume that a finite derivative $[dg(\tau)]/d\tau$ exists in the range $[0, 2\pi]$,

$$\lim_{\sigma \to \tau} Q(\sigma, \tau) = 2\frac{dg(\tau)}{d\tau}.$$

An approximate solution of equation (3.6.8) can be obtained by using efficient numerical methods.

§ 7. GALIN'S METHOD FOR POLYGONAL SECTION

1

Let us consider elastic–plastic torsion of a rod with a convex polygonal cross section Ω_n, bounded by the n-sided polygon Γ_n: $A_0 A_1 \cdots A_{\nu-1} A_\nu \cdots A_n$ (Fig. 3.9), where integer $n \geq 3$ and A_n coincides with A_0. The surface of plastic stresses $\Psi_p = \Psi_p(x, y)$ is a surface with a constant 45-deg slope passing through contour Γ_n. It is comprised of surfaces separated by edges. We shall now determine this surface. Let the normal equation of a polygon side $A_{\nu-1}A_\nu$ be

$$a_\nu x + b_\nu y - p_\nu = 0, \qquad \nu = 1, 2, \ldots n, \qquad a_\nu = \cos\theta_\nu, \qquad b_\nu = \sin\theta_\nu, \tag{3.7.1}$$

where θ_ν is an angle formed by the external normal to side $A_{\nu-1}A_\nu$ with the x axis. Hence, the plane generating surface $\Psi_p(x, y)$ and passing through $A_{\nu-1}A_\nu$ by virtue of (3.1.8) and (3.7.1) is defined by equation

$$\Psi_p = -(a_\nu x + b_\nu y - p_\nu). \tag{3.7.2}$$

Let us find an equation which describes the projection of the edge of surface Ψ_p passing through point $A_{\nu-1}$. The equation of surface Ψ_p, passing through side $A_{\nu-2}A_{\nu-1}$, has the form (3.7.2) with subscript $\nu - 1$. Therefore, the projection of the edge, passing through point $A_{\nu-1}$, will be

$$(a_\nu - a_{\nu-1})x + (b_\nu - b_{\nu-1})y - (p_\nu - p_{\nu-1}) = 0. \tag{3.7.3}$$

Stresses in the plastic region, corresponding to side $A_{\nu-1}A_\nu$, are equal to

$$\tau_{xz} = -\tau_s b_\nu, \qquad \tau_{yz} = \tau_s a_\nu. \tag{3.7.4}$$

Stresses τ_{xz} and τ_{yz} in the elastic zone can be expressed in terms of one harmonic function φ, which is proportional to warping W [see (3.1.2)]:

$$\tau_{xz} = G\alpha\left(\frac{\partial\varphi}{\partial x} - y\right), \qquad \tau_{yz} = G\alpha\left(\frac{\partial\varphi}{\partial y} + x\right). \tag{3.7.5}$$

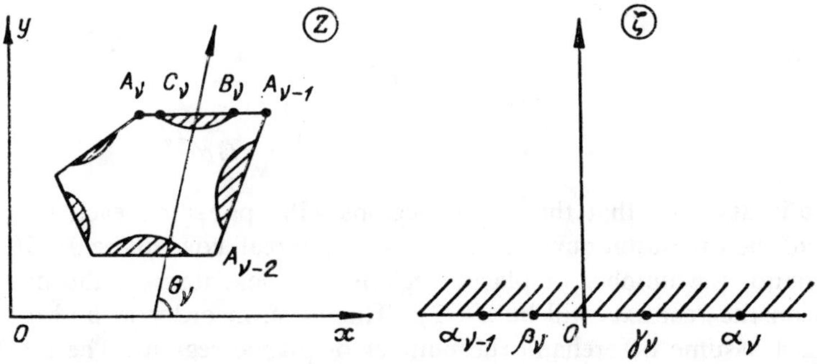

FIG. 3.9.

The boundaries of the elastic zone on sections $A_{\nu-1}B_\nu$ and $C_\nu A_\nu$ (Fig. 3.9) coincide with the external contour of a prismatic rod. Along these sections we have

$$\tau_{xz}\cos(\widehat{nx}) + \tau_{yz}\cos(\widehat{ny}) = 0,$$

$$\cos(\widehat{nx}) = \cos\theta_\nu, \qquad \cos(\widehat{ny}) = \sin\theta_\nu.$$

Accounting for expression (3.7.5), we obtain condition

$$a_\nu \frac{\partial \varphi}{\partial x} + b_\nu \frac{\partial \pi}{\partial y} + b_\nu x - a_\nu y = 0. \tag{3.7.6}$$

Due to the continuity of stresses τ_{xz} and τ_{yz} on the boundary between the elastic and plastic zones, we shall have

$$\tau_{xz} = -\tau_s b_\nu, \qquad \tau_{yz} = \tau_s a_\nu, \tag{3.7.7}$$

on the boundaries of the elastic core bordering the plastic deformation regions, which correspond to the side of the polygon having subscript ν.

Let us discuss two analytical functions

$$\begin{aligned} w_1 &= z^* = x + iy = u_1 + iv_1, \\ w_2 &= \frac{\partial \varphi}{\partial x} - i\frac{\partial \varphi}{\partial y} = u_2 + iv_2. \end{aligned} \tag{3.7.8}$$

We shall consider them as functions of a parametric complex variable ζ. Function $z^* = w_1(\zeta)$ performs conformal mapping of a half plane onto the elastic region; points $\alpha_{\nu-1}$, β_ν, γ_ν, and α_ν of the real axis (Fig. 3.9) correspond to $A_{\nu-1}$, B_ν, C_ν, A_ν. Point $\zeta = \infty$, in this instance, corresponds to A_0. The boundary conditions for determination of $w_1(\zeta)$ and $w_2(\zeta)$ on the real axis $\operatorname{Im}\zeta = 0$ will be as follows: On sections $\alpha_{\nu-1}\beta_\nu$ and $\gamma_\nu\alpha_\nu$,

$$\begin{aligned} a_\nu u_2 - b_\nu v_2 + b_\nu u_1 - a_\nu v_1 &= 0, \\ a_\nu u_1 + b_\nu v_1 &= p_\nu; \end{aligned} \tag{3.7.9}$$

on section $\beta_\nu\gamma_\nu$

$$v_1 - u_2 = \frac{\tau_s}{G\alpha} b_\nu, \qquad -v_2 + u_1 = \frac{\tau_s}{G\alpha} a_\nu. \tag{3.7.10}$$

If it is assumed that the plastic regions will appear on each side of the polygon, then the total number of points on the real axis will be $3n$. If, on the other hand, the number of plastic regions p is less than n, the number of points on the real axis will be $n + 2p$. Therefore, in order to find a solution one must assume beforehand the number of plastic regions. The location of points $\alpha_{\nu-1}$, β_ν, γ_ν, and α_ν shall be determined by using conditions [12], which follow from (3.7.3) and (3.7.4):

$$[(a_\nu - a_{\nu-1})u_1 + (b_\nu - b_{\nu-1})v_1 - (p_\nu - p_{\nu-1})]_{\zeta=\alpha_{\nu-1}} = 0,$$

$$[(a_{\nu+1} - a_\nu)u_1 + (b_{\nu+1} - b_\nu)v_1 - (p_{\nu-1} - p_\nu)]_{\zeta=\alpha_\nu} = 0,$$

$$(-b_\nu u_2 + b_\nu v_1 - a_\nu v_2 + a_\nu u_1)_{\zeta=\beta_\nu-0} = \tau_s/G\alpha,$$

$$(-b_\nu u_2 + b_\nu v_1 - a_\nu v_2 + a_\nu u_1)_{\zeta=\gamma_\nu+0} = \tau_s/G\alpha.$$

Up to now, we have assumed that part of the elastic zone contour coincides with the external boundary. This assumption was confirmed experimentally as well as theoretically [12].

Let us introduce functions [12]

$$W_1(\zeta) = \frac{dw_1(\zeta)}{d\zeta} = U_1 + iV_1,$$

$$W_2(\zeta) = \frac{dw_2(\zeta)}{d\zeta} = U_2 + iV_2.$$

Conditions (3.7.9) and (3.7.10) can be used to determine the introduced functions. After differentiating with respect to x, we have on sections $\alpha_{\nu-1}\beta_\nu$ and $\gamma_\nu\alpha_\nu$,

$$a_\nu U_2 - b_\nu V_2 + b_\nu U_1 - a_\nu V_1 = 0, \qquad a_\nu U_1 + b_\nu V_1 = 0,$$

and on section $\beta_\nu\gamma_\nu$,

$$U_2 - V_1 = 0, \qquad -V_2 + U_1 = 0.$$

Since $a_\nu + ib_\nu = \exp(i\theta_\nu)$, the boundary conditions for determination of functions $W_1(\zeta)$ and $W_2(\zeta)$ can finally be written in the following form: On sections $\alpha_{\nu-1}\beta_\nu$ and $\gamma_\nu\alpha_\nu$,

$$\mathrm{Im}[-\exp(i\theta_\nu)W_1 + i\exp(i\theta_\nu)W_2] = 0,$$

$$\mathrm{Im}[i\exp(-i\theta_\nu)W_1] = 0,$$

and on section $\beta_\nu\gamma_\nu$,

$$\mathrm{Im}[-W_1 - iW_2] = 0, \qquad \mathrm{Im}[iW_1 - W_2] = 0.$$

By assigning values from 1 to n to subscript ν, we obtain the conditions on the entire real axis.

The stated problem can be reduced to a search of a solution of a Fuchsian differential equation. The linear combinations of the linearly independent integrals are the following functions:

$$W_1(\zeta) = MW_3(\zeta) + NW_4(\zeta),$$

$$W_2(\zeta) = PW_3(\zeta) + QW_4(\zeta).$$

The Fuchsian differential equation has the form

$$W'' + \left\{ \sum_{k=1}^{3n-1} \frac{1 - \left(\rho_1^{(k)} + \rho_2^{(k)}\right)}{\zeta - h_k} - \sum_{k=1}^{r} \frac{1}{\zeta + l_k} \right\} W'$$

$$+ \frac{Q_{3n+r-3}(\zeta)}{\prod_{k=1}^{3n-1} (\zeta - h_k) \prod_{k=1}^{r} (\zeta - l_k)} W = 0. \tag{3.7.11}$$

Points h_k stand for

$$h_1 = \alpha_1, \qquad h_2 = \beta_1, \qquad h_3 = \gamma_1, \qquad h_4 = \alpha_2, \ldots;$$

points l_k correspond to the position of crack tips on the locus of ratio W_1/W_2; the prime denotes differentiation with respect to ζ; Q_{3n+r-3} denotes a polynomial to power $3n+r-3$ containing $3n+r-3$ constants, which must be determined. Constants ρ_1^k, ρ_2^k are determined in terms of h_k, l_k, and constants of polynomial Q_{3n+r-3}.

Following determination of $W_1(\zeta)$ and $W_2(\zeta)$, functions $w_1(\zeta)$ and $w_2(\zeta)$ are easily found. Function $w_1(\zeta)$ defines the shape of the elastic zone.

2

Let us present Galin's approximate solution for a convex polygonal section. This solution is obtained on the assumption that sections $A_{\nu-1}B_\nu$ and $C_\nu A_\nu$ (Fig. 3.9) are negligibly small and that condition (3.7.10) prevails on the entire real axis of plane ζ. We introduce functions

$$w_3(\zeta) = w_2(\zeta) + iw_1(\zeta), \qquad w_4(\zeta) = iw_2(\zeta) + w_1(\zeta). \tag{3.7.12}$$

For their solution, we consider the following problem. The real axis of plane ζ is divided by points $\alpha_1, \alpha_2, \ldots, \alpha_{n-1}$, which are to be determined, into intervals $(-\infty, \alpha), (\alpha_1, \alpha_2), \ldots, (\alpha_{n-1}, \infty)$. On intervals $(\alpha_{\nu-1}, \alpha_\nu)$, $\nu = 2, 3, \ldots, n-1$, the following conditions, which follow from equality (3.7.10), are met:

$$\begin{aligned} \operatorname{Re} w_3(\zeta) &= -\tau_s b_\nu / G\alpha \equiv \tau_{xz}^{(\nu)}, \\ \operatorname{Re} w_4(\zeta) &= \tau_s a_\nu / G\alpha \equiv \tau_{yz}^{(\nu)}, \end{aligned} \tag{3.7.13}$$

On intervals $(-\infty, \alpha_1)$ and (α_{n-1}, ∞), conditions (3.7.13) are satisfied for $\nu = 1$ and $\nu = n$, respectively. By using the Schwarz formula for a half plane, we find

$$w_3(\zeta) = \frac{1}{\pi i} \sum_{\nu=2}^{-n} \left(\tau_{xz}^{(\nu-1)} - \tau_{xz}^{(\nu)}\right) \ln(\alpha_{\nu-1} - \zeta) + \tau_{xz}^{(n)} + ic_1,$$

$$w_4(\zeta) = \frac{1}{\pi i} \sum_{\nu=2}^{n} \left(\tau_{yz}^{(\nu-1)} - \tau_{yz}^{(\nu)}\right) \ln(\alpha_{\nu-1} - \zeta) + \tau_{yz}^{(n)} + ic_2. \tag{3.7.14}$$

Here, c_1, c_2 are arbitrary real constants. From expressions (3.7.14) and (3.7.12) we find function

$$w_1(\zeta) = (w_4(\zeta) - iw_3(\zeta))/2, \tag{3.7.15}$$

which maps the upper half plane onto some region Ω, bounded by Γ. Since $w_1(\zeta)$ is unlimited, region Ω is infinite so that its boundary has asymptotes. The mapping function is dependent on $(n+1)$ constants $\alpha_1, \alpha_2, \ldots, \alpha_{n-1}, c_1, c_2$, which are determined from the following additional conditions of geometrical and mechanical character. Naturally, it is required that the asymptotes of curve Γ coincide with the projections on plane xy of the edges of surface $\Psi_p = \Psi_p(x, y)$, coming from the vertices of polygon Γ_n. If contour Γ_n is symmetrical with respect to axes x and y, the stresses at point $x = y = 0$ are equal to zero.

After finding all constants, functions $w_3(\zeta)$, $w_4(\zeta)$ and region Ω, bounded by curve Γ, are totally determined.

Let us take part of boundary Γ, lying inside region Ω_n, as the elastic–plastic boundary. The values of stresses in the elastic region $\Omega_n \cap \Omega$ are determined from function

$$w_2(\zeta) = (w_3(\zeta) - iw_4(\zeta))/2,$$

and in the plastic region from function $\Psi_p(x, y)$. In this case, the ratio $\text{mes}(\Omega \cap \Omega)/(\text{mes}\,\Omega_n)$ must be sufficiently small. The obtained solution will not satisfy the boundary condition (3.7.6) on the sides of Γ_n near the polygon vertices. A technique allowing satisfaction of condition (3.7.6) on a contour with rounded angles, close to contour Γ_n, was given by Galin [12].

It is important to note that mapping the elastic core onto the inside of a unit circle in the plane of complex variable ω is more convenient for determination of the constants compared to mapping onto a half plane. It can be done by assuming

$$\zeta = i\frac{1+\omega}{1-\omega}, \qquad \alpha_\nu = i\frac{1+\varepsilon_\nu}{1-\varepsilon_\nu}, \qquad \varepsilon_\nu = \exp(i\sigma_\nu), \tag{3.7.16}$$

$$\nu = 1, 2, \ldots, n-1.$$

Here, $\sigma_1 < \sigma_2 < \cdots < \sigma_{n-1}$ are constants which are sought instead of

$\alpha_1, \alpha_2, \dots, \alpha_{n-1}$. After substituting (3.7.16) into (3.7.14), we find

$$
\begin{aligned}
w_3(\omega) &= \sum_{\nu=2}^{n} \left(\tau_{xz}^{(\nu-1)} - \tau_{xz}^{(\nu)}\right) \ln \frac{\varepsilon_{\nu-1} - \omega}{1-\omega} + c_1^*, \\
w_4(\omega) &= \sum_{\nu=2}^{n} \left(\tau_{yz}^{(\nu-1)} - \tau_{yz}^{(\nu)}\right) \ln \frac{\varepsilon_{\nu-1} - \omega}{1-\omega} + c_2^*,
\end{aligned}
\tag{3.7.17}
$$

where c_1^*, c_2^* are constants.

Now, from equation (3.7.15) we find function $w_1 = w_1(\omega)$, which maps the interior of the circle onto region Ω. Points of the unit circle $1, \varepsilon_1, \dots, \varepsilon_{n-1}$ correspond to points $-\infty, \alpha_1, \alpha_2, \dots, \alpha_{n-1}$ of the real axis of plane ζ.

Let us discuss some examples:

a. Γ_n is a regular polygon inscribed in a circle of radius R. Let the origin of xy coordinates be at the center of the polygon and let axis x be directed perpendicular to a polygon side. The angle between the x axis and the normal to a side with subscript ν is equal to $\theta_\nu = 2\pi(\nu-1)/n$, $\nu = 1, 2, \dots, n$. The values of $\tau_{xz}^{(\nu)}$ and $\tau_{yz}^{(\nu)}$ are

$$
\tau_{xz}^{(\nu)} = -\frac{\tau_s}{G\alpha} \sin \frac{2\pi(\nu-1)}{n}, \qquad \tau_{yz}^{(\nu)} = \frac{\tau_s}{G\alpha} \cos \frac{2\pi}{n}(\nu-1).
$$

Constants σ_ν are selected from symmetry in the form

$$
\sigma_\nu = \frac{2\pi}{n}\nu, \qquad \nu = 1, 2, \dots, n-1,
$$

and constants c_1^* and c_2^* are selected such that $w_3(0) = w_4(0) = 0$.

Figure 3.10 shows, for a regular triangle ($n = 3$), how the plastic zone develops with increasing angle of twist [13].

b. Γ_n is a rectangle having sides $a < b$. The origin of xy coordinates is located at the center of the rectangle and the x axis is perpendicular to a side of length a. Thus, $\theta_\nu = \pi(\nu-1)/2$, $\nu = 1, 2, 3, 4$; $p_1 = p_3 = a/2$, $p_2 = p_4 = b/2$. From symmetry we select $\sigma_1 = \varkappa$, $\sigma_2 = \pi$, and $\sigma_3 = \pi + \varkappa$, where parameter $\varkappa$ is subject to determination. Constants c_1^* and c_2^* are chosen such that $w_3(0) =$

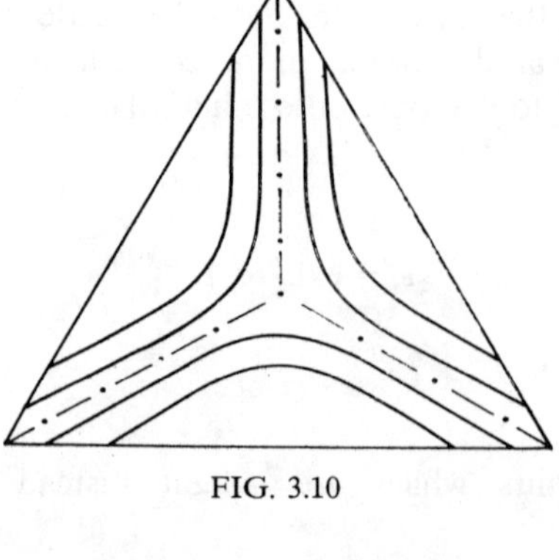

FIG. 3.10

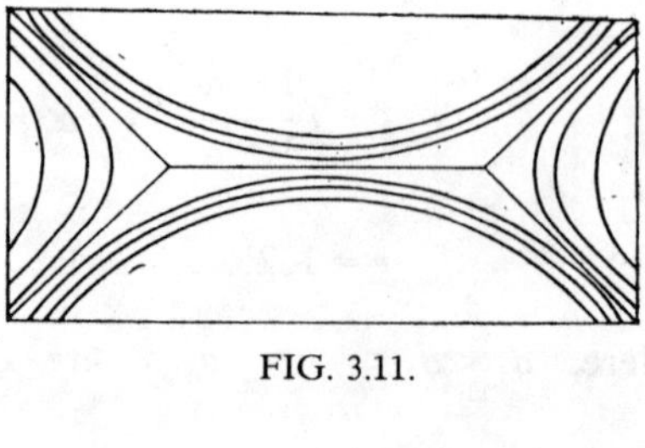

FIG. 3.11.

$w_4(0) = 0$. Therefore, we have

$$w_3(\omega) = \frac{\tau_s}{G\alpha}\left[\ln\frac{1-\omega}{1+\omega} + \ln\frac{e^{i\varkappa}-\omega}{e^{i\varkappa}+\omega}\right],$$

$$w_4(\omega) = \frac{\tau_s}{G\alpha}\left[-\ln\frac{1-\omega}{1+\omega} + \ln\frac{e^{i\varkappa}-\omega}{e^{i\varkappa}+\omega}\right].$$

Consequently, the mapping function is equal to

$$w_2(\omega) = \frac{\tau_s}{2G\alpha}\left[(i-1)\ln\frac{1-\omega}{1+\omega} + (i+1)\ln\frac{e^{i\varkappa}-\omega}{e^{i\varkappa}+\omega}\right].$$

The value $\varkappa = \arctan[(b-a)\pi G\alpha/2\tau_s]$ is determined from the condition of coincidence of projections of the crests with the asymptotes. Figure 3.11 depicts the progress of the plastic zone with increasing angle of twist [13].

§ 8. INITIAL DEVELOPMENT OF THE PLASTIC ZONES FOR RECTANGULAR SECTION

Let us present, in accordance with Annin and Sadovsky [7], an approximate solution of the torsion problem for a rod of rectangular section D with sides $a < b$ when the plastic zones develop only near one pair of sides (Fig. 3.12).

Let α_* and α^* be the values of twist at which plasticity condition (3.1.5) is reached at points $(\pm a/2, 0)$ and points $(\pm b/2, 0)$, respectively. It is known [9] that $\alpha_* = \tau_s(2G\alpha\eta)^{-1}$, with the dimensionless parameter η changing from 0.339 to 0.500 when b/a changes from 1 to infinity. The elastic–plastic boundary for $\alpha_* < \alpha < \alpha_*$ is comprised of two symmetrical, with respect to axis Ox, curves Γ^+ and Γ^-, which are symmetrical with respect to axis Oy (Fig. 3.12). Any straight line parallel to axis Ox intersects each of the curves Γ^+ and Γ^- at one point only.

Let us denote by B the open region defined as the difference of rectangle D and the regions bounded by Γ^+ and $x = a/2$, Γ^- and $x = -a/2$. Region B in Fig. 3.12 is shaded. Let us denote by Q the difference between regions D

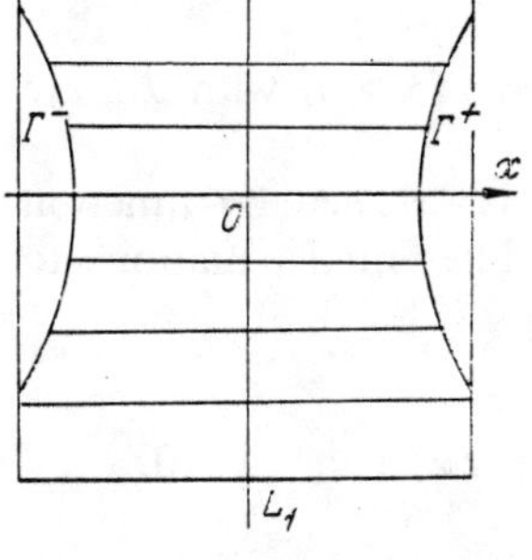

FIG. 3.12.

and B. Let L_1 and L_2 be the common parts of boundaries D and B. The elastic–plastic torsion problem under consideration becomes a matter of seeking curves Γ^+, Γ^- and function $\Psi = \Psi(x, y)$, which is determined and continuously differentiable on $B + L_1 + L_2 + \Gamma^+ + \Gamma^-$ and has second derivatives on B, according to the following conditions:

$$\left.\begin{aligned} &\frac{\partial^2\Psi}{\partial x^2} + \frac{\partial^2\Psi}{\partial y^2} = -2\lambda, \\ &\left(\frac{\partial\Psi}{\partial x}\right)^2 + \left(\frac{\partial\Psi}{\partial y}\right)^2 < 1 \end{aligned}\right\} \text{ in } B, \tag{3.8.1}$$

$$\lambda = G\alpha/\tau_s,$$

$$\begin{aligned} &\Psi = \frac{a}{2} - x, \qquad \frac{\partial\Psi}{\partial x} = -1 \quad \text{on } \Gamma^+, \\ &\Psi = \frac{a}{2} + x, \qquad \frac{\partial\Psi}{\partial x} = 1 \quad \text{on } \Gamma^-. \end{aligned} \tag{3.8.2}$$

The torque in this case is

$$M^{\vee} \equiv \frac{M}{\tau_s} = 2\iint_B \Psi\, dx\, dy + 2\iint_Q \left(\frac{a}{2} + x \operatorname{sign} x\right) dx\, dy. \tag{3.8.3}$$

It should be noted that condition $\Psi = a/2 \mp x$ is used in (3.8.2) instead of condition $\partial\Psi/\partial y = 0$ on Γ^+ and Γ^-.

We shall solve problems (3.8.1) and (3.8.2) using the method of straight lines. Let us define the harmonic function $u = u(x, y)$ on B in the form

$$u(x, y) = \Psi(x, y) + \lambda(y^2 - b^2/4). \tag{3.8.4}$$

Due to the symmetry of the problem,

$$u(x, y) = u(-x, y) = u(x, -y).$$

Let integer n be an odd number and

$$y_k = -b/2 + kh, \qquad k = 1, 2, \ldots, n,$$

$$h = b(n+1)^{-1}, \qquad u_k(x) \equiv (x, y_k).$$

The abscissas of the intersection points of rays $y = y_k$ $(x > 0)$ with L_1 are denoted by x_k.

In accordance with the Laplace equation, which is satisfied by function $u(x, y)$, we have the following system of ordinary differential equations to determine $u_k(x)$:

$$\frac{5}{6}\frac{d^2u_k}{dx^2} + \frac{1}{12}\left[\frac{d^2u_{k+1}}{dx^2} + \frac{d^2u_{k-1}}{dx^2}\right] + \frac{1}{h^2}[u_{k+1} - 2u_k + u_{k-1}] = 0.$$

By solving this system, we obtain

$$u_k(x) = \sum_{s=1}^{n} (-1)^{k+s} C_s \cosh(\beta_s x) \sin k\theta_s,$$

$$k = 1,2,\ldots,n, \qquad 0 \le x \le x_k \le a/2, \qquad C_s = \text{const}, \tag{3.8.5}$$

$$\beta_s^2 = 12(1+\cos\theta_s)h^{-2}(5-\cos\theta_s)^{-1}, \qquad \theta_s = \pi s(n+1)^{-1}.$$

For determination of constants C_k and x_k from (3.8.2) and (3.8.4), we obtain the following system of equations:

$$u_k(x_k) = \lambda\left(y_k^2 - \frac{b^2}{4}\right) + \frac{a}{2} - x_k, \tag{3.8.6}$$

$$\frac{du_k(x_k)}{dx} = -1, \quad \text{if } x_k < a/2, \tag{3.8.7}$$

and condition

$$\frac{du_k(x_k)}{dx} > -1 \quad \text{for } 0 \le x \le x_k \le a/2. \tag{3.8.8}$$

Equations (3.8.6) and (3.8.7) are solved by iterations. The superscript $i = 0,1,\ldots$ will denote the consecutive number of iteration. Assume $x_k^0 = a/2$, $k = 1,2,\ldots,n$. Let the values of x_k^i be known. By substituting them in (3.8.5) and (3.8.6) and solving the linear equations, we find C_s^i. Next we determine x_k^{i+1} using (3.8.7). The process comes to an end when

$$\max_k |x_k^{I+1} - x_k^I| < a\varepsilon \quad \text{for } i = I,$$

where ε is sufficiently small. We assume $x_k = x_k^{I+1}$ and $C_k = C_k^I$, and from formulas (3.8.4) and (3.8.5) we find approximate values of functions Ψ in B.

The computations on BESM-6 mainframe were conducted in the following manner. As a first step of the iterative process, $\lambda = 1$ was assumed and λ_* found from equation

$$\lambda_* = -\left[\frac{du_k(a/2)}{dx}\right]^{-1} \quad \text{for } k = \frac{n+1}{2}. \tag{3.8.9}$$

Next, for $\lambda = \lambda_* + l\delta$, $l = 0,1,\ldots,p-1$, approximate values of x_k and $M^\vee$ were determined by the algorithm. Numerical integration with respect to variable x in (3.8.3) was conducted by the Simpson rule. The system of linear equations was solved by the Gaussian method and the roots of transcendental equation (3.8.7) were found by the chord method. If a root of the equation did not exist for $0 \le x \le a/2$, then it was assumed that $x_k = a/2$. A specific value was given to δ and p was determined from the condition that quantity

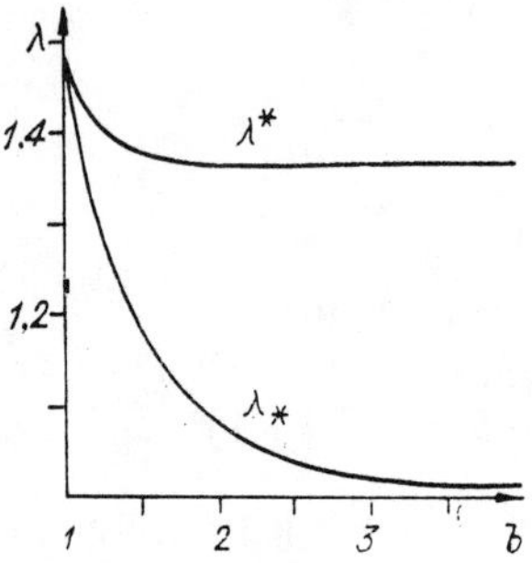

FIG. 3.13

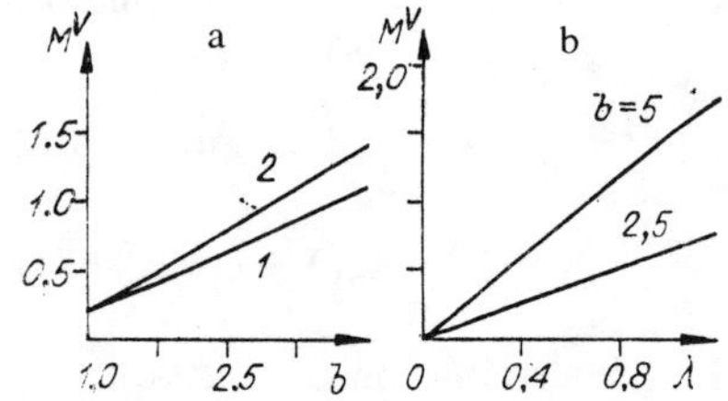

FIG. 3.14.

$T(\lambda) - 1$ changed its sign between $\lambda_* + (p-1)\delta$ and $\lambda_* + p\delta$. Here, $T(\lambda)$ is the difference approximation of the derivative at point $(0, -b/2)$, which was found in accordance with formula (3.8.4) from known $u_1(0), u_2(0), \ldots, u_m(0)$ with $m < n$. Finally, using the chord method, λ^* was obtained from equation $T(\lambda) = 1$ along with a corresponding value of $M^{\vee *}$.

Here are some results for $a = 1$, $1 \le b \le 5$, $m = 6$, $h = 0.005$, $\varepsilon = 10^{-5}$: The run time on BESM-6 did not exceed 5 min for the preceding parameters. Numerically obtained were relations between b and both λ_* and λ^* (Fig. 3.13), $M_*^{\vee}$ (curve 1) and $M^{\vee *}$ (curve 2) (Fig. 3.14a), and f^* (Fig. 3.15). Here and in the following $f = a/2 - x_k$ at $k = (n+1)/2$; the asterisks indicate that the values are taken at $\lambda = \lambda_*$ and λ^*, respectively. Graph $M^{\vee} = M^{\vee}(\lambda)$ for $b = 2$ and $b = 5$ is depicted in Fig. 3.14b. When analyzing the dependence $f = f(\lambda)$ for sufficiently large b (in Fig. 3.16, $b = 5$). Sokolovsky's analytical solution for $b = \infty$ can be used with a high degree of accuracy. This solution has the form

$$\Psi(x, y) = \begin{cases} \Psi_p(x) & \text{on } -a/2 \le x \le -x_k \text{ and } x_k \le x \le a/2, \\ -\lambda x^2 + a/2 - (4\lambda)^{-1} & \text{on } -a/2 \le x \le a/2. \end{cases}$$

Here

$$\Psi_p(x) = -|x| + a/2 \quad \text{for } -a/2 \le x \le a/2; \; x_k = (2\lambda)^{-1}.$$

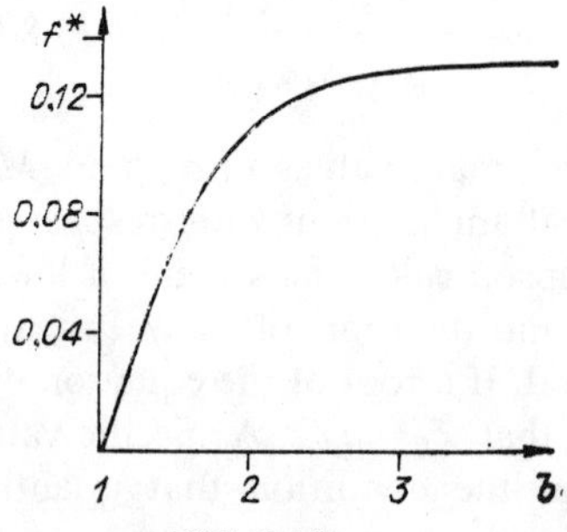

FIG. 3.15

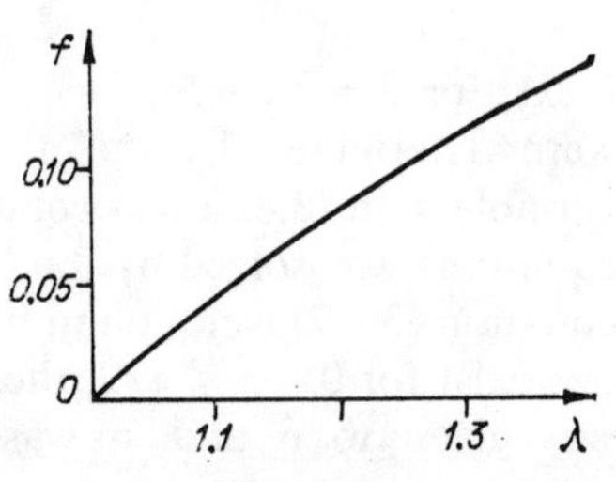

FIG. 3.16.

§ 9. SOLUTIONS WITH PLASTIC DISCONTINUITY LINES

Plastic deformation of bodies is accompanied by development of slip lines. If the stress gradient is negligible, the slip lines can be distributed uniformly across the entire volume of the body. Such behavior takes place in the case of advanced plastic deformation for work-hardening materials. For materials with clearly defined yield plateau (for metals such as mild steel, which are susceptible to yield delay) under nonuniform stress field with a large gradient, isolated slip lines appear in a volume of the body small compared to the elastic part. Consequently, the study of plastic deformations at the initial stages of their development can be reduced to discontinuous problems of the linear theory of elasticity. This fact was first noted and studied by Leonov and his co-workers [18, 19].

1. Elementary discontinuous deformation

The deformation of a body is described by a unique nonzero displacement $w(x, y)$. In plane $z=0$, displacements $w(x, y)$ are considered continuous everywhere with the exception of simple smooth lines L. Let the following condition be satisfied on line L:

$$w^{+}(t)-w^{-}(t)=\delta(s), \qquad s=s(t). \tag{3.9.1}$$

Here, $w^{+}(t)$ and $w^{-}(t)$ denote the values of w when it tends to point t of contour L from the left and right, respectively; $\delta(s)$ is a function of arc s on contour L satisfying, with its derivative $g(s) \equiv \delta'(s)$, the Hölder condition.

Assuming that the slip line L is composed of sections located along the y axis and recalling that function $w(x, y)$ satisfies the Laplace equation, we find from the boundary condition (3.9.1),

$$w(x, y)=\operatorname{Re}\left\{\frac{1}{2\pi i}\int_L \frac{\delta(s)\,dt}{t-\zeta}\right\}, \qquad t=is,\ \zeta=x+iy. \tag{3.9.2}$$

For stresses $\tau_{xz}=G(\partial w/\partial x)$ and $\tau_{yz}=G(\partial w/\partial y)$, accounting for (3.9.2) and condition $\delta(s)=0$ at the ends of the slip line L, we obtain

$$\tau_{xz}-i\tau_{yz}=\frac{G}{2\pi}\int_L \frac{g(s)\,dt}{\zeta-t}. \tag{3.9.3}$$

Let the ends a_k, b_k of line L have coordinates

$$a_k=i(mH-h), \qquad b_k=i(mH+h), \qquad m=0, \pm 1, \pm 2, \ldots, \pm 2n,$$

and let function $g(s)$ satisfy condition

$$\begin{aligned} g(2mH+\sigma) &= -g(2mH-\sigma)=g[(2m+1)H-\sigma] \\ &= -g[(2m+1)H+\sigma]=g(\sigma), \qquad -h \le \sigma \le h. \end{aligned} \tag{3.9.4}$$

With these assumptions, after passing to the limit $n \to \infty$, summation, and

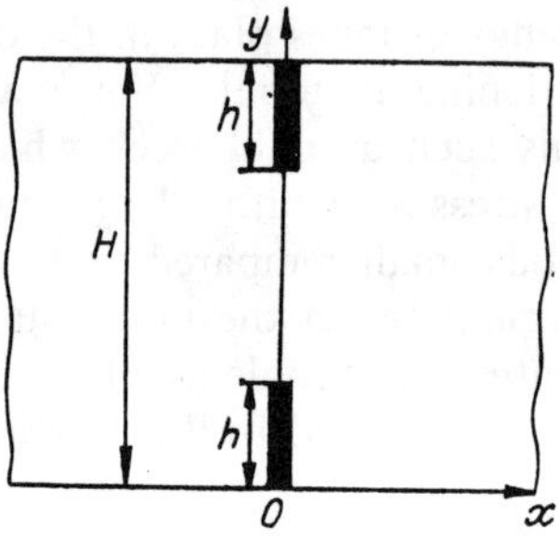

FIG. 3.17

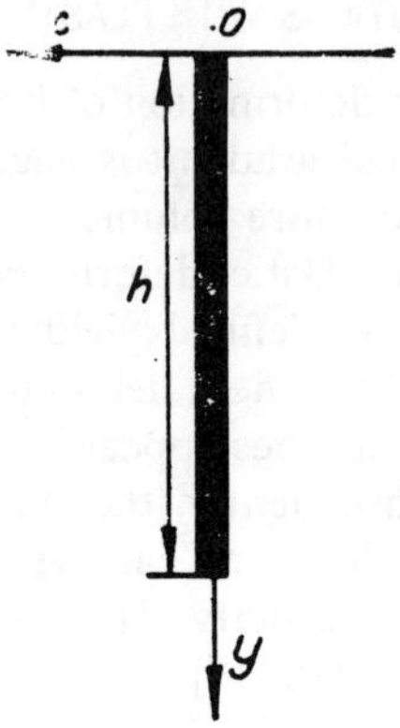

FIG. 3.18.

integration, function (3.9.3) can be written as

$$\tau_{xz} - i\tau_{yz} = \frac{G}{4H}\int_{-h}^{h}\left\{\cot\frac{\pi(\zeta - i\sigma)}{2iH} - \cot\frac{\pi[\zeta - i(H+\sigma)]}{2iH}\right\}g(\sigma)\,d\sigma$$

$$= \frac{G}{2H}\int_{-h}^{h}\frac{g(\sigma)\,d\sigma}{\sin[\pi(\zeta - i\sigma)/(iH)]}. \tag{3.9.5}$$

2. Plastic shear of a rectangular rod in torsion

Let us consider plastic slip lines in a rod with a rectangular cross section in torsion. The slip lines are normal to the contour at points located in the middle of the long sides of the cross section. Assuming the depth of the slip lines h is small compared with the length l of the cross section, we can model the cross section as an infinite strip (Fig. 3.17).

We seek a solution of the form

$$w(x, y) = w^0(x, y) + w^{(1)}(x, y),$$
$$\tau_{xz} = \tau_{xz}^0 + \tau_{xz}^{(1)}, \qquad \tau_{yz} = \tau_{yz}^0 + \tau_{yz}^{(1)}. \tag{3.9.6}$$

Here, $w^0(x, y)$ is warping and τ_{xz}^0, τ_{yz}^0 are shear stresses which do not account for plastic deformation while $w^{(1)}(x, y)$ is warping and $\tau_{xz}^{(1)}, \tau_{yz}^{(1)}$ are stresses caused by the plastic deformation.

Evidently, the following equality holds,

$$w^{(1)}(\pm 0, H - y) = w^{(1)}(\mp 0, y), \tag{3.9.7}$$

from which, for $2w^{(1)}(-0, y) = \delta(y)$, we obtain $\delta(H - y) = -\delta(y)$.

Function $g(y) = \delta'(y)$ satisfies condition

$$g(H - y) = g(y). \tag{3.9.8}$$

Stresses caused by the plastic deformations are determined from relation

(3.9.5) when condition (3.9.8) is satisfied. Function $g(\sigma)$ must be regarded as an odd function in order to satisfy the boundary conditions on the side surface of the strip $\tau_{yz}(x,0)=\tau_{yz}(x, H)=0$. The shear stresses on the slip lines are assumed to be opposite to the direction of shear and equal in magnitude to a constant (the lower yield limit), i.e.,

$$\tau_{xz}^{0}(0, y)+\tau_{xz}^{(1)}(0, y)=\begin{cases}\tau_c & \text{for } 0\le y\le h,\\ -\tau_c & \text{for } H-h\le y\le H.\end{cases} \tag{3.9.9}$$

It is convenient to introduce new variables

$$\eta=\tan \pi\zeta/iH, \qquad \xi=\tan \pi s/H. \tag{3.9.10}$$

Then formula (3.9.5) transforms to

$$\tau_{xz}^{(1)}-i\tau_{yz}^{(1)}=\frac{G}{2\pi}\sqrt{1+\eta^2}\int_{-\omega_1}^{\omega_1}\frac{g_*(\xi)\,d\xi}{\eta-\xi}, \tag{3.9.11}$$

where

$$\omega_1=\tan\frac{\pi h}{H}, \qquad g_*(\xi)=\frac{g[(H/\pi)\arctan \xi]}{\sqrt{1+\xi^2}}. \tag{3.9.12}$$

From (3.9.11) we find $\tau_{xz}^{(1)}(0, y)$ and substitute it in (3.9.9). As a result we obtain an integral equation for determination of function $g_*(\xi)$

$$\frac{1}{\pi}\int_{-\omega_1}^{\omega_1}\frac{g_*(\xi)\,d\xi}{\xi-\xi_1}=f_1(\xi_1). \tag{3.9.13}$$

Here

$$\xi_1=\tan\frac{\pi y}{H}, \qquad f_1(\xi_1)=\frac{2}{G\sqrt{1+\xi_1^2}}\left[\tau_{xz}^{0}\left(0,\frac{H}{\pi}\arctan \xi_1\right)-\tau_c\right], \tag{3.9.14}$$

$$0\le \xi_1\le \omega_1.$$

In addition, for function $f_1(\xi)$ it follows from (3.9.13) that $f_1(-\xi)=f_1(\xi)$. It can be shown that the solution of equation (3.9.13), in the class of totally bounded functions (stresses), has the form

$$g_*(\xi_1)=\frac{1}{\pi}\sqrt{\omega_1^2-\xi_1^2}\int_{-\omega_1}^{\omega_1}\frac{f_1(\xi)\,d\xi}{\sqrt{\omega_1^2-\xi^2}\,(\xi_1-\xi)}. \tag{3.9.15}$$

In this case the solvability condition

$$\int_{-\omega_1}^{\omega_1}\frac{f_1(\xi)\,d\xi}{\sqrt{\omega_1^2-\xi^2}}=0, \tag{3.9.16}$$

which determines the depth of the slip line h, must be satisfied.

If function $g_*(\xi)$ is known, then stresses $\tau_{xz}^{(1)}$ and $\tau_{yz}^{(1)}$ can be found from (3.9.11). We present the relationship linking stresses $\tau_{xz}^{(1)}$ and $\tau_{yz}^{(1)}$ in terms of $f_1(\xi)$ [18]:

$$\tau_{xz}^{(1)} - i\tau_{yz}^{(1)} = \frac{G}{2\pi}\sqrt{(1+\eta^2)(\eta^2-\xi_1^2)}\int_{-\omega_1}^{\omega_1}\frac{f_1(\xi)\,d\xi}{\sqrt{\omega_1^2-\xi^2}\,(\xi-\eta)}. \quad (3.9.17)$$

In the old variables ζ and σ, formulas (3.9.15)–(3.9.17) have the form

$$g(y) = \frac{1}{H}\frac{\sqrt{\cos 2\pi y/H - \cos 2\pi h/H}}{\cos \pi y/H} \times \int_{-h}^{h}\frac{(\cos \pi\sigma/H) f(\sigma)\,d\sigma}{\sqrt{\cos 2\pi\sigma/H - \cos 2\pi h/H}\cdot \sin\pi(y-s)/H}, \quad (3.9.18)$$

$$\int_{-h}^{h}\frac{f(\sigma)\,d\sigma}{\sqrt{\cos 2\pi\sigma/H - \cos 2\pi h/H}} = 0, \quad (3.9.19)$$

$$\tau_{xz}^{(1)} - i\tau_{yz}^{(1)} = \frac{\sqrt{\cos 2\pi\zeta/(iH) - \cos 2\pi h/H}}{iH\cos \pi\zeta/(iH)} \times \int_{-h}^{h}\frac{\cos \pi\sigma/H f(\sigma)\,d\sigma}{\sqrt{\cos 2\pi\sigma/H - \cos 2\pi h/H}\,\sin\pi(\zeta - i\sigma)/(iH)}. \quad (3.9.20)$$

Here

$$f(\sigma) = \frac{2}{G}\left[\tau_{xz}^{0}(0,\sigma) - \tau_c\right], \qquad 0 \le \sigma \le h,\ f(-\sigma) = f(\sigma). \quad (3.9.21)$$

3. The case of half space

Let the depth of the slip line h be small in comparison with the dimensions of the cross section. The rod can be considered as a half space (Fig. 3.18).

By passing to the limit $H \to \infty$ in formulas (3.9.18)–(3.9.21), we find

$$g(y) = \frac{1}{\pi}\sqrt{h^2-y^2}\int_{-h}^{h}\frac{f(\sigma)\,d\sigma}{\sqrt{h^2-\sigma^2}\,(y-\sigma)}, \quad (3.9.22)$$

$$\int_{-h}^{h}\frac{f(\sigma)\,d\sigma}{\sqrt{h^2-\sigma^2}} = 0, \quad (3.9.23)$$

$$\tau_{xz}^{(1)} - i\tau_{yz}^{(1)} = \frac{G}{2\pi}\sqrt{\zeta^2+h^2}\int_{-h}^{h}\frac{f(\sigma)\,d\sigma}{\sqrt{h^2-\sigma^2}\,(\zeta - i\sigma)}. \quad (3.9.24)$$

Stress $\tau_{xz}^0(0, y)$ for $0 \leq y \leq h$ can be expressed here with a sufficient accuracy in the form

$$\tau_{xz}^0(0, y) = \alpha\tau_m - by. \tag{3.9.25}$$

Here, α and b are constants that depend on the value of the torque and the shape of the cross section; τ_m is the upper yield limit.

According to (3.9.21), by using expression (3.9.25) we find

$$f(y) = \begin{cases} 2G^{-1}[(\alpha\tau_m - \tau_c) - by] & \text{for} \quad 0 \leq y \leq h, \\ 2G^{-1}[(\alpha\tau_m - \tau_c) + by] & \text{for} \quad -h \leq y \leq 0. \end{cases} \tag{3.9.26}$$

By accounting for (3.9.26) in relations (3.9.22)–(3.9.24) and computing the integrals, we finally obtain

$$g(y) = \frac{4by}{\pi G} \ln \left| \frac{h + \sqrt{h^2 - y^2}}{h} \right|, \tag{3.9.27}$$

$$h = \pi \frac{\alpha\tau_m - \tau_c}{2b}, \tag{3.9.28}$$

$$\tau_{xz}^{(1)} - i\tau_{yz}^{(1)} = -(\alpha\tau_m - \tau_c)\left[1 - \frac{\zeta}{2h} \ln \frac{\sqrt{\zeta^2 + h^2} + h}{\sqrt{\zeta^2 + h^2} - h}\right]. \tag{3.9.29}$$

The logarithm in formula (3.9.29) represents the analytical branch in plane $\zeta = x + iy$ with a slit along section $(-ih, ih)$, which assumes real values on the x axis. The formulas for stresses and depth of the plastic slip line are expressed in a closed form in terms of elementary functions including the case when $\tau_{xz}^0(0, y)$ for $0 \leq y \leq h$ is approximated by a polynomial of any order.

Let us find warping $w^{(1)}(x, y)$. Since $g(y) = \delta^{(1)}(y)$ and $\delta(h) = 0$, we obtain

$$\delta(y) = \int_h^y g(\sigma)\, d\sigma. \tag{3.9.30}$$

Computing integral (3.9.30) with $g(\sigma)$ defined by (3.9.27) gives us

$$\delta(y) = \frac{2b}{\pi G}\left[y^2 \ln \left| \frac{h + \sqrt{h^2 - y^2}}{y} \right| - h\sqrt{h^2 - y^2}\right]. \tag{3.9.31}$$

For $w^{(1)}(x, y)$, taking into account relation (3.9.31), we finally find from formula (3.9.2),

$$w^{(1)}(x, y) = \frac{b}{2\pi G} \operatorname{Re}\left\{\zeta^2 \ln \frac{\sqrt{\zeta^2 + h^2} + h}{\sqrt{\zeta^2 + h^2} - h} + 2h\sqrt{\zeta^2 + h^2} - 4h\right\}. \tag{3.9.32}$$

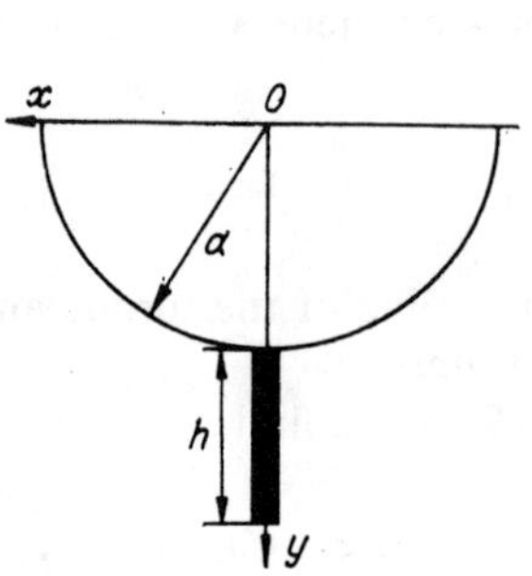

FIG. 3.19

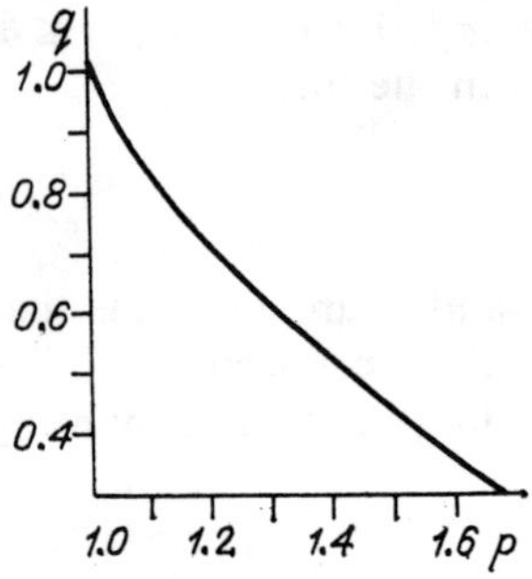

FIG. 3.20.

4. Plastic shear for prismatic rod with a small hemispherical groove in torsion

Let us consider torsion of a rod with a longitudinal groove on the surface. Assume that the diameter of the groove is small compared with the sectional dimensions of the rod, and the contour of the cross section in the vicinity of the groove is rectangular (Fig. 3.19).

We shall present the final results [19] as follows:

$$g(y)=\frac{\sqrt{(n^2-y^2)(y^2-m^2)}}{\pi}\int_L\frac{f(s)}{\sqrt{(n^2-s^2)(s^2-m^2)}}\frac{ds}{y-s},\tag{3.9.33}$$

$$\tau_{xz}^{(1)}-i\tau_{yz}^{(1)}=-\frac{G}{2\pi i}\sqrt{(\zeta^2+n^2)(\zeta^2+m^2)}\int_L\frac{f(s)\,ds}{\sqrt{(n^2-s^2)(s^2-m^2)}\,(\zeta-is)}.\tag{3.9.34}$$

The solvability condition of the problem, which determines the depth of the plastic slip line, has the form

$$\int_L\frac{s^kf(s)\,ds}{\sqrt{(n^2-s^2)(s^2-m^2)}}=0,\qquad k=0,1.\tag{3.9.35}$$

Contour L is composed of two sections $(-n,-m)$ and (m,n) symmetrical with respect to axis Ox, where $n=a+h$ and $m=a^2/(a+h)$; function $f(y)$ is determined from formula (3.9.21) on section $a\le y\le a+h$. On the remaining part of contour L, function $f(\sigma)$ is determined from relations

$$f\left(\frac{a^2}{y}\right)=\frac{y^2}{a^2}f(y),\qquad f(-y)=f(y).$$

In this case, using a solution for the Weber profile [19], we have

$$\tau_{xz}^0(0,y)=\frac{\alpha\tau_m}{2}\left(1+\frac{a_2}{y^2}\right),\qquad \alpha\ge1,\qquad a\le y\le a+h.\tag{3.9.36}$$

By finding function $f(y)$ in accordance with (3.9.21), substituting it in condition (3.9.35), and performing the integration, we find the following relation for the depth of plastic slip line h:

$$2q^2(p-1)\mathbf{F}(k,\varphi) - 2\mathbf{E}(k,\varphi) + p\mathbf{E}(k,\pi/2) = 0. \qquad (3.9.37)$$

Here, $p = \alpha\tau_m/\tau_c$ and $q = a/(a+h)$; **F**, **E** are elliptic integrals of the first and second order, $k = (1-q^2)^{1/2}$, and $\varphi = \arctan q$. Figure 3.20 depicts a plot of q as a function of p.

§ 10. NUMERICAL RESULTS

1. Rod with a square cross section

Let us analyze elastic–plastic torsion of a rod with $2a \times 2a$ square cross section [16] using a variational method. The shear stress $\tau = \tau(\gamma)$ is assumed to be bilinear:

$$\tau = \begin{cases} G\gamma & \text{for } \gamma \le 0.0025, \\ (1940 + 24\times 10^3\gamma)\ \text{kg/cm}^2 & \text{for } \gamma > 0.0025. \end{cases} \qquad (3.10.1)$$

The shear modulus $G = 8\times 10^5$ kg/cm^2. Dependence (3.10.1) describes, for example, the behavior of a nickel-alloyed steel.

The stress function is sought in the form

$$\Psi^{(k)} = c_1^{(k)}\Psi_1 + c_2^{(k)}\Psi_2, \qquad (3.10.2)$$

where $c_1^{(k)}$ and $c_2^{(k)}$ are constants, which are determined by an iterative process, k is the number of iteration,

$$\Psi_1 = (x^2/a^2 - 1)(y^2/a^2 - 1),$$
$$\Psi_2 = \Psi_1(x^2/a^2 + y^2/a^2). \qquad (3.10.3)$$

Computed coefficients $c_1^{(k)}$, $c_2^{(k)}$ are presented in Table III.4. Computations were conducted for $a\alpha = 0.015$ (α is the twist per unit length of rod). The

TABLE III.4

k	$c_1^{(k)} \times 10^{-5}$	$c_2^{(k)} \times 10^{-5}$
0	0.070135	0.014217
1	0.018023	−0.003382
2	0.017872	−0.003765
3	0.017728	−0.003802
4	0.017554	−0.003641
5	0.017551	−0.003740
6	0.017543	−0.003889
7	0.017461	−0.003757
8	0.017469	−0.003663
9	0.017400	−0.003693

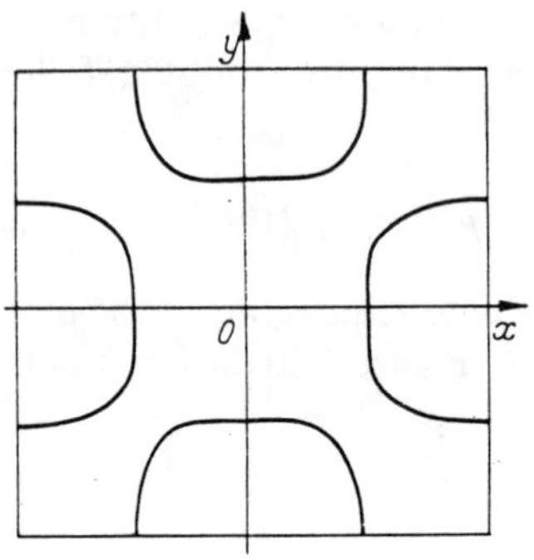

FIG. 3.21

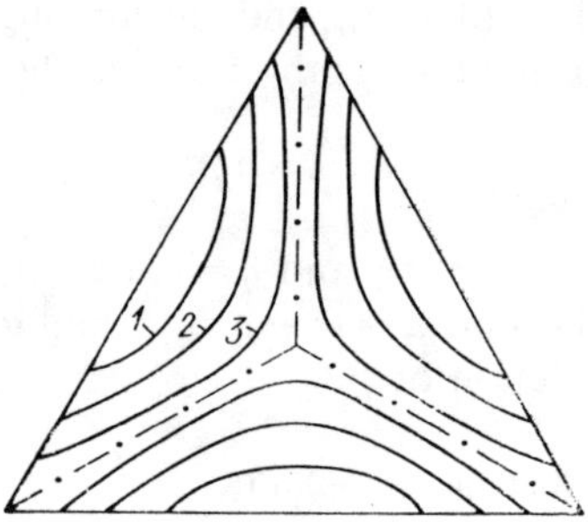

FIG. 3.22.

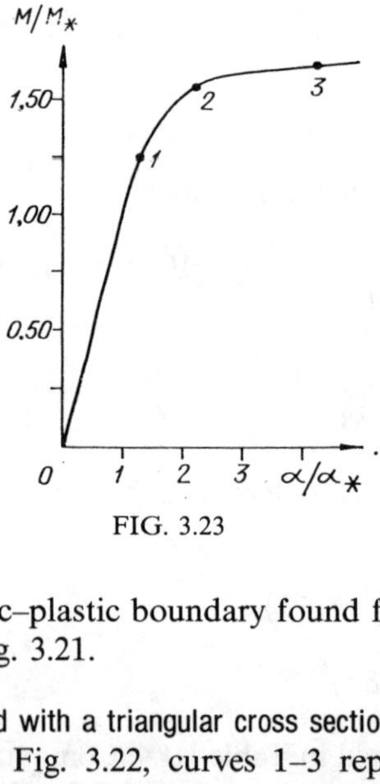

FIG. 3.23

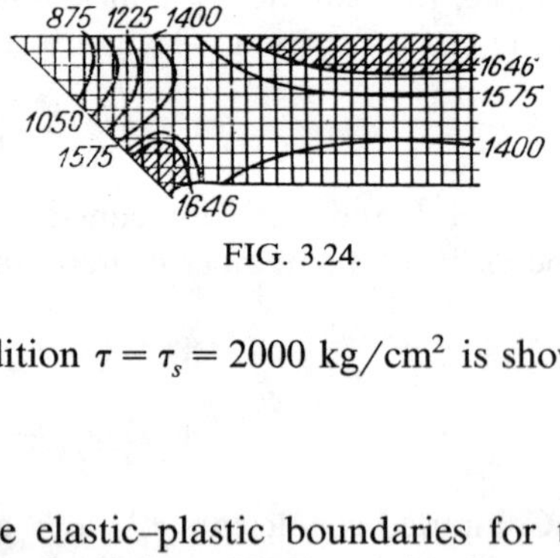

FIG. 3.24.

elastic–plastic boundary found from condition $\tau = \tau_s = 2000\ \mathrm{kg/cm^2}$ is shown in Fig. 3.21.

2. Rod with a triangular cross section

In Fig. 3.22, curves 1–3 represent the elastic–plastic boundaries for the following stages of torsion: $\alpha = 1.333\alpha_*$, $2\alpha_*$, $4\alpha_*$. Here, α_* is the twist corresponding to the onset of plastic deformations. The material of the rod is ideal elastic–plastic. The solution is obtained by the relaxation method [36]. Figure 3.23 shows the effect of the dimensionless angle of twist on the dimensionless torque.

3. Rod with a box-shaped rectangular cross section

The stress distribution and elastic–plastic boundary are shown in Fig. 3.24 (1/8 of the section is depicted) for $\alpha = 1.85\alpha_*$, where α_* is the twist at which plastic deformations first appear in the reentrant angles of the section. The material of the rod is considered ideal elastic–plastic. The solution is obtained by the relaxation method [35, 36].

4. Rod with a square cross section

Shown in Fig. 3.25 are the regions of plastic deformations in a rod of square cross section in torsion. The material of the rod is ideal elastic–plastic.

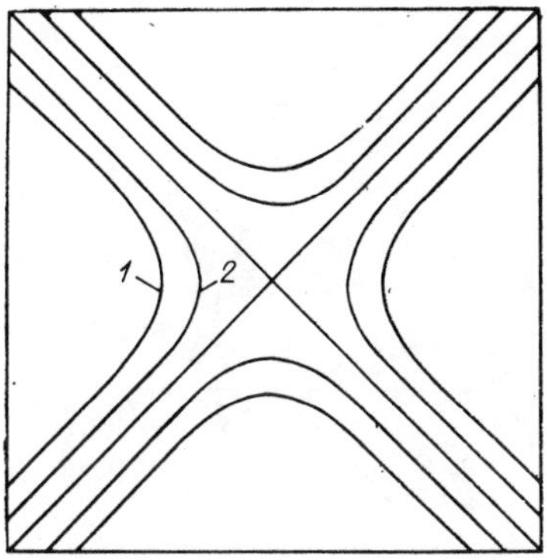

FIG. 3.25

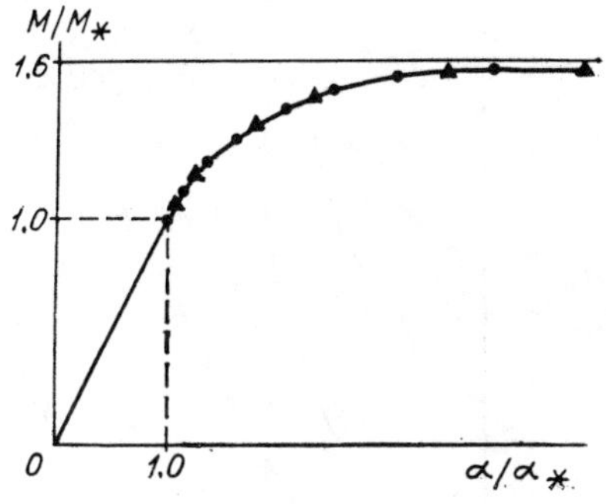

FIG. 3.26.

The solution is obtained by the relaxation method [23]. Curve 1 corresponds to torque $1.25M_*$ and curve 2 to torque $1.5M_*$. M_* is the maximum elastic torque corresponding to the onset of plastic deformation at the central point on a side of the cross section. The exact value of M_*^T, found by Timoshenko by series expansion (see [14], Chapter 2), is

$$M_*^T = 0.1406 G\alpha_* a^4, \tag{3.10.4}$$

where α_* is the twist corresponding to the onset of plastic deformations and a is a side of the square (see also § 8 of this chapter).

A similar result was given in [23]:

$$M_* = 0.1372 G\alpha_* a^4. \tag{3.10.5}$$

The effect of the dimensionless twist α/α_* on the torque M/M_* is shown in Fig. 3.26 [23, 28, 34].

5. Round rod weakened by six semicircular grooves

The material of the rod is considered ideal elastic–plastic and the solution is obtained by the relaxation method [38].

Curves 1–3 in Fig. 3.27 depict the elastic–plastic boundary for $\alpha = 1.448\alpha_*$, $2.173\alpha_*$, $4.345\alpha_*$.

The effect of the dimensionless twist on the dimensionless torque is shown in Fig. 3.28.

6. Sokolovsky's oval

When the plastic zone partially encloses the elastic core, an approximate solution of the elastic–plastic torsion problem of a rod with a cross section in the shape of Sokolovsky's oval (3.2.4) has the form [21]:

$$\begin{aligned} \tau_{xz} - i\tau_{yz} &= G\alpha\left[F'(z) - i\bar{z}\right], \\ F'(z) &= ia_1 z + ia_3 z^3 + ia_5 z^5 + \cdots, \end{aligned} \tag{3.10.6}$$

where $F(z)$ is a complex function of torsion.

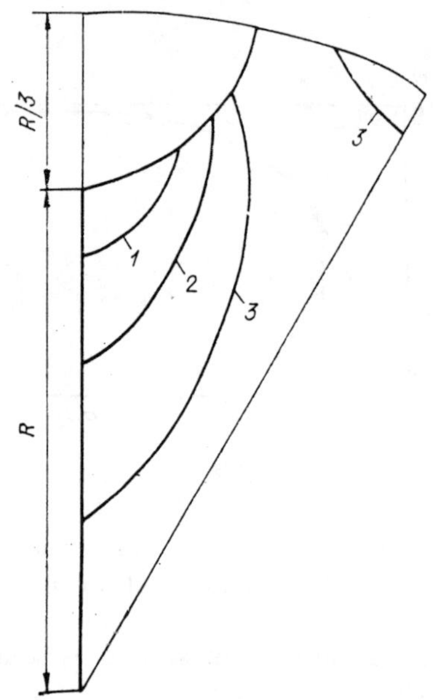

FIG. 3.27

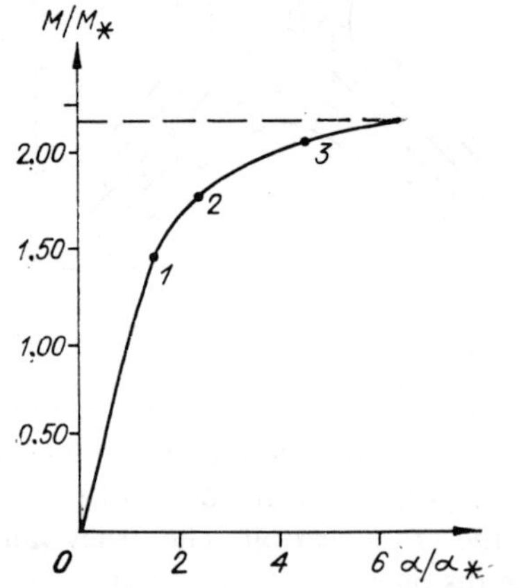

FIG. 3.28.

TABLE III.5

Point l	Intermediate points l_j: l_1	l_2	l_3	a_1	a_3	a_5	$\alpha\frac{G}{\tau_s}$
2.0	$1.96+i0.36$	—	—	0.296	−0.005	—	0.675
2.0	$1.96+i0.36$	$0.6+i1.25$	—	0.256	−0.013	−0.0006	0.61
2.0	$1.96+i0.36$	$0.6+i1.25$	$0.3+i1.42$	0.244	0.01	0.007	0.555

Table III.5 shows the values of coefficients a_1, a_3, a_5. The plastic deformations first appear at $\alpha_* = 0.545\tau_s/G$.

7. Cylindrical shafts with internal and external hyperbolic grooves.
a. Solid shaft. Calculations were conducted for three different grooves [26]:

$$v_0 = 1.20, \qquad a/\rho = 6.60, \qquad K_T = 1.63,$$
$$v_0 = 1.50, \qquad a/\rho = 199, \qquad K_T = 5.9,$$
$$v_0 = 1.55, \qquad a/\rho = 2311, \qquad K_T = 18.9,$$

where u, v, w are curvilinear coordinates linked to x, y, z by equations $x = \sinh u \cos v$, $y = \cosh u \sin v \cos w$, $z = \cosh u \sin v \sin w$; v_0 is a coordinate of the surface that is free from grooves, a is the radius of the shaft at the base of the groove, ρ is the radius of the groove at its base, and K_T is the elastic stress concentration factor. In all the calculations it was assumed that $\tau_s = 1050\ \text{kg/cm}^2$ and $G = 1.05 \times 10^6\ \text{kg/cm}^2$. Computations covered a range of values of the plastic region depth b/a and plastic shear modulus G^p.

TABLE III.6

Case number	K_T	G^p/G	b/a	τ_{max}	γ_{max}	K_σ^p	K_ε^p	K_σ^e	K_ε^e
1	1.63	0.200	0.50	2340	0.0086	1.40	1.83	1.21	4.60
2	1.63	0.200	0.75	1770	0.0049	1.38	2.25	1.22	3.62
3	1.63	0.033	0.50	1540	0.0146	1.22	2.18	0.96	9.71
4	1.63	0.033	0.75	1300	0.0083	1.19	3.92	1.01	6.83
5	1.63	0.033	0.85	1200	0.0053	1.11	2.98	1.02	4.77
6	1.63	0.033	0.95	1120	0.0031	1.05	2.37	1.21	3.48
7	1.63	0.000	0.75	1050	0.0148	1.00	6.91	0.83	12.33
8	5.9	0.200	0.50	9900	0.0512	4.63	7.17	4.07	22.3
9	5.9	0.200	0.75	6300	0.0307	4.49	10.36	3.82	19.7
10	5.9	0.033	0.50	4450	0.1009	3.17	8.93	2.55	61.0
11	5.9	0.033	0.75	3220	0.0645	2.90	25.00	2.35	49.7
12	5.9	0.033	0.85	2440	0.0415	2.15	22.40	2.02	36.4
13	5.9	0.033	0.95	1910	0.0259	1.79	19.88	2.43	35.0
14	18.9	0.200	0.50	32800	0.1825	14.65	23.41	12.97	75.8
15	18.9	0.200	0.75	20600	0.1127	14.20	34.25	12.08	69.4
16	18.9	0.033	0.50	14100	0.3855	9.73	30.31	7.88	226.8
17	18.9	0.033	0.75	9650	0.2547	8.66	88.76	6.92	192.0
18	18.9	0.033	0.85	67500	0.1685	6.22	88.37	5.51	145.0
19	18.9	0.033	0.95	4820	0.1117	4.52	85.13	6.11	149.4

Some results of the computations are presented in Table III.6, where $K_\sigma^p = \tau_{max}/\tau_3$, $K_\sigma^e = \tau_{max}/\tau_1$, $K_\varepsilon^p = \gamma_{max}/\gamma_3$, $K_\varepsilon^e = \gamma_{max}/\gamma_1$, τ_3 is the stress in the elastic–plastic shaft without grooves, τ_1 is the stress in the elastic shaft without grooves, and γ_3 and γ_1 are the strains, respectively.

Figure 3.29 illustrates the plastic zones in a solid shaft with a groove for various values of b/a and G^p/G. Curves 1, 3 correspond to $G^p/G = 0.20$, while curves 2, 4, 5, 6 correspond to $G^p/G = 0.033$. The following relation between stresses and strains was used in the analysis:

$$\gamma_{uw} = \frac{\tau_{uw}}{G} + \frac{\tau_{uw} - \tau_s}{G^p}. \tag{3.10.7}$$

It must be noted that some of the presented results correspond to unrealistic and extremely large values of the maximum stresses and strains.

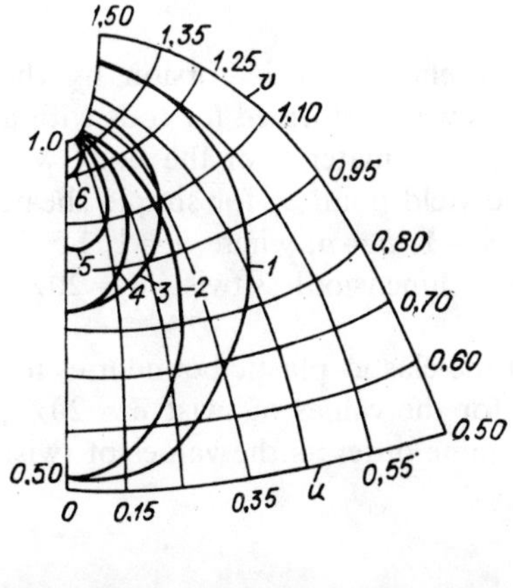

FIG. 3.29.

TABLE III.7

Groove	a/ρ	K_T	$\tau_{\max}$	$\gamma_{\max}$	K^e_σ	K^e_ε
External	6.6	2.14	2060	0.0067	1.99	6.82
	199	9.40	8 800	0.0450	7.34	39.6
	2311	31	29 500	0.163	24	140
Internal	6.6	2.14	1 550	0.0038	1.33	3.47
	199	9.40	6 550	0.0323	5.25	27.3
	2311	31	21 800	0.120	17.2	99

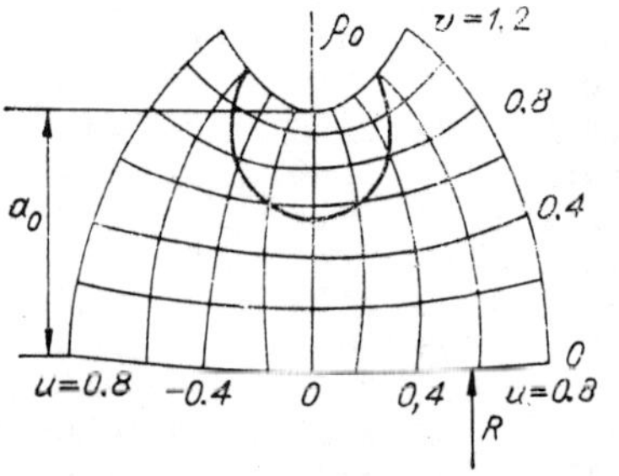

FIG. 3.30

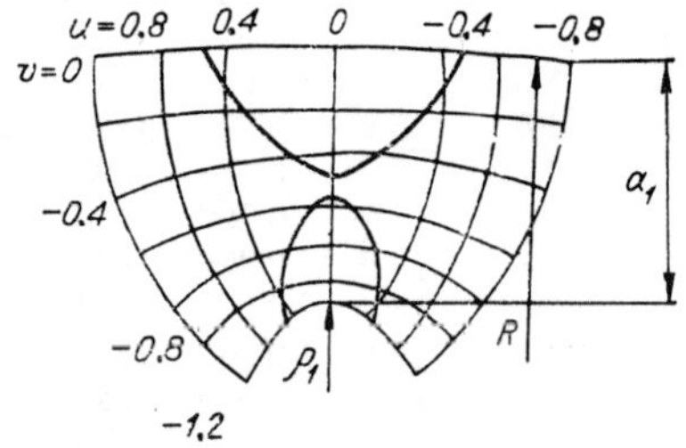

FIG. 3.31.

b. Hollow shaft. Computations were conducted for six cases:

$$v_0 = 1.2, \quad v_0 = 1.5, \quad v_0 = 1.55,$$
$$v_i = -1.2, \quad v_i = -1.5, \quad v_i = -1.55.$$

Here, v_0, v_i are the values of v on the external and internal surfaces of the shaft. The depth of the plastic zone is equal to 1.44 cm.

The results are presented in Table III.7 for $K = 1050$ kg/cm², $R = 250$ mm, $G = 1.05 \times 10^6$ kg/cm², and $G^p/G = 0.2$.

Again, some of the results correspond to unrealistic and very large values of stresses and strains.

Figures 3.30 and 3.31 illustrate the elastic–plastic boundaries for external and internal grooves when $a/\rho = 6.6$. In the last case, two plastic regions can develop as shown in Fig. 3.31.

8. Rods with square and polygonal cross section

We shall present the results obtained for elastic–plastic torsion by the method of local variations [16, 17]. The analysis was performed for rods with a rectangular cross section and some polygons. The material of the body was considered to be ideal elastic–plastic with the yield point τ_s for simple shear. Rods of rectangular cross section with sides $a = 1$, $b = n$, where $n = 1, 1.5, 2, 3, 5$ were analyzed at the following values of the dimensionless twist: $\bar{\alpha} = 20/3$, 10, 20, 40.

Curves 1, 2, 3 in Figs. 3.32 and 3.33 depict the elastic–plastic boundaries for rectangles with side ratios $n = 1, 1.5, 3$ and for the values of twist $\bar{\alpha} = 20/3$, 10, 20. The boundaries are numbered in the same order as the values of twist $\bar{\alpha}$.

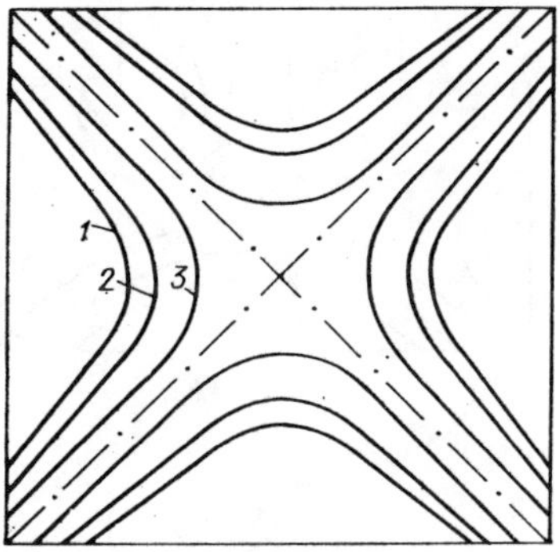

FIG. 3.32

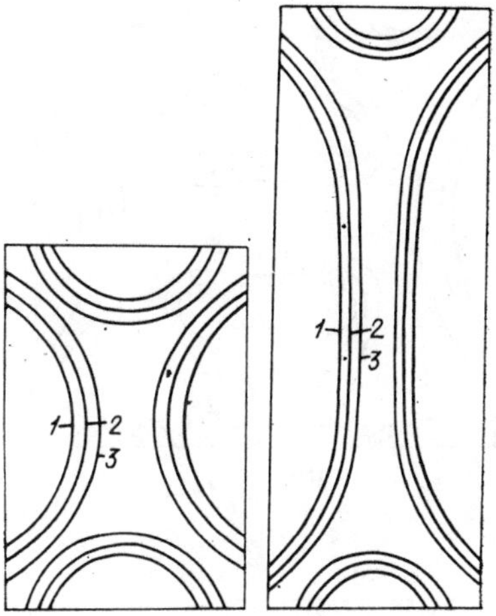

FIG. 3.33.

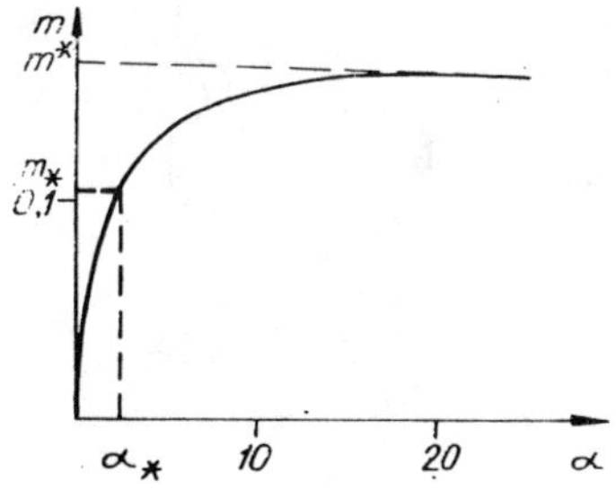

FIG. 3.34

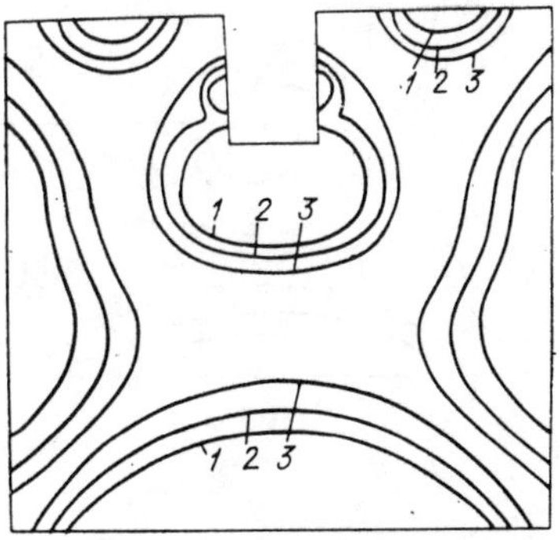

FIG. 3.35.

Figure 3.34 shows the effect of $\bar{\alpha}$ on the dimensionless torque m for a square cross section, where m^* is the limit torque.

Table III.8 gives the dimensionless torque for the rectangular regions with side ratios $n = 1, 1.5, 2, 3, 5$ obtained for various values of the twist $\bar{\alpha}$. Here, $\bar{\alpha} = 2G\alpha a/\tau_s$, $m = M/2\tau_s a^3$, $\bar{\alpha}_*$ is the dimensionless twist corresponding to the onset of plastic deformations, m_* is the dimensionless torque at which plastic deformations first appear, and $\bar{\alpha} = \infty$ corresponds to purely plastic torsion.

TABLE III.8

n	$\bar{\alpha}_*$	m_*	$\bar{\alpha} = 20/3$	$\bar{\alpha} = 10$	$\bar{\alpha} = 20$	$\bar{\alpha} = 40$	$\bar{\alpha} = 80$	$\bar{\alpha} = \infty$
1	2.9629	0.1040	0.1558	0.1620	0.1660	0.1662	0.1662	0.1667
1.5	2.3585	0.1155	0.2778	0.2855	0.2897	0.2904	0.2904	0.2917
2	2.1505	0.1230	0.3989	0.4096	0.4136	0.4150	0.4150	0.4166
3	2.0305	0.1335	0.6403	0.6548	0.6606	0.6627	0.6630	0.6667
5	2.0020	0.1455	1.1188	1.1375	1.1508	1.1560	1.1560	1.1667

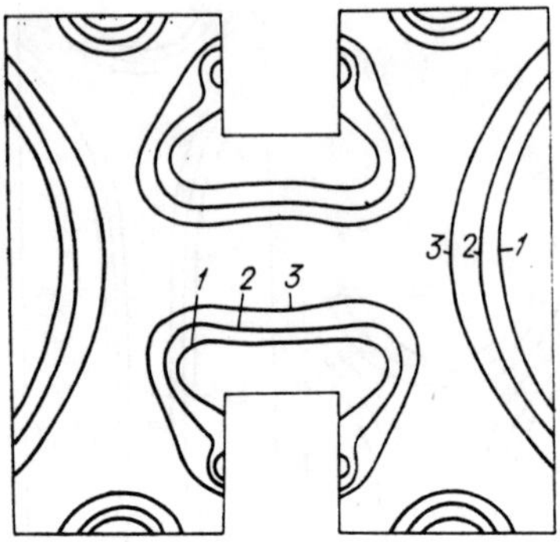

FIG. 3.36

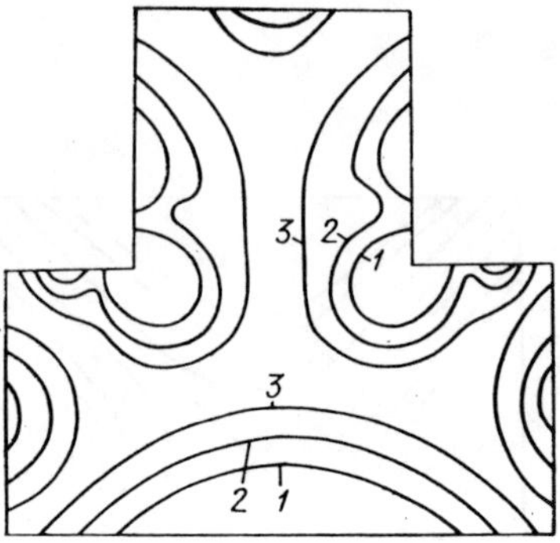

FIG. 3.37.

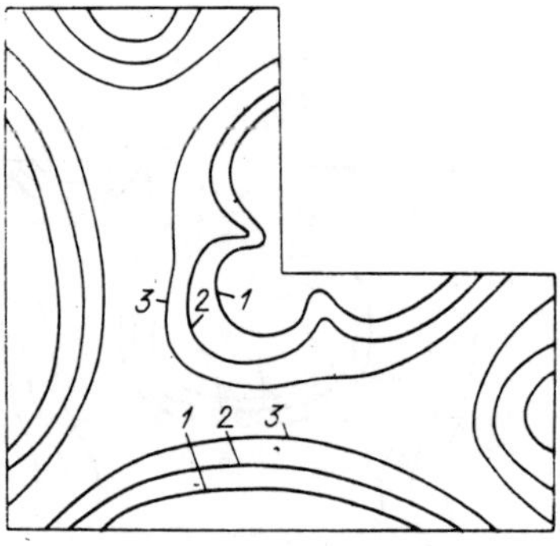

FIG. 3.38

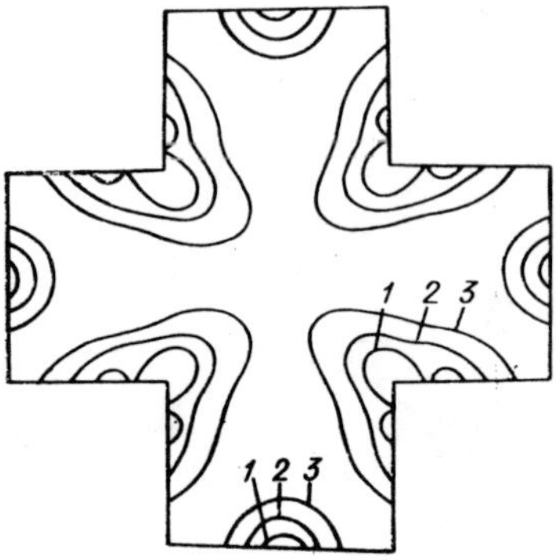

FIG. 3.39.

Curves 1, 2, 3 in Figs. 3.35–3.39 show the elastic–plastic boundaries for $\bar{\alpha} = 20/3$, 10, 20, respectively.

Figure 3.40 shows the effect of twist on the torque for polygons illustrated in Figs. 3.35–3.39; curves 1–5 correspond to Figs. 3.35–3.39.

Since unlimited stresses arise at the internal corner points due to any arbitrary small twist, we have $\bar{\alpha}_* = m_* = 0$ for these regions.

Function $m(\alpha)$ is generally nonlinear. However, for small α, it is practically linear since the plastic zones are negligibly small and their contribution to the torque is small.

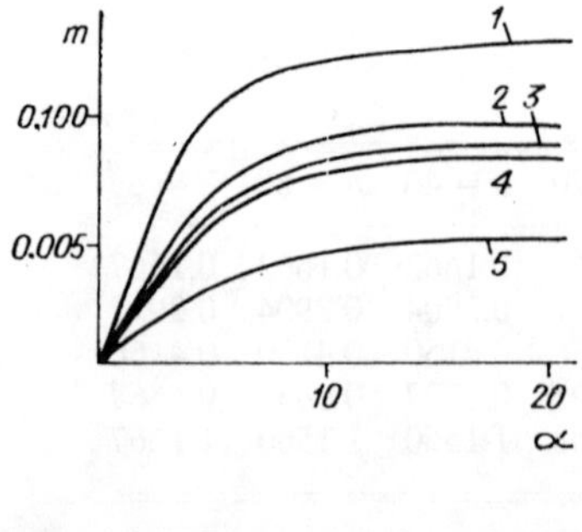

FIG. 3.40.

9. Composite rods

Let us consider a prismatic rod with a square section

$$D = D_1 + D_2, \qquad D_1 = \{0.4 \le x \le 1, 0 \le y \le 1\},$$

$$D_2 = \{0 \le x \le 0.4, 0 \le y \le 1\}.$$

The material of the rod is homogeneous in the elastic range: $G_1 = G_2 = G$, with a piecewise constant transverse distribution of plastic properties: $\tau_{s2}/\tau_{s1} = 0.65$. Here, τ_{si} is the yield point corresponding to subdomains D_i, $i = 1, 2$.

The analysis was performed for the following values of dimensionless twist: $\bar{\alpha} = 2.4$, 4, 10, 25. Here, $\bar{\alpha} = 2G\alpha/\tau_{s1}$ and $m = M/2\tau_{s1}$.

Purely elastic deformations occur prior to the moment when the shear stresses reach the lower yield point $\tau = \tau_{s2}$.

From the elastic solution, we can find the boundary values of the twist $\bar{\alpha}_* = 1.927$ and torque $m_* = 0.0676$. The values of the twist and torque, corresponding to the onset of plastic deformations in the region with the greater yield point $\tau_s = \tau_{s1}$, are equal to $\bar{\alpha}_*' = 2.963$ and $m_*' = 0.104$, respectively.

Thus, for twist $\bar{\alpha}$ in the range $\bar{\alpha}_* < \bar{\alpha} < \bar{\alpha}_*'$, plastic deformations appear only in materials with the lower yield point. For $\bar{\alpha}_*' < \bar{\alpha}$, plastic zones will appear in both materials.

The results for $\bar{\alpha} = 2.4$, 4, 10, 25, obtained by the method of local variations, are presented in Fig. 3.41.

Given in Table III.9 are the values of the torque at various $\bar{\alpha}$, and area D_p of the plastic regions in the section of the rod.

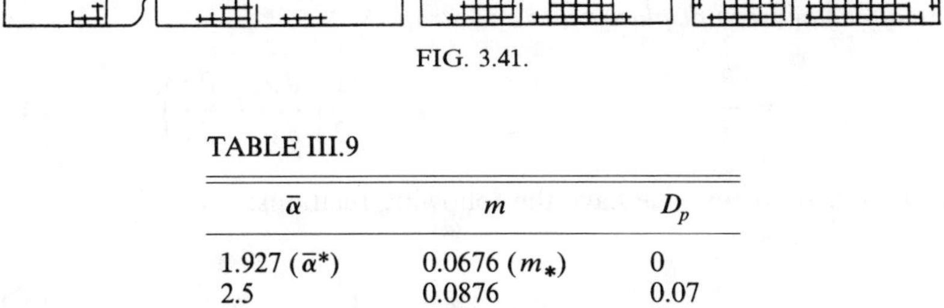

FIG. 3.41.

TABLE III.9

$\bar{\alpha}$	m	D_p
1.927 ($\bar{\alpha}^*$)	0.0676 (m_*)	0
2.5	0.0876	0.07
4	0.1145	0.20
10	0.1358	0.58
25	0.1387	0.83
∞	0.1390	1.00

4. PLANE STRAIN

§ 1. FUNDAMENTAL EQUATIONS. BRIEF REVIEW

In plane strain, the stresses and displacements are such that

$$u = u(x, y), \qquad v = v(x, y), \qquad w = 0,$$
$$\sigma_x = \sigma_x(x, y), \qquad \sigma_y = \sigma_y(x, y), \qquad \tau_{xy} = \tau_{xy}(x, y), \tag{4.1.1}$$
$$\sigma_z = \nu(\sigma_x + \sigma_y), \qquad 0 < \nu \le \tfrac{1}{2}, \qquad \tau_{xz} = \tau_{yz} = 0.$$

The components of the stress tensor under plane strain conditions in the absence of body forces satisfy two equilibrium equations

$$\frac{\partial \sigma_x}{\partial x} + \frac{\partial \tau_{xy}}{\partial y} = 0, \qquad \frac{\partial \tau_{xy}}{\partial x} + \frac{\partial \sigma_y}{\partial y} = 0. \tag{4.1.2}$$

In the elastic region, we have the following relations:
Hooke's law

$$\varepsilon_x = E^{-1}\left[(1-\nu^2)\sigma_x - \nu(1+\nu)\sigma_y\right],$$
$$\varepsilon_y = E^{-1}\left[(1-\nu^2)\sigma_y - \nu(1+\nu)\sigma_x\right], \tag{4.1.3}$$
$$\varepsilon_{xy} = (1+\nu)E^{-1}\tau_{xy}, \qquad \nu \ne \tfrac{1}{2};$$

The condition of strain compatibility that, expressed in terms of stresses, gives

$$\Delta(\sigma_x + \sigma_y) = 0; \tag{4.1.4}$$

The relations between displacements and strains

$$\varepsilon_x = \frac{\partial u}{\partial x}, \qquad \varepsilon_y = \frac{\partial v}{\partial y}, \qquad \varepsilon_{xy} = \frac{1}{2}\left(\frac{\partial u}{\partial y} + \frac{\partial v}{\partial x}\right). \tag{4.1.5}$$

In the plastic region, we have the following relations:
Plasticity condition

$$f(\sigma_x, \sigma_y, \tau_{xy}) - k = 0; \tag{4.1.6}$$

Associated law of plastic flow (1.3.6)

$$d\varepsilon_x^p = \lambda \frac{\partial f}{\partial \sigma_x}, \qquad d\varepsilon_y^p = \lambda \frac{\partial f}{\partial \sigma_y}, \qquad d\varepsilon_{xy}^p = \frac{\lambda}{2}\frac{\partial f}{\partial \tau_{xy}}; \tag{4.1.7}$$

Kinematic relations between displacements and strains

$$\varepsilon_x = \varepsilon_x^e + \varepsilon_x^p = \frac{\partial u}{\partial x}, \qquad \varepsilon_y = \varepsilon_y^e + \varepsilon_y^p = \frac{\partial v}{\partial y},$$
$$\varepsilon_{xy} = \varepsilon_{xy}^e + \varepsilon_{xy}^p = \frac{1}{2}\left(\frac{\partial u}{\partial y} + \frac{\partial v}{\partial x}\right). \tag{4.1.8}$$

Elastic strains in the plastic region are related to stresses by Hooke's law.

If the loads are given on the boundary of the body, then there exists a complete system of equations for the determination of stresses in the plastic region independently of strains, i.e., the problem is statically determinate.

We assume that the stresses and displacements on the elastic–plastic boundary are continuous.

In the elastic region, the Kolosov–Muskhelishvili formulas apply:

$$\sigma_x = \sigma_y = 4\,\mathrm{Re}\,\Phi(z),$$
$$\sigma_y - \sigma_x + 2i\tau_{xy} = 2[\bar{z}\Phi'(z) + \Psi(z)],$$
$$2G(u + iv) = \varkappa\varphi_*(z) - z\overline{\varphi_*'(z)} - \overline{\psi_*(z)}, \tag{4.1.9}$$
$$d\varphi_*(z)/dz \equiv \varphi_*'(z) = \Phi(z), \qquad \psi_*'(z) = \Psi(z), \qquad z = x + iy.$$

where $\varkappa = 3 - 4\nu$ for plane strain and $\varkappa = (3 - \nu)/(1 + \nu)$ for the plane stress $G = E2^{-1}(1 + \nu)^{-1}$.

Major progress in the solution of plane elastic–plastic problems was made in problems with a completely unknown elastic–plastic boundary. In this case, the corresponding mathematical problem is reduced to a boundary value problem for a biharmonic equation in a region whose boundary is unknown and is determined in the process of solving the problem from an additional boundary condition. Accounting for formulas (4.1.9) and mapping the region with an unknown boundary one-to-one and conformally onto a canonical region, this problem is reduced to a boundary value problem of the theory of complex-variable functions. The methods of solution for some of these problems are given by L. A. Galin, G. P. Cherepanov, and B. D. Annin.

In 1946, Galin gave an exact solution for the problem of stress distribution in the vicinity of a round hole in a body under plane strain conditions. Constant normal forces were applied to the contour, and the stresses at infinity were constant [12]. The solution was made possible due to the fact that the stress function in the plastic region is biharmonic. Displacements in the plastic region for this problem were determined by Ivlev [17], who used the perturbation method. An exact solution of the system of equations for displacements in the plastic zone for the Galin problem was obtained by Ostrosablin [45]. The Galin method was used for similar problems by Kuznetsov [24] in the case of a special discontinuity, and by Annin [4] in the case of exponentional plasticity condition approximating the limit state of rocks fairly well.

Cherepanov [66] found a set of exact solutions for a plane elastic–plastic problem subject to the following requirements: (a) the contour of the body is a polygon whose angles are multiples of $\pi/2$; (b) the tangential load over the entire contour is equal to zero; (c) part of the boundary of the polygon is loaded by constant normal stresses and is entirely enclosed by the plastic zone; (d) piecewise linear normal displacement is given on the remaining elastic part of the boundary. Ivlev, using the perturbation method, solved several plane elastic–plastic problems for an ideally plastic body. The shape of the hole was elliptical but near circular, and the stress state at infinity was close to hydrostatic compression [18, 19]. Ershov and Ivlev succeeded in solving an elastic–plastic problem for an elliptical tube under pressure using a similar method [16]. Kosmodamiansky [21] and Mirsalimov [31] discussed an elastic–plastic problem with an infinite row of single round holes. Annin [6] and Ostrosablin [43, 44] gave an approximate solution of an elastic–plastic problem for a plane surface weakened by a finite number of round holes. Kurshin and Suzdalnitsky [25] solved an elastic–plastic problem for a plane surface weakened by a double periodical system of circular holes.

Perlin, using his numerical method, solved a number of problems for circular and elliptical holes. He discussed also cases of the hole partially enclosed by the plastic zone and the case of a doubly connected region occupied by the body [50–52]. Sazhin used the same method to solve an elastic–plastic problem for a near-square hole; it was assumed that hydrostatic compression is applied at infinity and that the plastic region encloses the entire hole [59, 60]. Sazhin also discussed other interesting problems as applied to the development of rock pressure near openings of various shapes [61, 62].

Cherepanov ([33], in Chapter I) suggested a technique for the algebraization of an elastic–plastic problem in the case of a complete enclosure by the plastic zone with a statically determinate state of an arbitrary hole. Nottrot and Timman [85] and Nottrot [84] used this method for a numerical solution of some specific problems of this type.

Annin [5] obtained an integrodifferential equation for the case of a hole that is totally enclosed by the plastic zone for the boundary values of a function mapping the elastic region conformally onto a unit circle. He found the rationality condition of the mapping function.

Numerical methods are still rarely applied to plane elastic–plastic problems. In addition to the above mentioned papers by Perlin, Nottrot, and Timman, there are a few more publications that made use of numerical methods.

Southwell and de Allen discussed a strip with symmetrical semicircular and angular notches [88]. Teplitsky solved a plane problem of a rigid punch on elastic–plastic half space [63]. Banichuk, using the method of local variations, obtained a solution for a punch indented in an ideally elastic–plastic body [7]. The problem of a flat punch indented in ideally elastic–plastic medium was also discussed in papers [82, 89]. The solution in [89] was obtained by the relaxation method and in [82] by the finite-element method. Elastic–plastic problems for a crack were also solved numerically [23, 83]. Fomin [64] and Mirsalimov [30] discussed an elastic–plastic problem of a plane with a circular hole under a stationary temperature field when a biharmonic stress state and constant stresses are applied in the plastic zone and at infinity, respectively.

Effects of material anisotropy and inhomogeneities in the elastic–plastic problem were analyzed in [1, 2, 20]. Existence of a solution for plane elastic–plastic problems was discussed by Monakhov [40].

§ 2. METHODS OF SOLVING BOUNDARY VALUE PROBLEMS

Let us assume that the second derivatives of a biharmonic function, which are considered known functions of coordinates x and y, are given on an unknown closed contour L in the plane of the complex variable $z = x + iy$. It is required to determine boundary L and the biharmonic function. An elastic–plastic problem for a body under plane strain or plane stress conditions can be reduced to such a mathematical formulation, when the plastic zone entirely encloses the body contour since the stresses in the plastic region are, as a rule, determined directly from the boundary loads. Some problems of plate buckling and fracture of materials are reduced to a similar mathematical problem. When given boundary functions are the corresponding second derivatives of the biharmonic function, the problem can be solved by the Galin method [12].

Let us consider the methods of solving a class of problems in which the boundary functions might not satisfy the last condition [5, 67].

1. Formulation of the problem

Let a plane surface with a hole, bounded by sufficiently smooth contour γ, be subject to tangential and normal loads applied to contour γ. Stresses expressed as polynomial functions of the Cartesian coordinates x and y, are applied at infinity. Let us choose the origin of the coordinates inside the hole. The entire hole is enclosed by the plastic zone. We assume that the stresses in the plastic region are determined only by the shape of the hole contour and by the boundary load, and are independent of the stress state in the elastic region. In the elastic region, the components of the stress tensor $\sigma_x, \sigma_y, \tau_{xy}$ are determined by the Kolosov–Muskhelishvili formulas (4.1.9) in terms of functions $\Phi(z)$ and $\Psi(z)$. $\Phi(z)$ and $\Psi(z)$ are analytical functions of the variable $z = x + iy$, which, at an infinitely distant point, behave as follows:

$$\Phi(z) = \Phi_0(z) + O(z^{-1}), \qquad \Psi(z) = \Psi_0(z) + O(z^{-1}),$$
$$\Phi_0(z) = d_0 + d_1 z + \cdots + d_m z^m, \qquad \Psi_0(z) = e_0 + e_1 z + \cdots + e_n z^n, \tag{4.2.1}$$

where e_i, d_i are known constants.

Using the stress continuity on an unknown contour L separating the elastic and plastic regions and the Kolosov–Muskhelishvili formulas (4.1.9), we obtain the following boundary problem for the exterior of the contour:

$$\Phi(z) + \overline{\Phi(z)} = h(x, y) \quad \text{on } L,$$
$$\bar{z}\Phi'(z) + \Psi(z) = f(x, y) \quad \text{on } L. \tag{4.2.2}$$

Functions $h(x, y)$ and $f(x, y)$ are determined and continuously differentiable in an infinite region bounded by γ. These functions are known from solution of the corresponding problem in the theory of plasticity (the first of them is real and the second is complex).

Hence, it is required to find contour L and functions $\Phi(z)$ and $\Psi(z)$ based on the boundary conditions (4.2.1) and (4.2.2).

The internal plane elastic–plastic problem is formulated in exactly the same manner when the elastic region occupies the interior of the contour [there is no need in condition (4.2.1) in this case]. For the sake of brevity, only the external elastic–plastic problem will be considered.

In solving elastic–plastic problems, we adopt the usual assumptions [12]:

1. every characteristic originating from contour γ intersects the unknown contour L at one point;

2. under loading, the contours of the boundary between elastic and plastic zones successively enclose each other.

2. Method of functional equations [67]

Let us go over to a parametric plane of the complex variable ζ using transformation $z = \omega(\zeta)$. The analytical function $\omega(\zeta)$ conformally maps the exterior K^- of the unit circle in plane ζ onto the exterior of an unknown contour L in plane z with correspondence of infinitely distant points $\omega(\infty) = \infty$. This analytical function must be determined in the process of problem solution. Assume

$$\varphi(\zeta) \equiv \Phi[\omega(\zeta)], \qquad \psi(\zeta) \equiv \Psi[\omega(\zeta)],$$

$$\varphi_0(\zeta) \equiv \Phi_0[\omega(\zeta)], \qquad \psi_0(\zeta) \equiv \Psi_0[\omega(\zeta)],$$

$$F\left[\omega(\zeta),\overline{\omega(\zeta)}\right] = f\left[\frac{1}{2}\left(\omega(\zeta)+\overline{\omega(\zeta)}\right),\frac{1}{2i}\left(\omega(\zeta)-\overline{\omega(\zeta)}\right)\right], \tag{4.2.3}$$

$$H\left[\omega(\zeta),\overline{\omega(\zeta)}\right] = h\left[\frac{1}{2}\left(\omega(\zeta)+\overline{\omega(\zeta)}\right),\frac{1}{2i}\left(\omega(\zeta)-\overline{\omega(\zeta)}\right)\right].$$

From boundary condition (4.2.2) in plane ζ, the following boundary value problem is obtained, in the adopted notations, for determination of the analytical functions $\varphi(\zeta)$, $\psi(\zeta)$, and $\omega(\zeta)$:

$$\varphi(\zeta) + \overline{\varphi(\zeta)} = H\left[\omega(\zeta),\overline{\omega(\zeta)}\right] \quad \text{on}\, C\colon |\zeta| = 1,$$

$$\frac{\overline{\omega(\zeta)}}{\omega'(\zeta)}\varphi'(\zeta) + \psi(\zeta) = F\left[\omega(\zeta),\overline{\omega(\zeta)}\right]. \tag{4.2.4}$$

The prime denotes differentiation with respect to ζ. At an infinitely distant point, functions $\varphi(\zeta)$, $\psi(\zeta)$, and $\omega(\zeta)$, based on (4.2.1), behave in the following fashion:

$$\varphi(\zeta) = \varphi_0(\zeta) + O(\zeta^{-1}) = O(\zeta^m), \qquad \omega(\zeta) = O(\zeta),$$

$$\psi(\zeta) = \psi_0(\zeta) + O(\zeta^{-1}) = O(\zeta^n). \tag{4.2.5}$$

Let us discuss the second boundary condition (4.2.4). It represents a finite equation with respect to $\overline{\omega(\zeta)}$.

Proposition 1. *Let the second boundary condition* (4.2.4) *be solvable with respect to*

$$\overline{\omega(\zeta)} = \Gamma[\omega(\zeta), \psi(\zeta), \varphi'(\zeta)/\omega'(\zeta)] \tag{4.2.6}$$

and function

$$\chi(\zeta) = \Gamma[\omega(\zeta), \psi(\zeta), \varphi'(\zeta)/\omega'(\zeta)] \tag{4.2.7}$$

be analytic outside the unit circle $|\zeta| > 1$, *with the exception of, perhaps, a finite number of isolated singular points of a unique character. Then the boundary value problem* (4.2.4) *is solved in a closed form.*

Indeed, let functions $\omega(\zeta)$ *and* $\chi(\zeta)$ *at* $|\zeta| \to \infty$ *have the form*

$$\begin{aligned} \omega(\zeta) &= c\zeta + \sum_{k=0}^{\infty} c_k \zeta^{-k}, \\ \chi(\zeta) - \chi_0(\zeta) &= O(\zeta^{\nu}), \qquad \nu \geq 0. \end{aligned} \tag{4.2.8}$$

Here, $\chi_0(\zeta)$ *is a function that is analytic in plane* ζ, *with the exception of singular points of function* $\chi(\zeta)$, *at which it has singularities coinciding with the singularities of function* $\chi(\zeta)$. *We assume that function* $\chi_0(\zeta)$ *is known with an accuracy, perhaps, up to undefined constants. The solution of the boundary value problem* (4.2.6) *can, evidently, be written as*

$$\begin{aligned} \omega(\zeta) &= c\zeta + \overline{P}_{\nu}(1/\zeta) + \overline{\chi_0}(1/\zeta), \\ \chi(\zeta) &= \bar{c}/\zeta + P_{\nu}(\zeta) + \chi_0(\zeta). \end{aligned} \tag{4.2.9}$$

$P_{\nu}(\zeta)$ *is a polynomial of* ν*th order with, as yet, undefined coefficients. Now function* $\varphi(\zeta)$ *is found from the solution of the Dirichlet problem* (4.2.4). *The unknown constants are determined from finite equations obtained by expanding functions* $\varphi(\zeta)$, $\psi(\zeta)$, *and* $\omega(\zeta)$ *in the vicinity of the singular points of function* $\chi(\zeta)$ *and at infinity, and by using function* (4.2.7).

In the assumption just referred to, the analyticity of function $\chi(\zeta)$ *outside the unit circle is established a posteriori. We shall indicate a technique, which is convenient to use when the preceding assumption is used. Let us consider the functional equation* ($|\zeta| > 1$)

$$\frac{\varphi'(\zeta)}{\omega'(\zeta)}\overline{\omega}\left(\frac{1}{\zeta}\right) + \psi(\zeta) = F\left[\omega(\zeta), \overline{\omega}\left(\frac{1}{\zeta}\right)\right]. \tag{4.2.10}$$

Function $\overline{\omega}(1/\zeta)$ *is evidently analytic inside the unit circle* $|\zeta| < 1$, *with the exception of the origin of the coordinates, at which it has the first order pole.*

Proposition 2. *Let function* $\omega(\zeta)$ *be analytic in the entire plane* ζ, *except for an infinitely distant point at which it has the first order pole and, probably, for a* *finite number of unique isolated points located inside the unit circle* $|\zeta| < 1$. *Then the second boundary condition* (4.2.4) *allows the analytical extension outside the unit circle* $|\zeta| < 1$ *with the help of the functional equation* (4.2.10).

Indeed, function $\omega(\zeta)$ under the preceding assumptions can be written in the form (4.2.9). As the examples that follow show, function $\chi_0(\zeta)$ can be identically equal to zero. Therefore, evidently,

$$\bar{\omega}(1/\zeta) = \bar{c}/\zeta + P_\nu(\zeta) + \chi_0(\zeta), \tag{4.2.11}$$

and the left-hand side of the functional equation (4.2.10) is analytic outside the unit circle, except for an infinitely distant point and, probably, for a finite number of unique singular points. The right-hand side of the functional equation (4.2.10), i.e. function $F[\omega(\zeta), \bar{\omega}(1/\zeta)]$, must also be analytic outside the unit circle, except for an infinitely distant point and, probably for a finite number of singular points where the singularities coincide with the singularity of the left-hand part. The second boundary condition (4.2.4) in this case will, evidently, be satisfied.

Thus, the necessary condition that function $\omega(\zeta)$ have the form of (4.2.9) is the analyticity of the right-hand side of the functional equation (4.2.10) outside the unit circle, with the exception of a finite number of unique singular points. This condition is also sufficient if the singularities of the right- and left-hand sides of the functional equation (4.2.10) can be chosen so that they coincide. The above mentioned condition sometimes allows finding closed-form solutions of the boundary value problem (4.2.10) easily even when it is unclear whether function $F[\omega(\zeta), \bar{\omega}(1/\zeta)]$ is analytic or not for $|\zeta| > 1$. It can be done by formally substituting expressions (4.2.9) and (4.2.11) for functions $\omega(\zeta)$ and $\bar{\omega}(1/\zeta)$ in equation (4.2.10), setting as a prerequisite the analyticity of function $F[\omega(\zeta), \bar{\omega}(1/\zeta)]$ essentially everywhere in $|\zeta| > 1$, and equating the singularities of the left- and right-hand sides of the functional equation (4.2.10).

If a certain choice of the undefined coefficients and functions can satisfy these requirements, then the existence of a solution (4.2.9) is established.

The preceding considerations are valid not only for the second boundary condition (4.2.4), but also for a boundary condition of a more general form.

Let the following boundary condition be given on the contour of a unit circle $|\zeta| = 1$:

$$g\left[\omega(\zeta), \overline{\omega(\zeta)}, a_1(\zeta), \ldots, a_n(\zeta)\right] = 0. \tag{4.2.12}$$

Here, $\omega(\zeta)$ is a function that is analytic outside the unit circle, with the exception of, probably, an infinitely distant point at which it can have a pole: $a_1(\zeta), \ldots, a_n(\zeta)$ are functions some of which may not be known beforehand; g is a given function of its variables. Consider the functional equation

$$g[\omega(\zeta), \bar{\omega}(1/\zeta), a_1(\zeta), \ldots, a_n(\zeta)] = 0. \tag{4.2.13}$$

Let us find out to what extent and under which conditions the functional equation (4.2.13) can replace the initial boundary condition (4.2.12). The following theorem applies [68].

Theorem 1. *Let the coefficients $a_i(\zeta)$ and function $\omega(\zeta)$ be functions that are analytic in the entire plane ζ, with the exception of a finite number of isolated singular points, and $g(\omega, \bar{\omega}, a_1, \ldots, a_n)$ be an analytical function of all its arguments, with the exception of a finite number of isolated singular point. Then (i) the functional equation* (4.2.13) *is valid in the expanded plane ζ; (ii) any solution of the functional equation* (4.2.13) *in the indicated class of functions satisfies the boundary condition* (4.2.12) *and vice versa.*

It is not difficult to prove the theorem by using an analytical extension and the theorem of uniqueness of analytical functions. It is also not difficult to generalize it for a class of functions, whose natural boundary is different from the isolated points and can be regarded as a region. An analogous theorem is true for a system of boundary conditions and several desired functions.

For the solution of the functional equation, it is natural to expand all analytical functions into series in the vicinity of a point and reduce the problem to the solution of the infinite system of equations with respect to the unknown coefficients. In some cases the infinite system degenerates into a finite system, thereby allowing solution of the initial problem in a closed form.

Specific examples of solving various functional equations of type (4.2.12) are presented in this chapter and also in Chapter 5.

3. Integrodifferential equations for boundary values of the mapping function [5]

a. Let us introduce an alternative route to the solution of the boundary value problem (4.2.4). We shall assume that in (4.2.1),

$$d_k = e_l = 0, \qquad k = 0, 1, \ldots, m, \quad l = 0, 1, \ldots, n. \tag{4.2.14}$$

This assumption does not result in any loss of generality. In case conditions (4.2.14) are not satisfied, the following functions should be considered:

$$\Phi^*(z) = \Phi(z) - \sum_{k=0}^{m} d_k z^k,$$

$$\Psi^*(z) = \Psi(z) - \sum_{l=0}^{n} e_l z^l.$$

For these functions, condition (4.2.14) has already been satisfied. Determination of functions $\Phi^*(z)$ and $\Psi^*(z)$, in this case, is reduced to a problem of type (4.2.2) with somewhat modified functions $h(x, y)$ and $f(x, y)$. If the main force vector, applied to L, is equal to zero, the following representations hold in the vicinity of an infinitely distant point of plane ζ accounting for (4.2.14):

$$\begin{aligned} \varphi(\zeta) &= \alpha_2/\zeta^2 + \alpha_3/\zeta^3 + \cdots, \\ \psi(\zeta) &= \beta_2/\zeta^2 + \beta_3/\zeta^3 + \cdots, \\ \omega(\zeta) &= c\zeta + c_0 + c_1/\zeta^1 + c_2/\zeta^2 + \cdots, \qquad c > 0. \end{aligned} \tag{4.2.15}$$

Using the Schwarz formula for the outside of the circle, we obtain from the first of conditions (4.2.4),

$$\varphi(\zeta)=\frac{1}{2\pi i}\int_C\frac{H\big(\omega(\tau),\overline{\omega(\tau)}\big)}{\zeta-\tau}d\tau, \tag{4.2.16}$$

$$\zeta\in K^-,\qquad \tau\in C,\qquad \tau=\exp(i\theta),\qquad 0\le\theta\le 2\pi.$$

Here and henceforth, integration over contour C is performed so that region K^-: $|\zeta|>1$ remains on the right.

In accordance with the mean value theorem for the harmonic function $\operatorname{Re}\varphi(\zeta)$, we have

$$\int_C\frac{H\big(\omega(\tau),\overline{\omega(\tau)}\big)}{\tau}d\tau=0. \tag{4.2.17}$$

By virtue of (4.2.15), the coefficient of ζ^{-1} in expansion $\varphi(\zeta)$ in the vicinity of point $\zeta=\infty$ is equal to zero; therefore,

$$\int_C H\big(\omega(\tau),\overline{\omega(\tau)}\big)\,d\tau=0. \tag{4.2.18}$$

From (4.2.16) we find

$$\varphi'(\zeta)=\frac{1}{2\pi i}\int_C\frac{\mu(\tau)}{\zeta-\tau}d\tau,\qquad \zeta\in K^-,$$

$$\mu(\tau)=\frac{dH\big(\omega(\tau),\overline{\omega(\tau)}\big)}{d\tau}=-i\exp(-i\theta)\frac{dH}{d\theta},$$

$$\mu(\tau)=A\big(\omega(\tau),\overline{\omega(\tau)}\big)\omega'(\tau)-\tau^{-2}\overline{A\big(\omega(\tau),\overline{\omega(\tau)}\big)\omega'(\tau)},$$

$$A\big(\omega(\tau),\overline{\mu(\tau)}\big)\equiv h_x\left[\frac{\omega(\tau)+\overline{\omega(\tau)}}{2},\frac{\omega(\tau)-\overline{\omega(\tau)}}{2i}\right]$$

$$-ih_y\left[\frac{\omega(\tau)+\overline{\omega(\tau)}}{2},\frac{\omega(\tau)-\overline{\omega(\tau)}}{2i}\right],$$

$$h_x(x,y)=\frac{\partial h(x,y)}{\partial x},\qquad h_y(x,y)=\frac{\partial h(x,y)}{\partial y}.$$

In proceeding to the limit $\zeta \to \tau \in C$, we obtain

$$\varphi'(\tau) = \frac{1}{2}\mu(\tau) - \frac{1}{2\pi i}\int_C \frac{\mu(t)}{t-\tau}\,dt. \tag{4.2.19}$$

In accordance with (4.2.15), the coefficients of ζ^{-1} and ζ^{-2} in expansion $\varphi'(\zeta)$ in the vicinity $\zeta = \infty$ are equal to zero; therefore,

$$\int_C \mu(t)\,dt = 0, \qquad \int_C \mu(t)\,t\,dt = 0. \tag{4.2.20}$$

By going over to the conjugate value in the second of expressions (4.2.20), and assuming that $\mu(t)\,dt$ is real, we obtain

$$\int_C \frac{\mu(t)}{t}\,dt = 0. \tag{4.2.21}$$

Let us rewrite the second of conditions (4.2.4) in the form

$$\psi(\tau)\omega'(\tau) = -\overline{\omega(\tau)}\,\varphi'(\tau) + F\Big(\omega(\tau), \overline{\omega(\tau)}\Big)\omega'(\tau). \tag{4.2.22}$$

Function $\psi(\zeta)\omega'(\zeta)$ is analytical in K^-; consequently,

$$\frac{1}{2}\psi(\tau)\omega'(\tau) + \frac{1}{2\pi i}\int_C \frac{\psi(t)\omega'(t)}{t-\tau}\,dt = 0, \qquad t, \tau \in C. \tag{4.2.23}$$

Function $\omega(\zeta)$ is analytical in K^-, with the exception of point $\zeta = \infty$, at which it has the first order pole with the principal part $c\zeta + c_0$. Therefore, for any $\tau \in C$ the following condition is met:

$$\int_C \frac{\overline{\omega(t)}}{t-\tau}\,dt = \pi i\overline{\omega(\tau)} - 2c\pi i\bar{\tau}. \tag{4.2.24}$$

By substituting (4.2.22) in (4.2.23), taking into account (4.2.19) and (4.2.24), and using the Poincaré–Bertrand formula, we obtain a singular integro-differential equation for the boundary values of the mapping function

$$\tau V^+(\tau) = -\frac{1}{2\pi}\int_C \frac{\overline{\omega(t)} - \overline{\omega(\tau)}}{\bar{t}-\bar{\tau}}\mu(t)\,d\sigma,$$

$$V^+(\tau) \equiv \frac{1}{2}F\Big(\omega(\tau), \overline{\omega(\tau)}\Big)\omega'(\tau) + \frac{1}{2\pi i}\int_C \frac{F\Big(\omega(t), \overline{\omega(t)}\Big)\omega'(t)}{t-\tau}\,dt, \tag{4.2.25}$$

$$t, \tau \in C, \qquad t = \exp(i\sigma), \qquad 0 \le \sigma \le 2\pi.$$

A functional equation, equivalent to equation (4.2.25), was obtained by Ostrosablin [47] for the mapping function $\omega(\zeta)$ in the form

$$\frac{1}{2\pi i}\int_C \frac{F\left(\omega(t),\overline{\omega(t)}\right)\omega'(t)-\overline{\omega(t)}\;dH\left(\omega(t),\overline{\omega(t)}\right)/dt}{t-\zeta}dt$$

$$+\bar{\omega}\left(\frac{1}{\zeta}\right)\frac{1}{2\pi i}\int_C \frac{dH\left(\omega(t),\overline{\omega(t)}\right)/dt}{t-\zeta}dt=0, \qquad |\zeta|<1. \quad (4.2.26)$$

In proceeding to the limit at $\zeta \to \tau \in C$ in this equation, we obtain equation (4.2.25). Equation (4.2.26) is sometimes more convenient for determination of function $\omega(\zeta)$.

Let $\omega_*(t)$ be a solution of equation (4.2.25) subject to conditions (4.2.17), (4.2.18), (4.2.20), and (4.2.24) and let function

$$z=\omega_*(\zeta)=-\frac{1}{2\pi i}\int_C \frac{\omega_*(t)}{t-\zeta}dt+c\zeta+c_0,$$

where constant c_0 maps K^- one-to-one and conformally onto some infinite region D with boundary L enclosing curve γ. Then, using (4.2.15), (4.2.16), and (4.2.22) we determine functions $\varphi(\zeta)$ and $\psi(\zeta)$,

$$\varphi(\zeta)=\frac{1}{2\pi i}\int_C \frac{H\left(\omega_*(t),\overline{\omega_*(t)}\right)}{\zeta-t}dt,$$

$$\psi(\zeta)=\frac{[\omega_*'(\zeta)]^{-1}}{2\pi i}\int_C \frac{F\left(\omega_*(t)\overline{\omega_*(t)}\right)\omega_*'(t)}{\zeta-t}dt \qquad (4.2.27)$$

$$-\frac{[\omega_*'(\zeta)]^{-1}}{2\pi i}\int_C \frac{\overline{\omega_*(t)}\,\varphi'(t)}{\zeta-t}dt, \qquad \zeta\in K^-,$$

which are analytic in K^- and provide, when combined with $\omega_*(\zeta)$, the solution of problem (4.2.4).[1] If curve L is symmetrical with respect to axes x and y, it can be assumed that $c_0=0$.

Sometimes an exact solution of equation (4.2.25) can be found. Let, for example,

$$H\left(\omega(\tau),\overline{\omega(\tau)}\right)=k\left[\ln\left(\omega(\tau)\overline{\omega(\tau)}\right)+1\right]-2d_0,$$

$$F\left(\omega(\tau),\overline{\omega(\tau)}\right)=k\overline{\omega(\tau)}/\omega(\tau)-e_0, \qquad (4.2.28)$$

$$d_0=\bar{d}_0, \qquad e_0=\bar{e}_0.$$

[1] In this case, it is assumed that the coefficient of ζ^{-1} in expansion $\psi(\zeta)$ in the vicinity of point $\zeta=\infty$ is equal to zero.

This occurs in the Galin problem, which will be discussed in detail. In this case

$$\mu(\tau)=k\omega'(\tau)/\omega(\tau)-k\overline{\omega'(\tau)\tau^2}/\overline{\omega(\tau)}$$

and the integral equation (4.2.25) has the form

$$\frac{e_0c}{k}=\frac{1}{2\pi i}\int_C\frac{\overline{\omega(t)}-\overline{\omega(\tau)}}{t-\tau}\frac{\overline{\omega'(t)t^2}}{\overline{\omega(t)}}dt. \tag{4.2.29}$$

Based on equalities of type (4.2.24), which hold true for functions $\omega'(\zeta)\zeta^2-c\zeta^2$ and $\zeta^2\omega'(\zeta)/\omega(\zeta)-\zeta$ and are equal to zero at infinity, it follows from (4.2.29) that

$$\omega(\tau)=c(\tau+e_0/\tau k). \tag{4.2.30}$$

The constant c_0 in this instance is equal to zero. Constant c is determined from (4.2.17),

$$c=R\exp(d_0/k-1/2). \tag{4.2.31}$$

Thus

$$\omega(\zeta)=c(\zeta+e_0/\zeta k),\qquad \zeta\in K^-,$$

and functions $\varphi(\zeta)$ and $\psi(\zeta)$ are determined from relations (4.2.27).

b. Let us consider a case of a rational mapping function. Let there be such a solution of the boundary value problem (4.2.4) that $\omega(\zeta)$ has the form

$$\omega(\zeta)=c\zeta+c_0+P_r(\zeta)/Q_s(\zeta), \tag{4.2.32}$$

where $P_r(\zeta)=p_0+p_1\zeta+\cdots+p_r\zeta^r$, $p_r\neq0$, $Q_s(\zeta)=q_0+q_1\zeta+\cdots+q_s\zeta^s$, $q_s\neq0$, are mutually prime polynomials with respect to ζ, $r<s$, and the zeros of the polynomial $Q_s(\zeta)$ are located inside the unit circle, i.e., inside region K^+. Let us find the required conditions of representability of $\omega(\zeta)$ in the form of (4.2.32). Substituting (4.2.32) in (4.2.25), we shall have

$$V^+(\tau)=\bar\tau\Gamma_m(\bar\tau)/\overline{Q_s(\tau)},\qquad |\tau|=1, \tag{4.2.33}$$

where $\Gamma_m(\zeta)$ is a polynomial of order $m\le s-1$, the coefficients of which depend on c_0, c, and on the coefficients of polynomials $P_r(\zeta)$, $Q_s(\zeta)$. Equality (4.2.33) indicates that

$$\omega'(\tau)F\big(\omega(\tau),\overline{\omega(\tau)}\big)=N_*^-(\tau)+\Gamma_m(\bar\tau)\bar\tau/\overline{Q_s(\tau)}, \tag{4.2.34}$$

where $N_*^-(\tau)$ is a boundary value on C of the analytical function $N_*(\zeta)$ in K^-, which tends to zero at infinity. From (4.2.34), the required rationality condition of $\omega(\zeta)$ is

$$F\big(\omega(\tau),\overline{\omega(\tau)}\big)=N^-(\tau)+\overline{\omega(\tau)}\,T^-(\tau), \tag{4.2.35}$$

where $N^{-}(\tau)$ and $T^{-}(\tau)$ are boundary values on C of the analytical functions $N(\infty) = \text{const}$ and $T(\infty) = \text{const}$ in K^{-}.

If it is assumed that $\omega(\zeta)$ is analytical in $K^{-} + C$, i.e., L is an analytical curve, condition (4.2.35) becomes a sufficient rationality condition of $\omega(\zeta)$. In this case, the following functional equation exists in $K^{-} + C$:

$$\bar{\omega}(\zeta^{-1})[\Phi'(\zeta)/\omega'(\zeta) - T(\zeta)] + \Psi(\zeta) - N(\zeta) = 0. \qquad (4.2.36)$$

The required rationality condition (4.2.35) will be fulfilled, in particular, if the stress function in the plastic region satisfies a biharmonic function. This is equivalent to the case when the following overdetermined system has a solution:

$$\frac{\partial \sigma_x}{\partial x} + \frac{\partial \tau_{xy}}{\partial y} = 0, \qquad \frac{\partial \tau_{xy}}{\partial x} + \frac{\partial \sigma_y}{\partial y} = 0, \qquad (4.2.37)$$

$$f^*(\sigma_x, \sigma_y, \tau_{xy}) = 0, \qquad (4.2.38)$$

$$\frac{\partial^2(\sigma_x + \sigma_y)}{\partial x^2} + \frac{\partial^2(\sigma_x + \sigma_y)}{\partial y^2} = 0, \qquad (4.2.39)$$

where $f^* = 0$ is the plasticity condition.

Since stresses in the plastic region satisfy the equilibrium equations and plasticity condition, only the last equation, i.e., (4.2.39) needs to be checked when specific solutions are discussed. Let us consider a few examples (see [1], Chapter 1).

1. In the case of plasticity condition

$$f^* = (\sigma_y - \sigma_x)^2 + 4\tau_{xy}^2 - 4k^2 = 0,$$

system (4.2.37)–(4.2.39) has only a solution of the form

$$\sigma_x = \sigma - k\cos 2\theta^*, \qquad \sigma_y = \sigma + k\cos 2\theta^*, \qquad \tau_{xy} = k\sin 2\theta^*,$$

$$\sigma = k\ln\left[(x-a)^2 + (y-b)^2\right] + n, \qquad \theta^* = -\arctan[(y-b)/(x-a)] + m,$$

where a, b, m, n are arbitrary constants. This solution is used later.

2. Consider the axisymmetric solution in the plastic zone near a circular hole of radius R when the following exponential yield condition is satisfied:

$$f^* = (\sigma_y - \sigma_x)^2 + 4\tau_{xy}^2 - 4k^2\left[1 - \exp\left(-\sigma_0/k + (\sigma_x + \sigma_y)/2k\right)\right]^2 = 0,$$

where σ_0, k are constants. It has the form

$$\begin{aligned} \sigma_r &= \sigma_0 - k - 2k\ln(dr/R) - kR^2/d^2e^2r^2, \\ \sigma_\theta &= \sigma_0 - 3k - 2k\ln(dr/R) + kR^2/d^2e^2r^2, \qquad \tau_{r\theta} = 0. \end{aligned} \qquad (4.2.40)$$

Here, r, θ are polar coordinates, $r > R, 0 \le \theta < 2\pi$; d is the root of equation

$$-p/k + \sigma_0/k = 1 + 2\ln d + e^{-2}d^{-2}, \qquad d > e^{-1};$$

e is the base of natural logarithms; p is a constant. It is readily seen that expression

$$\sigma_x + \sigma_y = \sigma_r + \sigma_\theta = 2\sigma_0 - 2k - 4k \ln(dr/R)$$

satisfies the Laplace equation. Therefore, the stresses, determined by equalities (4.2.40), satisfy equation (4.2.39) and, hence, the required rationality condition (4.2.35) is met.

3. Consider a problem of the plastic state in a plate with a hole bounded by a smooth convex contour Γ (Fig. 3.1). Both shear and normal stresses on the contour are equal to zero. We shall use the Tresca–Saint Venant plasticity condition

$$f^* = (\sigma_y - \sigma_x)^2 + 4\tau_{xy}^2 - 4(\sigma_s - |\sigma_x + \sigma_y|/2)^2 = 0, \qquad \sigma_x\sigma_y \geq \tau_{xy}^2, \quad (4.2.41)$$

where σ_s is the yield point under pure tension. We shall define the equation of Γ in the form of (3.2.1) and introduce the orthogonal λ, β coordinates, which are related to x, y coordinates by

$$x = \frac{dM(\beta)}{d\beta}\cos\beta + (M(\beta) + \lambda)\sin\beta,$$

$$y = \frac{dM(\beta)}{d\beta}\sin\beta - (M(\beta) + \lambda)\cos\beta,$$

$$\frac{\pi}{2} \leq \beta < \frac{5\pi}{2}, \qquad 0 \leq \lambda < \infty.$$

Here, $M(\beta)$ is the support function of contour Γ and λ is the distance of point (x, y) along the normal to Γ.

Solving the equilibrium equations (4.2.37) and plasticity condition (4.2.41), expressed in λ, β coordinates, and accounting for the absence of external stresses on Γ, we find

$$\begin{aligned} &\sigma_\lambda = \sigma_s[1 - \rho(\beta)/(\rho(\beta) + \lambda)], \\ &\sigma_\beta = \sigma_s, \qquad \tau_{\lambda\beta} = 0, \qquad \rho(\beta) = M(\beta) + d^2M(\beta)/d\beta^2. \end{aligned} \quad (4.2.42)$$

Here, $\rho(\beta)$ is the radius of curvature of Γ as a function of β; σ_λ, σ_β, $\tau_{\lambda\beta}$ are the components of the stress tensor in λ, β coordinates. Substituting expression

$$\sigma_x + \sigma_y = \sigma_\lambda + \sigma_\beta = \sigma_s[2 - \rho(\beta)/(\rho(\beta) + \lambda)]$$

into the Laplace equation, written in λ, β coordinates, we come to the conclusion that it is satisfied only for $\rho(\beta) = 0$.

Thus, the stress function in the plastic region will not be biharmonic in this case; however, the required rationality condition can be met.

Let, for example, Γ be a circle of radius R; then $\beta = \theta + \varphi/2$, $\rho(\beta) = R$, and $R + \lambda = r$, where r, θ are polar coordinates. From (4.2.42), it follows that

$$\sigma_r = \sigma_s(1 - R/r), \qquad \sigma_\theta = \sigma_s, \qquad \tau_{r\theta} = 0.$$

In this case,

$$H\left(\omega(\tau), \overline{\omega(\tau)}\right) = \sigma_s/2 - \sigma_s R\left[\omega(\tau)\overline{\omega(\tau)}\right]^{-1/2}/4 - 2d_0,$$

$$F\left(\omega(\tau), \overline{\omega(\tau)}\right) = \sigma_s R\left[\overline{\omega(\tau)}\right]^{1/2}[\omega(\tau)]^{-3/2}/2 - e_0.$$

If $\omega(\tau) = c\tau + c_1\tau^{-1} + c_3\tau^{-3}$, and $c_1^2 = 4c_3c$, then $F[\omega(\tau), \overline{\omega(\tau)}]$ will be a boundary value on C of the analytical function, tending to a constant at infinity, in K^-. The functional equation (4.2.36) exists in region K^-. A complete solution of this elastic–plastic problem is given in Chapter 5.

§ 3. GALIN PROBLEM

Let an infinite body, under plane strain, with a circular hole of radius R be in tension by constant stresses applied at infinity

$$\sigma_x = \sigma_x^\infty, \qquad \sigma_y = \sigma_y^\infty, \qquad \tau_{xy} = 0. \tag{4.3.1}$$

Constant external stresses

$$\sigma_r = p, \qquad \tau_{r\theta} = \tau. \tag{4.3.2}$$

are applied to the contour of the hole. When the body is in tension, stress concentration occurs in the vicinity of the hole. If the stresses $\sigma_x^\infty, \sigma_y^\infty$ are sufficiently large, plastic regions appear. Let us assume that the plastic zone encloses the circular hole entirely (Fig. 4.1). In this case, stresses in the plastic region are directly determined from the boundary conditions, equilibrium equations, and plasticity condition.

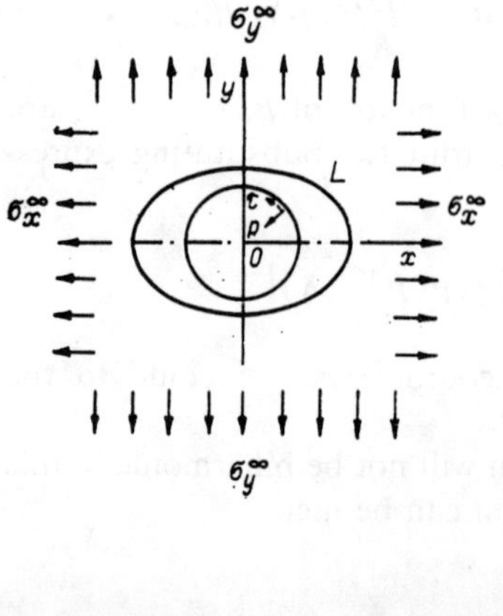

FIG. 4.1.

Stresses in the plastic region under the Tresca–Saint Venant condition are expressed by [39]

$$
\begin{aligned}
\sigma_r &= \varepsilon k\left[2\ln\left(\sqrt{r^2-A}+\sqrt{r^2+A}\right)-r^{-2}\sqrt{r^4-A^2}\right]+Bk,\\
\sigma_\theta &= \varepsilon k\left[2\ln\left(\sqrt{r^2-A}+\sqrt{r^2+A}\right)+r^{-2}\sqrt{r^4-A^2}\right]+Bk,\\
\tau_{r\theta} &= Ak/r^2, \qquad A=\tau R^2/k,\\
B &= p/k-\varepsilon\left[2\ln\left(\sqrt{R^2+A}+\sqrt{R^2-A}\right)-R^{-2}\sqrt{R^4-A^2}\right],
\end{aligned}
\tag{4.3.3}
$$

where r, θ are polar coordinates, k is the plasticity constant, and $\varepsilon = \pm 1$.

On the basis of relations (4.1.9) and formulas (4.3.3), the boundary conditions on an unknown contour L, separating the elastic and plastic regions, can be expressed as

$$
4\,\mathrm{Re}\,\Phi(z) = 4\varepsilon k\ln\left(\sqrt{r^2-A}+\sqrt{r^2+A}\right)+2Bk \quad \text{on } L,
$$

$$
\bar{z}\Phi'(z)+\Psi(z)=\left(\frac{\varepsilon k}{r^2}\sqrt{r^4-A^2}+iA\frac{k}{r^2}\right)e^{-2i\theta} \quad \text{on } L;
$$

for $|z|\to\infty$,

$$
\Phi(z)=\tfrac{1}{4}\left(\sigma_x^\infty+\sigma_y^\infty\right)+O(z^{-2}),
$$

$$
\Psi(z)=\tfrac{1}{2}\left(\sigma_y^\infty-\sigma_x^\infty\right)+O(z^{-2}).
$$

Solution of the boundary value problem

Let us go over to the parametric plane of the complex variable ζ using transformation $z=\omega(\zeta)$. The analytical function $\omega(\zeta)$ conformally maps the exterior of the unit circle in plane ζ onto the exterior of an unknown contour in plane z with correspondence of infinitely distant points $\omega(\infty)=\infty$; this function must be determined in the process of problem solving. Let

$$
\varphi(\zeta)=\Phi[\omega(\zeta)], \qquad \psi(\zeta)=\Psi[\omega(\zeta)].
$$

The following boundary value problem, in the adopted notations, is obtained on plane ζ from the boundary condition on L, for determination of the three analytical functions $\varphi(\zeta)$, $\psi(\zeta)$, and $\omega(\zeta)$:

$$
\begin{aligned}
&\varphi(\zeta)+\overline{\varphi(\zeta)} = Bk+2\varepsilon k\ln\left[\sqrt{\omega(\zeta)\overline{\omega(\zeta)}-A}+\sqrt{\omega(\zeta)\overline{\omega(\zeta)}+A}\right],\\
&\frac{\overline{\omega(\zeta)}}{\omega'(\zeta)}\varphi'(\zeta)+\psi(\zeta)=k\frac{Ai+\varepsilon\sqrt{\left[\omega(\zeta)\overline{\omega(\zeta)}\right]^2-A^2}}{[\omega(\zeta)]^2}, \qquad |\zeta|=1;
\end{aligned}
\tag{4.3.4}
$$

for $|\zeta| \to \infty$,

$$\varphi(\zeta) = \tfrac{1}{4}(\sigma_x^\infty + \sigma_y^\infty) + O(\zeta^{-2}),$$

$$\psi(\zeta) = \tfrac{1}{2}(\sigma_y^\infty - \sigma_x^\infty) + O(\zeta^{-2}),$$

$$\omega(\zeta) = O(\zeta).$$

Let us consider the functional equation in the expanded plane

$$\frac{\phi'(\zeta)}{\omega'(\zeta)}\bar{\omega}\left(\frac{1}{\zeta}\right) + \psi(\zeta) = k\frac{Ai + \varepsilon\sqrt{[\omega(\zeta)\bar{\omega}(1/\zeta)]^2 - A^2}}{[\omega(\zeta)]^2}. \tag{4.3.5}$$

We seek a solution of the functional equation (4.3.5) in the form

$$\omega(\zeta) = c\zeta + \bar{P}_\nu(1/\zeta). \tag{4.3.6}$$

Here, $P_\nu(\zeta)$ is a polynomial of νth order with, as yet, undetermined coefficients. By substituting expression (4.3.6) in the functional equation (4.3.5) and expanding all functions into a series in the vicinity of an infinitely distant point, it can readily be seen that $\nu = 1$. Thus, if function $\omega(\zeta)$ has the form (4.3.6), then it must be equal to

$$\omega(\zeta) = c\zeta + c_1/\zeta. \tag{4.3.7}$$

Here, c, c_1 are undetermined constants (from the symmetry conditions they turn out to be real). Constants c and c_1 must be selected such that the right-hand side of the functional equation (4.3.5) is an analytical function outside the unit circle. For this purpose, it is necessary and sufficient that the roots of equation

$$(c\zeta + c_1/\zeta)^2(c/\zeta + c_1\zeta)^2 - A^2 = 0 \tag{4.3.8}$$

coincide in pairs or are all located inside a unit circle $|\zeta| < 1$.

Analysis of equation (4.3.8) shows that it has roots, which coincide in pairs when and only when $A = 0$. By solving the Dirichlet problem (4.3.4) at $A = 0$ for function $\varphi(\zeta)$ and using the conditions at infinity, we obtain

$$\begin{aligned}
\varphi(\zeta) &= \frac{p + k\varepsilon}{2} + \varepsilon k \ln\frac{\omega(\zeta)}{R\zeta}, \\
\psi(\zeta) &= \frac{k\varepsilon}{\zeta\omega'(\zeta)}\bar{\omega}\left(\frac{1}{\zeta}\right), \\
\omega(\zeta) &= c\left(\zeta + \frac{\sigma_y^\infty - \sigma_x^\infty}{2k\varepsilon\zeta}\right), \\
c &= R\exp\left[\frac{\sigma_x^\infty + \sigma_y^\infty - 2p}{4\varepsilon k} - \frac{1}{2}\right].
\end{aligned} \tag{4.3.9}$$

This is the classical Galin solution [12]. Annin ([1], Chapter 1) proved the uniqueness of this solution in the Hölder classes of functions $\varphi(\zeta)$, $\psi(\zeta)$, and $\omega(\zeta)$.

The boundary between the elastic and plastic regions will be an ellipse,

$$\frac{x^2}{c^2(1+\beta)^2}+\frac{y^2}{c^2(1-\beta)^2}=1, \tag{4.3.10}$$

where $\beta=(\sigma_y^\infty-\sigma_x^\infty)/2k\varepsilon$, $\varepsilon=\pm 1$. In the special case when $\sigma_x^\infty=\sigma_y^\infty$, the boundary between the elastic and plastic regions will be a circle of radius

$$c=R\exp((\sigma_x^\infty-p-\varepsilon k)/2\varepsilon k). \tag{4.3.11}$$

Stresses and displacements in the elastic region are determined from formulas (4.1.9) and the stresses in the plastic region are determined from formulas (4.3.3).

It can be shown that, when $A\neq 0$, there always exists a root of equation (4.3.8), which is located outside the unit circle, so that the analyticity condition for the right-hand side of the functional equation (4.3.5) cannot be satisfied by function $\omega(\zeta)$ of type (4.3.7).

Let us consider the boundaries of the obtained solution. The solution of this problem holds true only for the loading paths, for which the contours separating the elastic and plastic regions enclose each other in succession or, at least, come into contact over some sections. We ascribe subscript 1 to the values at the time of preceding loading and subscript 2 to the values at the time of the following moment. Then

$$c_2(1+\beta_2)\geq c_1(1+\beta_1), \qquad c_2(1-\beta_2)\geq c_1(1-\beta_1). \tag{4.3.12}$$

Inequalities (4.3.12) and the relations for c and β determine the required limits of stress variations for σ_x^∞, σ_y^∞, and p. It is significant that the plastic region should enclose circle $r=R$. For this to happen, it is necessary to satisfy the following condition:

$$\left(1-\frac{|\sigma_x^\infty-\sigma_y^\infty|}{2k}\right)\exp\left(\frac{\sigma_x^\infty+\sigma_y^\infty}{4\varepsilon k}-\frac{p}{2\varepsilon k}-\frac{1}{2}\right)\geq 1. \tag{4.3.13}$$

For the solution to hold true, it is necessary that any point in the plastic region could be connected with the contour of the hole by two slip lines, which are located completely in the plastic zone. This condition takes place when the following inequality [12][2] is satisfied:

$$|\beta|\leq 0.171. \tag{4.3.14}$$

When determining displacements in the plastic region, we make use of the associated law of plastic flow (4.1.7) and the Tresca–Saint Venant condition

[2] Ostrosablin [45] has shown that $|\beta|<\sqrt{2}-1$.

(1.3.11). In the case under consideration

$$d\varepsilon_r^p = \lambda, \qquad d\varepsilon_\theta^p = -\lambda, \qquad d\gamma_{r\theta}^p = 0. \tag{4.3.15}$$

Displacements in the plastic region can be obtained from the following system of equations:

$$\varepsilon_r + \varepsilon_0 = 0, \qquad \gamma_{r\theta} = 0, \tag{4.3.16}$$

which can also be expressed in terms of displacement vector components in the polar system of coordinates

$$\begin{aligned} \frac{\partial u_r}{\partial r} + \frac{u_r}{r} + \frac{1}{r}\frac{\partial u_\theta}{\partial \theta} &= 0, \\ \frac{\partial u_\theta}{\partial r} - \frac{u_\theta}{r} + \frac{1}{r}\frac{\partial u_r}{\partial \theta} &= 0. \end{aligned} \tag{4.3.17}$$

It can be shown that the system of equations (4.3.17) is of hyperbolic type. Equations (4.3.17) can be solved numerically. This system of equations was solved by Ivlev [17] using the perturbation method, which resulted in approximate analytial expressions for the displacements:

$$\begin{aligned} \frac{4Gu_r^e}{kc} &= \frac{2}{\rho} - 2\beta\left(\rho + \frac{2}{\rho} - \frac{2}{\rho^3}\right)\cos 2\theta + \beta^2\left[\frac{2}{\rho} + 4\left(-\frac{1}{\rho^3} + \frac{1}{\rho^5}\right)\cos 4\theta\right] \\ &\quad + \beta^3\left[\frac{2}{\rho^3}\cos 2\theta - 2\left(\frac{4}{\rho^5} - \frac{5}{\rho^7}\right)\cos 6\theta\right] + \cdots, \\ \frac{4Gu_\theta^e}{kc} &= 2\beta\left(\rho + \frac{1}{\rho^3}\right)\sin 2\theta + 2\beta^2\left(-\frac{1}{\rho^3} + \frac{1}{\rho^5}\right)\sin 4\theta \\ &\quad + \beta^2\left[\frac{2}{\rho^3}\sin 2\theta + 2\left(-\frac{8}{3\rho^5} + \frac{5}{\rho^7}\right)\sin 6\theta\right] + \cdots, \end{aligned} \tag{4.3.18}$$

$$\begin{aligned} \frac{4Gu_r^p}{kc} &= \frac{2}{\rho} - 4\beta\left[\cos(\sqrt{3}\ln\rho) + \frac{1}{\sqrt{3}}\sin(\sqrt{3}\ln\rho)\right]\cos 2\theta \\ &\quad + \frac{2\beta^2}{\rho} - 2\beta^3\left\{\left[\cos(\sqrt{3}\ln\rho) - \frac{3}{\sqrt{3}}\sin(\sqrt{3}\ln\rho)\right]\cos 2\theta\right. \\ &\quad \left. + \left[\cos(\sqrt{35}\ln\rho) - \frac{3}{\sqrt{35}}\sin(\sqrt{35}\ln\rho)\right]\cos 6\theta\right\} + \cdots, \end{aligned} \tag{4.3.19}$$

$$\frac{4Gu_\theta^p}{kc} = 2\beta\left[2\cos(\sqrt{3}\ln\rho) - \frac{2}{\sqrt{3}}\sin(\sqrt{3}\ln\rho)\right]\sin 2\theta \qquad (4.3.19, \text{cont})$$

$$+ 2\beta^2\left[\cos(\sqrt{15}\ln\rho) + \frac{1}{\sqrt{15}}\sin(\sqrt{15}\ln\rho)\right]\sin 4\theta$$

$$- \beta^3\Big\{\left[2\cos(\sqrt{3}\ln\rho) + 2\sqrt{3}\sin(\sqrt{3}\ln\rho)\right]\sin 2\theta$$

$$+ \left[\frac{2}{3}\cos(\sqrt{35}\ln\rho) + \left(\frac{1}{\sqrt{35}} + \frac{\sqrt{35}}{3}\right)\sin(\sqrt{35}\ln\rho)\right]\sin 6\theta\Big\} + \cdots$$

Here, superscript e denotes the elastic region and superscript p denotes the plastic region, $\rho = r/c$.

These approximations show that the contour of the hole increases under biaxial tension gaining some elongation in the direction of the larger acting force.

Another approximate solution for displacements was obtained by Erlikhman [74].

Exact formulas for displacements in the plastic zone for the Galin problem, taking into account compressibility of the material, were obtained by Ostrosablin [45] on the basis of the Haar–Karman theory

$$\frac{2Gu_r}{(\varkappa+1)kRc} = \frac{2(1-2\nu)}{\varkappa+1}\rho_0\left[\frac{p}{2k} + \ln(\rho_0 c)\right] + \frac{1}{2}[f(\theta_1) + f(\theta_2)]$$

$$+ \frac{1}{2}\int_{\theta_1}^{\theta_2}\Big\{f(\theta)I_0^*[\lambda(\theta)]\left[\ln\frac{\rho(\theta)}{\rho_0} - (\theta-\theta_0)\frac{\rho'(\theta)}{\rho(\theta)}\right]$$

$$+ I_0[\lambda(\theta)]\left[g'(\theta) + f(\theta) - g(\theta)\frac{\rho'(\theta)}{\rho(\theta)}\right]\Big\}\,d\theta,$$

$$\frac{2Gu_\theta}{(\varkappa+1)kRc} = \frac{1}{2}[g(\theta_1) + g(\theta_2)]$$

$$+ \frac{1}{2}\int_{\theta_1}^{\theta_2}\Big\{g(\theta)I_0^*[\lambda(\theta)]\left[\ln\frac{\rho(\theta)}{\rho_0} - (\theta-\theta_0)\frac{\rho'(\theta)}{\rho(\theta)}\right] \qquad (4.3.20)$$

$$+ I_0[\lambda(\theta)]\left[f'(\theta) - g(\theta) + f(\theta)\frac{\rho'(\theta)}{\rho(\theta)}\right]\Big\}\,d\theta,$$

where the prime denotes a derivative with respect to θ, $\varkappa = 3 - 4\nu$,

$$\rho = \frac{r}{Rc}, \qquad c = \exp\left[-\frac{1}{2}\left(1 + \frac{p}{k} - \frac{\sigma_x^\infty + \sigma_y^\infty}{2k}\right)\right],$$

$$\rho(\theta) = \frac{1 - \beta^2}{\sqrt{1 + \beta^2 - 2\beta\cos 2\theta}}, \qquad \beta = \frac{\sigma_y^\infty - \sigma_x^\infty}{2k},$$

$$f(\theta) = \frac{1 - \beta^2}{\rho(\theta)} - \frac{1}{2}\rho(\theta),$$

$$g(\theta) = \rho(\theta)\left[\arctan\left(\frac{-\beta\sin 2\theta}{1 - \beta\cos 2\theta}\right) + \frac{2\beta}{1 - \beta^2}\sin 2\theta\right],$$

$$\lambda(\theta) = \sqrt{\ln^2\frac{\rho(\theta)}{\rho_0} - (\theta - \theta_0)^2},$$

$I_0(\lambda) = \Sigma_{n=0}^{\infty}(1/n!\,n!)(\lambda/2)^{2n}$ is the Bessel function of zero order $I_0^*(\lambda) = (1/\lambda)dI_0/d\lambda$, and θ_1, θ_2 are determined from equations

$$\theta_1 = \theta_0 - \ln\frac{\rho(\theta_1)}{\rho_0}, \qquad \theta_2 = \theta_0 + \ln\frac{\rho(\theta_2)}{\rho_0}.$$

These formulas determine displacements at point ρ_0, θ_0 when the plastic zone completely encloses the hole and the characteristics intersect boundary L at one point, i.e., when the following inequality is satisfied:

$$|\beta| \le 1 - \exp\left[\frac{1}{2}\left(1 + \frac{p}{k} - \frac{\sigma_x^\infty + \sigma_y^\infty}{2k}\right)\right] \equiv f_1, \qquad |\beta| < \sqrt{2} - 1. \quad (4.3.21)$$

The solution of system (4.3.17), which satisfies the continuity conditions for the displacements on the elastic–plastic boundary L is obtained from the formulas when $\nu = \frac{1}{2}$.

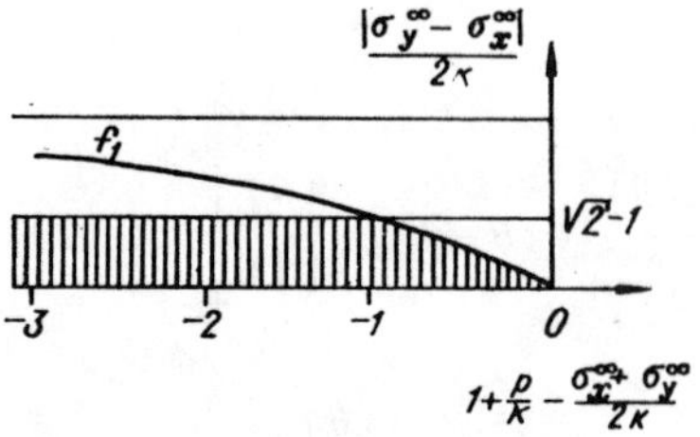

FIG. 4.2

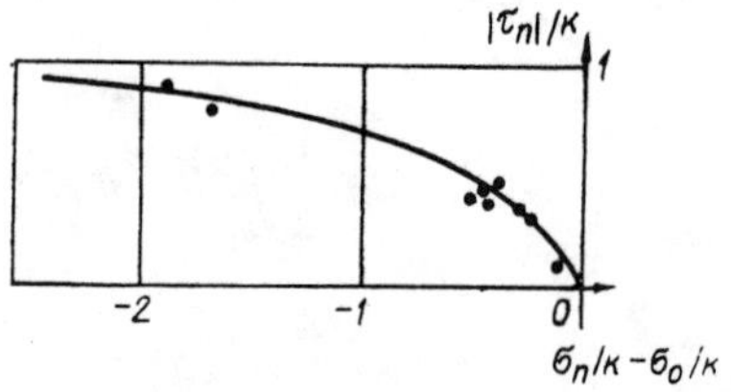

FIG. 4.3.

The region of parameters (4.3.21), for which solution (4.3.20) exists, is shown in Fig. 4.2 (shaded).

In conclusion, we note that the Galin problem has an interesting mechanical analogy with the problem of plate bending [13]. This analogy leads to a number of variational formulations, which are equivalent to a variational inequality.[3]

§ 4. SOME GENERALIZATIONS OF THE GALIN PROBLEM

The problem of elastic–plastic deformation of a plane with a circular hole by constant forces at infinity (see § 3 above) was generalized in a number of directions: discussed were other plasticity conditions, polynomial conditions at infinity, and systems of holes. In the present section some of these generalizations are presented.

1

Let an infinite body with a circular hole of radius R be subjected, under conditions of plane strain, to constant stresses (4.3.1) at infinity. Let also the constant stresses (4.3.2) be applied to the contour of the hole so that $\tau = 0$ (Fig. 4.1). Let the following exponential plasticity condition be satisfied in the plastic zone appearing around the hole:

$$(\sigma_x - \sigma_y)^2 + 4\tau_{xy}^2 = 4k^2\left[1 - \exp\left(-\frac{\sigma_0}{k} + \frac{\sigma_x + \sigma_y}{2k}\right)\right]^2, \qquad \sigma_x + \sigma_y < 2\sigma_0, \tag{4.4.1}$$

This condition was first introduced by Annin [4]. Here, k, σ_0 are positive constants, having the dimensions of stress. The equation of the limit curve on the Mohr plane $\sigma_n - \tau_n$ for condition (4.4.1) has the form

$$\sigma_n - \sigma_0 = k\left[\ln \sin t + (1 - \sin t)\sin t\right],$$

$$|\tau_n| = k(1 - \sin t)\cos t, \qquad 0 < t \leq \pi/2.$$

Figure 4.3 depicts this curve for $k = 70.4$ kg/cm^2, $\sigma_0 = 3.2$ kg/cm^2, and the experimental limit points for coal. Based on the assumption that $-\infty < p < \sigma_0$, we consider two possible cases.

1. In the entire plastic region $\sigma_r - \sigma_\theta \geq 0$ [4]. By using (4.1.2), (4.4.1), (4.3.1), and (4.3.2), we find that the stresses in the plastic region are determined by relations (4.2.40). The sum $\sigma_r + \sigma_\theta$ satisfies the Laplace equation; therefore, the required rationality condition of the mapping function (see § 2) is satisfied.

[3] See references [4, 20, 27] to Chapter 3.

Let us present the final result. The mapping function has the form

$$z=\omega(\zeta)=c\left(\zeta-\frac{\beta}{\zeta}\right),\qquad |\zeta|\ge 1,$$

$$c=R\exp\left(-\frac{\sigma_x^\infty+\sigma_y^\infty}{4k}-\frac{1}{2}+\frac{1}{2d^2e^2}\right),\tag{4.4.2}$$

$$\beta=\frac{\sigma_y^\infty-\sigma_x^\infty}{2k}.$$

The elastic–plastic boundary is an ellipse

$$x^2a^{-2}+y^2b^{-2}=1,\qquad a=c(1-\beta),\qquad b=c(1+\beta).$$

Stresses in the elastic region are equal to

$$\sigma_y-\sigma_x+2i\tau=\sigma_y^\infty-\sigma_x^\infty+2k\left[\frac{\zeta\bar{z}-c(1+\beta^2)}{(\zeta^2+\beta)c}-\frac{\bar{z}}{z}\right]+\frac{2kR^2}{z^2d^2e^2},$$

$$\sigma_y+\sigma_x=\sigma_x^\infty+\sigma_y^\infty+4k\,\mathrm{Re}\left[\ln\left(\frac{1}{2}+\sqrt{\frac{1}{4}+\beta c^2z^{-2}}\right)\right],$$

$$\zeta=\frac{z}{2c}+\sqrt{\frac{z^2}{4c^2}+\beta}.$$

The condition that circle $r=R$ lie inside the ellipse imposes the following restriction on stresses σ_x^∞, σ_y^∞, and p:

$$\left(1+\frac{\sigma_x^\infty-\sigma_y^\infty}{2k}\right)\exp\left(-\frac{\sigma_x^\infty+\sigma_y^\infty}{4k}-\frac{1}{2}+\frac{1}{2d^2e^2}\right)\ge 1.$$

In addition

$$|\sigma_y^\infty-\sigma_x^\infty|<2k\left[1-\exp\left(-\frac{\sigma_0}{k}+\frac{\sigma_x^\infty+\sigma_y^\infty}{2k}\right)\right],\qquad \sigma_x^\infty+\sigma_y^\infty<2\sigma_0.$$

Finally, one more restriction on the parameters of the problem σ_x^∞, σ_y^∞, and p follows from the fact that every point in the plastic region must be connected with the contour of the hole by two characteristics of the systems (4.1.2), (4.4.1), which are located entirely inside the plastic zone.

2. In the entire plastic region $\sigma_r-\sigma_\theta\le 0$ [46]. By using (4.1.2), (4.4.1), (4.3.1), and (4.3.2), we find an expression for the stresses in the plastic region:

$$\sigma_r=\sigma_0+k\ln\left[1+QR^2r^{-2}-\sqrt{(1+QR^2r^{-2})^2-1}\right.$$

$$\left.+QkR^2r^{-2}-k\sqrt{(1+QR^2r^{-2})^2-1}\right.,\qquad \tau_{r\theta}=0,$$

$$\sigma_\theta=2\sigma_0+2k\ln\left[(1+QR^2r^{-2})^2-1\right]-\sigma_r,\qquad Q=d^{-1}(1-d)^2/2.$$

Constant d is the root of equation

$$\ln d - 1 + d = (p - \sigma_0)/k, \qquad 0 < d \leq 1.$$

The boundary value problem (4.2.4) for determination of functions $\varphi(\zeta)$, $\psi(\zeta)$, and $\omega(\zeta)$ has the form ($R = 1$, $k = 1$)

$$\varphi(\zeta) + \overline{\varphi(\zeta)} = 2q + \ln\frac{\sqrt{\omega(\zeta)\overline{\omega(\zeta)} + Q/2} - \sqrt{Q/2}}{\sqrt{\omega(\zeta)\overline{\omega(\zeta)} + Q/2} + \sqrt{Q/2}}, \tag{4.4.3}$$

$$\frac{\overline{\omega(\zeta)}}{\omega'(\zeta)}\varphi'(\zeta) + \psi(\zeta) = \beta + \frac{\sqrt{2Q}}{\omega^2(\zeta)}\left(\sqrt{\omega(\zeta)\overline{\omega(\zeta)} + Q/2} - \sqrt{Q/2}\right), \tag{4.4.4}$$

$$q = \frac{\sigma_0}{2k} - \frac{\sigma_x^\infty + \sigma_x^\infty}{4k}, \qquad \beta = \frac{\sigma_x^\infty - \sigma_y^\infty}{2k}.$$

If we assume that

$$\omega(\zeta) = c\zeta + c_1/\zeta + c_3/\zeta^3 \tag{4.4.5}$$

and that constants c, c_1, c_3 are related by (see [1] in Chapter 1)

$$(c_1^2 - 4cc_3)(c - c_3)^2 - 2cc_3Q = 0,$$

then the right-hand side of equality (4.4.4) will be a limit value on C: $|\zeta| = 1$ of the analytical function in K^-: $|\zeta| > 1$, which tends to a constant at infinity. Consequently, the required rationality condition of the mapping function is fulfilled (see § 2). By substituting (4.4.5) in (4.4.3) and (4.4.4), we find the required functions as

$$z = \frac{R\sqrt{Q/2}}{(1 - a^2)\sinh q}\omega_1(\zeta),$$

$$\omega_1(\zeta) = \zeta + \frac{[2a\cosh q]}{\zeta} + \frac{a^2}{\zeta^3},$$

$$\Phi(\zeta) = \frac{\sigma_x^\infty + \sigma_y^\infty}{4} + k\ln\frac{1 + ae^q/\zeta^2}{1 + ae^{-q}/\zeta^2}, \tag{4.4.6}$$

$$\Psi(\zeta) = 2k\sinh q\left\{\frac{(1 - a^2)\left[a(\zeta^2 + \zeta^{-2}) + e^{-q} + a^2e^q\right]}{\omega_1^2(\zeta)} + \frac{2a\omega_1(\zeta^{-1})}{\zeta^2\omega_1(\zeta)\omega_1'(\zeta)}\right\}.$$

Here

$$c=\left(\sqrt{\frac{\beta^2}{16\sinh^2 q}+\frac{1}{27}}-\frac{\beta}{4\sinh q}\right)^{1/3}-\left(\sqrt{\frac{\beta^2}{16\sinh^2 q}+\frac{1}{27}}+\frac{\beta}{4\sinh q}\right)^{1/3}.$$

The solution of (4.4.6) is meaningful when $|\omega(\zeta)|\geq R$ for $|\zeta|=1$ and function $\omega'(\zeta)$ has no zeros for $|\zeta|\geq 1$. These conditions will be satisfied if

$$\frac{(1-a^2)\sinh q}{1-2|a|\cosh q+a^2}\leq\sqrt{Q/2}\,,\qquad 0<1-2|a|\cosh q-3a^2.$$

The elastic–plastic boundary L in parametric form is defined by

$$\frac{x}{R}=\frac{\sqrt{Q/2}}{(1-a^2)\sinh q}\left[(1+2a\cosh q)\cos t+a^2\cos 3t\right],$$

$$\frac{y}{R}=\frac{\sqrt{Q/2}}{(1-a^2)\sinh q}\left[(1-2a\cosh q)\sin t-a^2\sin 3t\right], \tag{4.4.7}$$

$$0\leq t\leq 2\pi.$$

We note, in conclusion, that the use of other yield conditions [24, 41, 42, 55] also leads to rational mapping functions of type (4.4.2), (4.4.5).

2

Let an infinite body with a round hole of radius $R=1$ be loaded with a constant force (4.3.2) with $\tau=0$, and let the stresses at infinity be determined by the Kolosov–Muskhelishvili polynomial potentials

$$\Phi_0(z)=d_0+d_1z+\cdots+d_nz^n,$$

$$\Psi_0(z)=e_0+e_1z+\cdots+e_mz^m,$$

where $d_0, d_1,\ldots, d_n, e_0, e_1,\ldots, e_m$ are given constants. We assume that the Guber–Mises plasticity condition is satisfied in the plastic zone forming around the hole

$$(\sigma_y-\sigma_x)^2+4\tau_{xy}=4k^2. \tag{4.4.8}$$

The solution in the plastic region has the form of (4.3.3) with $\tau=0$.

Let us consider functions (see § 2)

$$\Phi^*(z)=\Phi(z)-\Phi_0(z),\qquad \Psi^*(z)=\Psi(z)-\Psi_0(z). \tag{4.4.9}$$

The boundary value problem (4.2.4) of determining functions $\varphi(\zeta) \equiv \Phi^*(\omega(\zeta))$, $\Psi = \Psi^*(\omega(\zeta))$, and $\omega(\zeta)$ of form (4.2.5) is ($|\zeta| = 1$)

$$\varphi(\zeta) + \overline{\varphi(\zeta)} = p + \varepsilon k \ln\left(\omega(\zeta)\overline{\omega(\zeta)}\right) + \varepsilon k - \sum_{j=0}^{n} d_j \omega^j(\zeta) - \sum_{j=0}^{n} \bar{d}_j \overline{\omega^j(\zeta)}, \qquad \varepsilon = \pm 1, \quad (4.4.10)$$

$$\frac{\overline{\omega(\zeta)}}{\omega'(\zeta)}\varphi'(\zeta) + \psi(\zeta) = \varepsilon k \frac{\overline{\omega(\zeta)}}{\omega(\zeta)} - \left[\overline{\omega(\zeta)} \sum_{j=1}^{n} j d_j \omega^{j-1}(\zeta) + \sum_{j=0}^{m} e_j \omega^j(\zeta)\right]. \quad (4.4.11)$$

According to (4.2.15), when $|\zeta| \to \infty$ we have

$$\omega^j(\zeta) = A_0^{(j)} + A_1^{(j)}\zeta + \cdots + A_j^{(j)}\zeta^j + O(1/\zeta).$$

The coefficients $A_0^{(j)}, A_1^{(j)}, \ldots$ are determined from the known relations [48] in terms of $c, c_1, c_2, \ldots$.

Let $s = \max(n-1, m)$. Divide both sides of equality (4.4.11) by ζ^s. It is easy to see that the rationality condition for the mapping function (4.2.35) is satisfied. Therefore, the solution to the boundary value problem (4.4.10), (4.4.11) can be sought in the form of (4.2.32) from the functional equation (4.2.36). This, however, is convenient to do for specific values of n, m ([12, 29, 56–58, 65] and [2], Chapter 1). It must be noted that function $\omega(\zeta)$, in this case, is explicitly determined from integrodifferential equations (4.2.25).

3

Let us consider an elastic–plastic distribution of stresses in a plane weakened by two circular holes [6].

Let an infinite body with two circular holes of the same radius $R = 1$ separated by a distance $2a$ (Fig. 4.4) be subjected to uniform stresses (4.3.1) at infinity. Uniform pressure (4.3.2) is applied to the contour of each hole and $\tau = 0$. We assume that the plasticity condition (4.4.8) is satisfied in the plastic zones appearing around the holes. The solution in each plastic zone has the form of (4.3.3) at $\tau = 0$, where r, θ are the polar coordinates with a pole at point O_i, $i = 1, 2$. Let us introduce the following notations: L_1, L_2 are components of the elastic–plastic boundary; D^- is an infinite doubly connected region bounded by L_1 and L_2; D_i^- is an infinite singly connected region bounded by L_i, $i = 1, 2$. In the doubly connected region D^-, bounded by the required curves L_1 and L_2, the material is in an elastic state. The Kolosov–Muskhelishvili potentials $\Phi(z), \Psi(z)$, which define this state, are analytical functions z in region D^-. By accounting for force and geometrical symmetry with respect to axes x and y, the potentials can be presented in the

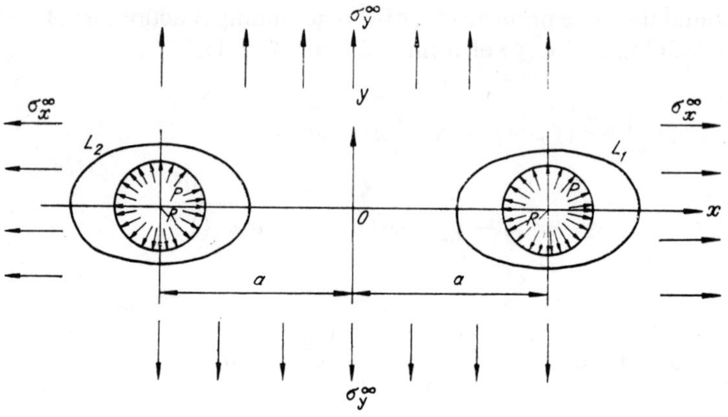

FIG. 4.4.

form

$$
\begin{aligned}
&\Phi(z)=\alpha_0+\Phi_1(z)+\Phi_2(z), \qquad \Psi(z)=\beta_0+\Psi_1(z)+\Psi_2(z),\\
&\Phi_1(z)=\sum_{k=1}^{\infty}\alpha_k(z-a)^{-k}, \qquad \Phi_2(z)=\Phi_1(-z),\\
&\Psi_1(z)=\sum_{k=1}^{\infty}\beta_k(z-a)^{-k}, \qquad \Psi_2(z)=\Psi_1(-z),\\
&\quad z\in D^-+L_1+L_2,\\
&\operatorname{Im}\alpha_k=\operatorname{Im}\beta_k=0, \qquad k=0,1,\ldots,\\
&\quad \alpha_0=(\sigma_x^\infty+\sigma_y^\infty)/4, \qquad \beta_0=(\sigma_y^\infty-\sigma_x^\infty)/2,\\
&\Phi(\zeta)=\alpha_0+O(\zeta^{-2}), \qquad \Psi(\zeta)=\beta_0+O(\zeta^{-2}) \quad \text{for } |\zeta|\to\infty.
\end{aligned}
\tag{4.4.12}
$$

Curve L_2 is obtained by imaging L_1 relative to the y axis; curves L_1 and L_2 are symmetrical with respect to the x axis.

Let us assume $d=\max_{x\in L_1}\{(x-a)^2+y^2\}/2$. We suppose that $d/a\ll 1$ and find an approximate solution for the problem by assuming that the following relations are satisfied:

$$
\begin{aligned}
&\Phi_2(z)=\sum_{m=0}^{n}A_m(z-a)^m, \qquad \Psi_2(z)=\sum_{m=0}^{n}B_m(z-a)^m \quad \text{on } L_1,\\
&\Phi_1(z)=\sum_{m=0}^{n}A_m(-1)^m(z+a)^m,\\
&\Psi_1(z)=\sum_{m=0}^{n}B_m(-1)^m(z+a)^m \quad \text{on } L_2,\\
&A_m=\frac{(-1)^m}{m!}\left(\frac{d^m}{dz^m}\Phi_1(z)\right)_{z=-a}, \qquad B_m=\frac{(-1)^m}{m!}\left(\frac{d^m}{dz^m}\Psi_1(z)\right)_{z=-a}
\end{aligned}
\tag{4.4.13}
$$

Here, n is considered fixed. Let the function

$$z - a = \omega(\zeta) = c\zeta + c_0 + c_1/\zeta + c_2/\zeta^2 + \cdots, \qquad c > 0, \qquad (4.4.14)$$

conformally map the outside of the unit circle K^-: $|\zeta| > 1$ onto D_1^-. We shall assume that

$$\begin{aligned} \varphi(\zeta) &= \Phi_1(a + \omega(\zeta)) - \frac{\alpha_1}{\omega(\zeta)}, \\ \psi(\zeta) &= \Psi_1(a + \omega(\zeta)) - \frac{\beta_1}{\omega(\zeta)} + a\frac{\varphi'(\zeta)}{\omega'(\zeta)}, \\ f(\omega(\zeta)) &= \alpha_0 + \frac{\alpha_1}{\omega(\zeta)} + \sum_{m=0}^{n} A_m[\omega(\zeta)]^m, \\ g(\omega(\zeta)) &= \beta_0 + \frac{\beta_1}{\omega(\zeta)} + \sum_{m=0}^{n} B_m[\omega(\zeta)]^m. \end{aligned} \qquad (4.4.15)$$

Functions $\varphi(\zeta)$, $\psi(\zeta)$, which are analytic in K^- and equal to zero at infinity, satisfy the following conditions on circle C: $|\zeta| = 1$:

$$\varphi(\zeta) + \overline{\varphi(\zeta)} = p + \varepsilon k \ln\left(\omega(\zeta)\overline{\omega(\zeta)}\right) + \varepsilon k - \frac{\left[f(\omega(\zeta)) + \overline{f(\omega(\zeta))}\right]}{2},$$

$$\begin{aligned} \frac{\overline{\omega(\zeta)}}{\omega'(\zeta)}\varphi'(\zeta) + \psi(\zeta) \underset{\varepsilon = \pm 1}{=} \overline{\omega(\zeta)}\left[\frac{\varepsilon k}{\omega(\zeta)} - \frac{df(\omega(\zeta))}{d\omega}\right] \\ - a\frac{df(\omega(\zeta))}{d\omega} - g(\omega(\zeta)). \end{aligned} \qquad (4.4.16)$$

The boundary value problem (4.4.16) is solved similarly to the boundary value problem (4.4.10), (4.4.11). The obtained functions $\varphi(\zeta)$, $\psi(\zeta)$, and $\omega(\zeta)$ will depend on the unknown constants A_m and B_m, $m = 0, 1, \ldots, n$. The nonlinear system of $2(n+1)$ equations for their determination follows from relations (4.4.13).

The solution to this system can be sought by the perturbation method. Assuming $c_0 = 0$ and the perturbation $\varepsilon_* = 1/2a$, we present the final expressions $\omega(\zeta)$ for $n = 2$ [43, 44] accurate to ε_*^3 ($\delta = \text{sign } A_1$):

$$z - a = \omega(\zeta) = c\zeta + \frac{l_1}{\zeta} + \frac{l_2}{\zeta - \zeta_0}, \qquad |\zeta_0| < 1,$$

$$c = \exp\left(-\frac{N}{2} + A_0\right), \qquad l_1 = -\frac{cB_1}{A_1},$$

$$l_2 = -\frac{\zeta_0}{A_1}\left(M - B_0 - \frac{A_1}{2\varepsilon_*} - \frac{B_1}{A_1}\right), \qquad \zeta_0 = \frac{1}{2A_1c} - \delta\sqrt{\frac{1}{(2A_1c)^2} - 1},$$

$$N = -\frac{1}{2}\left(\frac{\sigma_x^\infty + \sigma_y^\infty}{k}\right) - \frac{p}{k} + 1, \qquad M = -\frac{\beta_0}{2k},$$

$$A_0 = -e^{-N} M \varepsilon_*^2, \qquad A_1 = 2e^{-N} M \varepsilon_*^3, \qquad A_2 = 0,$$

$$B_0 = e^{-N}\left(1 + \frac{M}{2}\right)\varepsilon_*^2, \qquad B_1 = -2e^{-N}\left(1 + \frac{1}{2}M\right)\varepsilon_*^3, \qquad B_2 = 0.$$

§ 5. THE PERIODIC PROBLEM

1. Single-periodic problem

Let us consider the elastic–plastic problem for an infinite plane xy, weakened by an infinite row of identical circular holes of radius $\lambda < 1$. The centers of the holes are located at points

$$P_m = m\omega_0, \qquad m = 0, \pm 1, \pm 2, \ldots, \omega_0 = 2.$$

Normal pressure $p < 0$ applied to the contours of the holes Γ_m is such that the resulting plastic zones completely enclose the holes, but do not merge with each other. We assume that the Tresca–Saint Venant plasticity condition is satisfied in the plastic state. The stresses at infinity are assumed equal to zero.

By virtue of the geometrical and force symmetry, the elastic–plastic boundaries L_m are congruent. Let us denote the exterior of contours L_m by D_z^1. The boundary conditions on the unknown contour L_0 have the form

$$\operatorname{Re}\Phi(z) = (k+p)/2 + k\ln(|z|/\lambda), \qquad (4.5.1)$$
$$\bar{z}\Phi'(z) + \Psi(z) = k\bar{z}/z, \qquad z = x + iy.$$

Let us go over to the parametric plane of the complex variable ζ using transformation $z = \omega(\zeta)$. Function $\zeta = \omega^{-1}(z)$ carries out the conformal mapping of region D_z^1 onto region D_ζ^1 in plane ζ, which is the exterior of the circular holes l_m of radius μ with the centers at points P_m.

Let

$$\varphi(\zeta) = \Phi(\omega(\zeta)), \qquad \psi(\zeta) = \Psi(\omega(\zeta)).$$

Then the boundary value problem (4.5.1) on the parametric plane ζ for determination of the three analytical functions $\varphi(\zeta)$, $\psi(\zeta)$, and $\omega(\zeta)$ is written as

$$\operatorname{Re}\varphi(\zeta) = \frac{k+p}{2} + \frac{k}{2}\ln\frac{\omega(\zeta)\overline{\omega(\zeta)}}{\lambda^2}, \qquad \text{on } l_0. \qquad (4.5.2)$$
$$\frac{\overline{\omega(\zeta)}}{\omega'(\zeta)}\varphi'(\zeta) + \psi(\zeta) = k\frac{\overline{\omega(\zeta)}}{\omega(\zeta)}.$$

We seek a solution of this boundary value problem in the form [36]

$$\begin{aligned}
\varphi(\zeta) &= \alpha_0 + \sum_{k_1=0}^{\infty} \alpha_{2k_1+2} \frac{\mu^{2k_1+2}\rho^{(2k_1)}(\zeta)}{(2k_1+1)!}, \\
\psi(\zeta) &= \sum_{k_1=0}^{\infty} \beta_{2k_1+2} \frac{\mu^{2k_1+2}\rho^{(2k_1)}(\zeta)}{(2k_1+1)!} - \sum_{k_1=0}^{\infty} \alpha_{2k_1+2} \frac{\mu^{2k_1+2}s^{(2k_1)}(\zeta)}{(2k_1+1)!}, \qquad (4.5.3) \\
\omega(\zeta) &= \zeta + \sum_{k_1=0}^{\infty} A_{2k_1+2} \frac{\mu^{2k_1+2}\rho^{(2k_1-1)}(\zeta)}{(2k_1+1)!},
\end{aligned}$$

where

$$\begin{aligned}
\rho(\zeta) &= \left(\frac{\pi}{\omega}\right)^2 \frac{1}{\sin^2(\pi\zeta/\omega_0)} - \frac{1}{3}\left(\frac{\pi}{\omega_0}\right)^2, \\
s(\zeta) &= \sum_m{}' \left[\frac{P_m}{(\zeta - P_m)^2} - \frac{2\zeta}{P_m^2} - \frac{1}{P_m}\right].
\end{aligned} \qquad (4.5.4)$$

The prime by the sum sign means that case $m = 0$ is excluded during the summation. From the symmetry it follows that

$$\operatorname{Im} \alpha_{2k_1+2} = \operatorname{Im} \beta_{2k_1+2} = \operatorname{Im} A_{2k_1+2} = 0, \qquad k_1 = 0, 1, \ldots .$$

From the conditions of equality to zero of the main vector of forces acting on the arc joining two congruent points in D_ζ, it follows that $\alpha_0 = \pi^2\beta_2\mu^2/24$. By expanding functions $\varphi(\zeta)$, $\psi(\zeta)$, and $\omega(\zeta)$ into a Laurent series in the vicinity of points $\zeta = 0$, substituting them in (4.5.2), and comparing the coefficients of $\operatorname{Re}\zeta^{k_1}$ and $\operatorname{Im}\zeta^{k_1}$, $k_1 = 0, \mp 1, \ldots$, we obtain an infinite system of nonlinear algebraic equations with respect to the coefficients of the Laurent series.

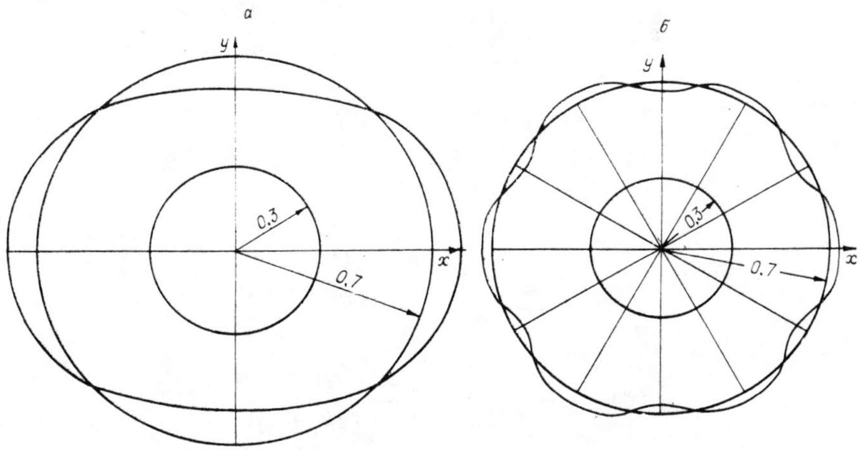

FIG. 4.5.

To a first approximation, the equation of the elastic–plastic boundary in plane xy has the following form in the polar system:

$$r = \mu(d + d_1 \cos 2\theta)^{1/2}.$$

In particular, for $\lambda = 0.3$, $p/k = 2.12$, it is found [36] that $\mu = 0.7$, $d = 1.013$, and $d_1 = 0.326$. The elastic–plastic boundary is shown in Fig. 4.5*a*.

2. Double-periodic problem

Let us consider the elastic–plastic problem for an infinite plane xy weakened by a triangular network of circular holes of equal radius $\lambda < 1$. The centers of the holes are located at points

$$P_{mn} = m\omega_1 + n\omega_2, \qquad \omega_1 = 2, \qquad \omega_2 = 2\exp(i\pi/3), \quad m, n = 0, \pm 1, \ldots .$$

Normal pressure $p < 0$, applied to the contour of holes Γ_{mn}, is such that the resulting plastic zones completely enclose the holes without merging with each other. We consider that the Tresca–Saint Venant condition is satisfied in the plastic state. By virtue of force and geometrical symmetry, the elastic–plastic boundaries are congruent. The boundary conditions on an unknown contour L_{00} have the form (4.5.1). Let us denote the exterior of contours Γ_{mn} by D_z^2. We go over to the parametric plane of the complex variable ζ using transformation $z = \omega(\zeta)$. Function $\zeta = \omega^{-1}(z)$ carries out the conformal mapping of region D_z^2 onto region D_ζ^2 in plane ζ. The latter is the exterior of the circular holes l_{mn} of radius μ with the centers at points P_{mn}. The boundary value problem on the parametric plane ζ for determination of the three functions $\varphi(\zeta) = \Phi(\omega(\zeta))$, $\psi(\zeta) = \Psi(\omega(\zeta))$, and $\omega(\zeta)$ is written in the form of equations (4.5.2), which hold true on l_{00}.

The solution in the parametric plane ζ is sought in the form [25]

$$\begin{aligned}
\varphi(\zeta) &= \alpha_0 + \sum_{k_1=0}^{\infty} \alpha_{2k_1+2} \frac{\mu^{2k_1+2}\wp^{(2k_1)}(\zeta)}{(2k_1+1)!}, \\
\psi(\zeta) &= \beta_0 + \sum_{k_1=0}^{\infty} \beta_{2k_1+2} \frac{\mu^{2k_1+2}\wp^{(2k_1)}(\zeta)}{(2k_1+1)!} \\
&\quad - \sum_{k_1=0}^{\infty} \alpha_{2k_1+2} \frac{\mu^{2k_1+2}Q^{(2k_1+1)}(\zeta)}{(2k_1+1)!}, \\
\omega(\zeta) &= \zeta + \sum_{k_1=0}^{\infty} A_{2k_1+2} \frac{\mu^{2k_1+2}\wp^{(2k_1-1)}(\zeta)}{(2k_1+1)!}.
\end{aligned} \tag{4.5.5}$$

Here, $\wp(\zeta)$ is the elliptical Weierstrass function and $Q(\zeta)$ is a special meromorphic function

$$\wp(\zeta) = \frac{1}{\zeta^2} + \sum_{m,n}{}' \left[\frac{1}{(\zeta - P_{mn})^2} - \frac{1}{P_{mn}^2} \right],$$

$$Q(\zeta) = \sum_{m,n}{}' \left[\frac{\overline{P}_{mn}}{(\zeta - P_{mn})^2} - 2\zeta \frac{\overline{P}_{mn}}{P_{mn}^3} - \frac{\overline{P}_{mn}}{P_{mn}^2} \right], \tag{4.5.6}$$

$$m, n = 0, \pm 1, \pm 2, \ldots .$$

The prime by the sum sign means that the case $m = n = 0$ is excluded. From the condition of equality to zero of the main vector of forces, which act on the arc connecting two congruent points in D_ζ^2, it follows that $\alpha_0 = 3^{-1/2}\pi\beta_2\mu^2$ and $\beta_0 = \sqrt{3}\,\pi\alpha_2\mu^2/6$. The symmetry conditions lead to equalities $\alpha_{6k_1+2} = \beta_{6k_1\mp 2} = A_{6k_1+2} = 0$. By expanding functions $\varphi(\zeta)$, $\psi(\zeta)$, and $\omega(\zeta)$ into a Laurent series in the vicinity of point $\zeta = 0$, substituting them in (4.5.2), and comparing the coefficients of $\operatorname{Re}\zeta^{k_1}$ and $\operatorname{Im}\zeta^{k_1}$, $k_1 = 0, \pm 1, \ldots$, we obtain an infinite system of nonlinear algebraic equations with respect to the coefficients of the Laurent series.

As a first approximation, the elastic–plastic boundary in plane xy has the form $r = \mu(d + d_1\cos 6\theta)^{1/2}$ in polar coordinates. In particular, for $\lambda = 0.3$, $p/k = 2.24$, it is found [25] that $\mu = 0.7$, $d = 0.998$, and $d_1 = 0.090$. The elastic–plastic boundary is shown in Fig. 4.5*b*.

§ 6. THE PERTURBATION METHOD

The perturbation method is an extremely useful tool for solving elastic–plastic problems. It allows finding a solution that is close to the already known exact one; both the shape of the body and the boundary conditions can be subjected to perturbation. The method and some results for plane and three-dimensional elastic–plastic problems are presented in sufficient detail in the book by Ivlev and Ershov ([7], Chapter 1).

Let us discuss an application, other than the Ivlev–Ershov scheme, of the perturbation method to a problem of triaxial tension by force $p^\infty > 0$ of plane xy, which is weakened by a hole. The hole is a singly connected region bounded by a convex curve γ that has continuous curvature and is close to a circle. The normal stress on the boundary of the hole is $p < 0$, and the shear stress is equal to zero.

Equation of γ has the form (3.2.1) and

$$M(\beta) = R(1 + \varepsilon m(\beta)), \qquad \pi/2 \le \beta < \tfrac{5}{2}\pi,$$

$$m(\beta) = \sum_{n=2}^{\infty} b_n \cos(n\beta + a_n), \tag{4.6.1}$$

where ε is a perturbation, $\varepsilon \ll 1$, a_n and b_n are constants, $m(\beta)$ is a periodical (with period 2π) and continuous function, and $m(\pi/2) = m(5\pi/2)$. The radius of curvature for this contour, which is equal to

$$\rho_\gamma(\beta) = R + R\varepsilon\big(m(\beta) + d^2 m(\beta)/d\beta^2\big), \tag{4.6.2}$$

differs insignificantly from constant R, and the contour γ approaches a circle. For example, for an ellipse with semiaxes $R(1+\varepsilon)$ and $R(1-\varepsilon)$ we get $M(\beta)=R(1+\varepsilon\cos 2\beta)$. Suppose that the plastic zone entirely encloses the hole. The components of the stress tensor in the plastic region are determined by[4]

$$\begin{aligned} \sigma_x &= k+p+2k\chi+k\cos 2\lambda, \\ \sigma_y &= k+p+2k\chi-k\cos 2\lambda, \\ \tau_{xy} &= k\sin 2\lambda, \end{aligned} \tag{4.6.3}$$

where parameters χ, λ are expressed in terms of x, y from equations

$$\begin{aligned} -X(\chi,\lambda)\sin\lambda + Y(\chi,\lambda)\cos\lambda &= x, \\ X(\chi,\lambda)\cos\lambda + Y(\chi,\lambda)\sin\lambda &= y, \end{aligned} \tag{4.6.4}$$

$$\begin{gathered} X(\chi,\lambda)=X_0+\varepsilon X_1, \qquad Y(\chi,\lambda)=Y_0+\varepsilon Y_1, \\ X_0=R\exp\chi, \qquad Y_0=0, \\ X_1=-R\sum_{n=2}^{\infty} nb_n\sin(N\chi-f_n)\cos(n\lambda+a_n), \\ Y_1=R\sum_{n=2}^{\infty} nb_n\cos(N\chi)\sin(n\lambda+a_n), \\ f_n=\arcsin(1/n), \qquad N=(n^2-1)^{1/2}. \end{gathered} \tag{4.6.5}$$

This is an exact solution of the equations of the theory of plasticity. We shall seek functions $\chi(x, y)$ and $\lambda(x, y)$, which satisfy system (4.6.4), in the form

$$\chi=\chi_0+\varepsilon\chi_1+O(\varepsilon^2), \qquad \lambda=\lambda_0+\varepsilon\lambda_1+O(\varepsilon^2). \tag{4.6.6}$$

Let r, θ be polar coordinates in plane x, y; from (4.6.4) we have

$$r^2=X^2+Y^2, \qquad X\cos(\lambda-\theta)+Y\sin(\lambda-\theta)=0. \tag{4.6.7}$$

Accounting for (4.6.5) and (4.6.6), we obtain

$$\begin{gathered} \chi_0=\ln\left(\frac{r}{R}\right), \qquad \lambda_0=\theta+\frac{\pi}{2}, \\ \chi_1=\chi_1(r,\theta)=\frac{R}{r}\sum_{n=2}^{\infty} nb_n\sin(N\chi_0-f_n)\cos(n\lambda_0+a_n), \\ \lambda_1=\lambda_1(r,\theta)=\frac{R}{r}\sum_{n=2}^{\infty} nb_n\cos(N\chi_0)\sin(n\lambda_0+a_n). \end{gathered} \tag{4.6.8}$$

---

[4] See the monograph by Sokolovsky ([27], Chapter 1).

Thus, in the plastic zone the following conditions are satisfied:

$$\sigma_x^p + \sigma_y^p = 2(k+p) + 4k\ln(r/R) + 4k\varepsilon\chi_1(r,\theta) + O(\varepsilon^2),$$
$$\sigma_y^p - \sigma_x^p + 2i\tau_{xy}^p = 2k\exp(-2\theta i)(1-2\varepsilon i\lambda_1(r,\theta)) + O(\varepsilon^2). \quad (4.6.9)$$

By conformally mapping the exterior of the unit circle K^-: $|\zeta|>1$ onto the elastic region using function

$$z = \omega(\zeta) = c\zeta + c_0 + c_1\zeta^{-1} + c_2\zeta^{-2} + \cdots,$$

we obtain the following boundary value problem for the Kolosov–Muskhelishvili potentials and function $\omega(\zeta)$:

$$\varphi(\zeta) + \overline{\varphi(\zeta)} = k + p + 2k\ln\left(\frac{r}{R}\right) + 2k\varepsilon\chi_1(r,\theta),$$

$$\frac{\overline{\omega(\zeta)}}{\omega'(\zeta)}\varphi'(\zeta) + \psi(\zeta) = k\exp(-2i\theta)(1-2\varepsilon i\lambda_1(r,\theta)), \quad (4.6.10)$$

$$r^2 = \omega\bar{\omega}, \qquad \frac{\bar{\omega}}{\omega} = \exp(-2i\theta);$$

for $|\zeta|\to\infty$,

$$\varphi(\zeta) = \frac{p^\infty}{2} + O(\zeta^{-2}), \qquad \psi(\zeta) = O(\zeta^{-2}).$$

For $\varepsilon = 0$, an axisymmetric solution is obtained:

$$\omega_0(\zeta) = R_*\zeta, \qquad \varphi_0(\zeta) = p^\infty/2, \qquad \psi(\zeta) = k/\zeta^2,$$
$$R_* = R\exp((p^\infty - p - k)/2k). \quad (4.6.11)$$

Let us assume that when $\varepsilon \neq 0$ the elastic–plastic boundary approaches a circle of radius R_* and its equation is of the form

$$r = R_*(1+\varepsilon\delta(\theta)),$$
$$\delta(\theta) = \alpha_0 + \alpha_1\cos\theta + \beta_1\sin\theta + \alpha_2\cos 2\theta + \beta_2\sin 2\theta + \cdots, \quad (4.6.12)$$

where $\alpha_0, \alpha_1, \beta_1, \alpha_2, \beta_2, \ldots$ are the constants to be determined. We seek an approximate solution of the boundary value problem (4.6.10) in the form

$$\omega(\zeta) = R_*\zeta + \varepsilon\omega_1(\zeta) + O(\varepsilon^2),$$
$$\varphi(\zeta) = p^\infty/2 + \varepsilon\varphi_1(\zeta) + O(\varepsilon^2),$$
$$\psi(\zeta) = k/\zeta^2 + \varepsilon\psi_1(\zeta) + O(\varepsilon^2), \quad (4.6.13)$$
$$\varphi_1(\zeta) = O(\zeta^{-2}), \qquad \psi_1(\zeta) = O(\zeta^{-2}) \text{ when } |\zeta|\to\infty.$$

Let ρ, τ be the polar coordinates in plane ζ. The following relations[5] hold true:

$$\omega(\zeta) = R_*\zeta - \varepsilon R_*\zeta\left[\alpha_0 + \frac{\alpha_1 + i\beta_1}{\zeta} + \frac{\alpha_2 + i\beta_2}{\zeta^2} + \cdots\right],$$

$$\omega'(\zeta) = R_* - \varepsilon R_*\left[\alpha_0 - \frac{\alpha_2 + i\beta_2}{\zeta^2} + \cdots\right], \tag{4.6.14}$$

$$\theta = \tau + \varepsilon\Delta(\tau), \qquad 0 \le \tau \le 2\pi,$$

where

$$\Delta(\tau) = \frac{1}{2\pi}\int_0^{2\pi}\delta(s)\cot\frac{s-\tau}{2}$$

$$= -\alpha_1\sin\tau + \beta_1\cos\tau - \alpha_2\sin 2\tau + \beta_2\cos 2\tau + \cdots. \tag{4.6.15}$$

From (4.6.12) and (4.6.13) it follows that

$$\ln r = \ln R_* + \varepsilon\delta(\theta) + O(\varepsilon^2) = \ln R_* + \varepsilon\delta(\tau) + O(\varepsilon^2),$$
$$\exp(-2i\theta) = \exp(-2i\tau)(1 - 2i\varepsilon\Delta(\tau)). \tag{4.6.16}$$

By substituting (4.6.13) in (4.6.10) and accounting for (4.6.15), (4.6.16), we obtain, with an accuracy up to ε^2, the boundary value problem

$$\varphi_1(\zeta) + \overline{\varphi_1(\zeta)} = 2k(\chi_1(R_*, \tau) + \delta(\tau)),$$
$$\zeta R_*\varphi_1' + \zeta^2\psi_1 = -2ki(\lambda_1(R_*, \tau) + \Delta(\tau)). \tag{4.6.17}$$

The real part of the function $\zeta R_*\varphi_1' + \zeta^2\psi_1$, which is analytic in K^-, is equal to zero on circumference C; therefore, for all $|\zeta| \ge 1$,

$$\zeta R_*\varphi_1' + \zeta^2\psi_1 = i2kA,$$

where A is a real constant. From this relation and from (4.6.15) we obtain an equation for $\delta(\tau)$:

$$\lambda_1(R_*, \tau) + \frac{1}{2\pi}\int_0^{2\pi}\delta(s)\cot\frac{s-\tau}{2}ds + A = 0. \tag{4.6.18}$$

The solvability condition for this equation has the form [14]

$$\frac{1}{2\pi}\int_0^{2\pi}\lambda_1(R_*, \tau)\,d\tau + A = 0. \tag{4.6.19}$$

[5]See the book by Lavrentev and Shabat ([13], to Chapter 1).

As evident from (4.6.8), the free term in the expansion of function $\lambda_1(R_*, \tau)$ into a Fourier series is equal to zero; hence, $A = 0$. By using the Hilbert transformation formula we obtain from (4.6.18)

$$\delta(\theta) = -\frac{1}{2\pi}\int_0^{2\pi}\lambda_1(R_*, \tau)\cot\frac{\theta-\tau}{2}\,d\tau + B,$$

where B is an arbitrary constant.

Since $\varphi_1(\infty) = 0$, it follows from the first of equations (4.6.17) that

$$\int_0^{2\pi}(\lambda_1(R_*, \tau) + \delta(\tau))\,d\tau = 0;$$

therefore, $B = 0$ as well. The final expression for $\delta(\theta)$ has the form

$$\delta(\theta) = -\frac{R}{R_*}\sum_{n=2}^{\infty} nb_n\cos\left(N\ln\left(\frac{R_*}{R}\right)\right)\cos\left(n\theta + \frac{n\pi}{2} + a_n\right).$$

Functions $\varphi_1(\zeta)$ and $\psi_1(\zeta)$ are determined as follows:

$$\varphi_1(\zeta) = -\frac{1}{2\pi i}\int_C\frac{\chi_1(R_*, \tau) + \delta(\tau)}{\tau - \zeta}dt, \qquad \psi_1(\zeta) = -\frac{R_*\varphi_1'}{\zeta}, \qquad t = \exp(i\tau).$$

In particular, for an ellipse with semiaxes $R(1+\varepsilon)$ and $R(1-\varepsilon)$,

$$\delta(\theta) = \frac{2R}{R_*}\cos\left(\sqrt{3}\ln\left(\frac{R_*}{R}\right)\right)\cos 2\theta.$$

This coincides with the result presented in the book by Ivlev and Ershov ([5], Chapter I).

§ 7. SOLUTIONS WITH PLASTIC DISCONTINUITY LINES

1. Formulation of the problem

Let a homogeneous isotropic body from an ideally elastic–plastic material contain normal (mode I) cracks which satisfy the condition of local symmetry. Let us also assume that the following condition holds:

$$L \gg d \gg \rho. \tag{4.7.1}$$

Here, L is a characteristic linear dimension of the body, d is characteristic dimension of the plastic region, and ρ is the radius of curvature of the crack tip. Condition (4.7.1) is satisfied, for example, for many metallic structures in a wide range of dimensions, temperatures, strain rates, and initial cracks (see [8], Chapter 2). The concept of the stress intensity factor applies in this case, and the problem of the fine structure of a crack tip [70] (i.e., distribution of stresses and strains at distances satisfying the condition $L \gg r \gg \rho$) can be stated as follows. Let us examine the vicinity of the crack tip, which is small compared with the characteristic linear dimension of the body, but larger compared with the characteristic dimension of the plastic zone. Thus, a crack on plane xy can be modeled as a semiinfinite slit along $y = 0$, $-\infty < x < 0$, which is free from external loads. In the vicinity of the origin a plastic zone exists, which needs to be determined, and at an infinitely distant point the following asymptotic condition (at $y = 0$, $x \to \infty$) is known:

$$\lim\left(\sigma_y\sqrt{2\pi x}\right) = K_{\mathrm{I}}. \tag{4.7.2}$$

Here, K_{I} is the stress intensity factor.

We assume that the plastic deformation is localized along several slip lines coming from the crack tip. Experience shows a general trend toward the formation of plastic regions at the start of loading in the form of narrow slip bands. This is typical, in particular, for mild steels that are susceptible to delayed yielding. The slip lines can evidently be only straight lines. Indeed, the preceding problem is self-similar, the role of time being played by parameter $K_{\mathrm{I}}/(2\pi G^2)$; the corresponding self-similar dimensionless variables are:

$$x_* = \frac{2\pi x G^2}{K_{\mathrm{I}}}, \qquad y_* = \frac{2\pi y G^2}{K_{\mathrm{I}}},$$
$$\sigma_x^* = \frac{\sigma_x}{G}, \qquad \sigma_y^* = \frac{\sigma_y}{G}, \qquad \tau_{xy}^* = \frac{\tau_{xy}}{G}, \tag{4.7.3}$$
$$\vec{u}_* = \frac{2\pi \vec{u} G^2}{K_{\mathrm{I}}}, \qquad \tau = \frac{\tau_s}{G}.$$

Here, $\vec{u}$ is a displacement vector and τ_s is the yield limit under shear. Henceforth, the asterisk will be omitted for the sake of simplicity.

Experimental data on plane strain indicate that the initial plastic deformation in the vicinity of the crack tip is localized along narrow bands making an angle of 45° with the direction of the crack [80, 83]. The experimental results for thin plates are more contradictory [22, 54, 76, 79].

In view of the preceding discussion, the formulation and solution of the following problem of plane theory of elasticity [28] is of interest:

$$\sigma_y = \tau_{xy} = 0 \quad \text{for } y = 0,\ -\infty < x < 0,$$
$$\sigma_y = 2\tau, \qquad \tau_{xy} = 0 \quad \text{for } y = 0, 0 < x < a,$$
$$\tau_{r\theta} = \tau, \qquad [\sigma_\theta] = 0, \tag{4.7.4}$$
$$[u_\theta] = 0 \quad \text{for } y = \pm x, 0 < x < b/\sqrt{2},$$
$$\lim\left(\tau_y\sqrt{2\pi x}\right) = 1 \quad \text{for } y = 0,\ x \to \infty,$$

where r, θ are polar coordinates, the square bracket denotes a jump during transition through the discontinuity line, and a and b are dimensionless parameters that need to be determined (Fig. 4.6).

In particular, case $a = 0$ corresponds to plane strain, and when $b = 0$ the well-known Dugdale formulation is obtained. The normal conditions on the discontinuity surface of the shear displacement in the ideally elastic–plastic

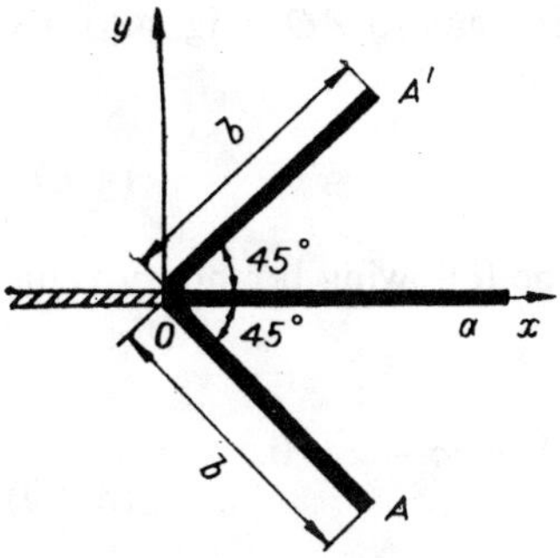

FIG. 4.6.

body are given in the second and third lines. Case $a \neq 0$ is physically possible in thin plates (neck) only and corresponds to the slip plane at angle 45° to the plane of the plate [81].

2. Solution of the boundary value problem

To start with, we obtain an exact solution of the following plane problem of the theory of elasticity:

$$\begin{aligned} &\sigma_y = \tau_{xy} = 0 \quad \text{for } y = 0,\ -\infty < x < 0, \\ &\sigma_y = 2\tau, \qquad \tau_{xy} = 0 \quad \text{for } y = 0, 0 < x < a, \\ &\sigma_\theta - i\tau_{r\theta} = 0, \qquad [u_r + iu_\theta] = f(r) \quad \text{for } y = \pm x, 0 < x < b/\sqrt{2}, \\ &\lim\left(\sigma_y\sqrt{2\pi x}\right) = 1 \quad \text{for } y = 0,\ x \to \infty. \end{aligned} \tag{4.7.5}$$

Let us agree to follow the plastic line in the direction from A to A' (Fig. 4.6); the jump in the value of X is equal to $[X] = X^+ - X^-$.

We shall use the Kolosov–Muskhelishvili relationship

$$\begin{aligned} &\sigma_x + \sigma_y = \sigma_\theta + \sigma_r = 4\,\mathrm{Re}\,\Phi(z), \qquad z = x + iy, \\ &\sigma_y - \sigma_x + 2i\tau_{xy} = e^{-2i\theta}(\sigma_\theta - \sigma_r + 2i\tau_{r\theta}) = 2[\bar{z}\Phi'(z) + \Psi(z)], \\ &2(u + iv) = 2e^{i\theta}(u_r + iu_\theta) = \varkappa\varphi(z) - z\overline{\varphi'(z)} - \overline{\psi(z)}, \\ &\varphi'(z) = \Phi(z), \qquad \psi'(z) = \Psi(z). \end{aligned} \tag{4.7.6}$$

Using representations (4.7.6), the boundary problem (4.7.5) can be written as

$$\begin{aligned} &\Phi(z) + \overline{\Phi(z)} + \bar{z}\Phi'(z) + \Psi(z) = 0 \quad \text{for } y = 0,\ -\infty < x < 0, \\ &\Phi(z) + \overline{\Phi(z)} + \bar{z}\Phi'(z) + \Psi(z) = 2\tau \quad \text{for } y = 0, 0 < x < a, \\ &\left[\Phi(z) + \overline{\Phi(z)} \pm i\bar{z}\Phi'(z) \pm i\Psi(z)\right] = 0, \\ &\left[\varkappa\varphi(z) - z\overline{\varphi'(z)} - \overline{\psi'(z)}\right] = 2f(r)e^{\pm i\pi/4} \quad \text{for } y = \pm x, 0 < x < b/\sqrt{2}, \\ &\lim\left\{\sqrt{2\pi z}\,\Phi(z)\right\} = \tfrac{1}{2}, \qquad \lim\left\{\sqrt{2\pi z}\,\Psi(z)\right\} = \tfrac{1}{4} \quad \text{for } z \to \infty. \end{aligned} \tag{4.7.7}$$

The upper signs correspond to OA' and the lower signs to AO (Fig. 4.6). By introducing a supplementary analytical function

$$\Omega(z) = z\Phi'(z) + \Psi(z), \tag{4.7.8}$$

the boundary problem (4.7.7) can be reduced to the following boundary value problem of the theory of analytical functions:

$$\begin{aligned} &\Phi(z) + \overline{\Phi(z)} + \Omega(z) = 0 \quad \text{for } y = 0,\ -\infty < x < 0, \\ &\Phi(z) + \overline{\Phi(z)} + \Omega(z) = 2\tau \quad \text{for } y = 0,\ 0 < x < a, \end{aligned} \tag{4.7.9}$$

$$[\Phi(z)] = g_1(t), \qquad [\Omega(z)] = g_2(t) \quad \text{for } y = \pm x,\ 0 < x < b/\sqrt{2},$$

$$\lim\{\sqrt{2\pi z}\,\Phi(z)\} = \tfrac{1}{2}, \qquad \lim\{\sqrt{2\pi z}\,\Omega(z)\} = 0 \quad \text{for } z \to \infty. \tag{4.7.10}$$

Here

$$g_1(t) = 2f'(r)/(\varkappa + 1), \qquad t = re^{\pm \pi i/4},$$

$$g_2(t) = \frac{2}{\varkappa + 1}\{rf''(r) \pm 2if'(r) \pm rif''(r)\}.$$

The upper and lower signs correspond to OA' and AO, respectively (Fig. 4.6). We shall seek a solution of the boundary value problem (4.7.9) in the form

$$\begin{aligned} \Phi(z) &= \frac{1}{2\pi i}\int_L \frac{g_1(t)\,dt}{t - z} + \Phi_0(z) + \frac{c_1}{z - z_0} + \frac{\overline{c_1}}{z - \overline{z_0}}, \\ \Omega(z) &= \frac{1}{2\pi i}\int_L \frac{g_2(t)\,dt}{t - z} + \Omega_0(z) + \frac{c_2}{z - z_0} + \frac{\overline{c_2}}{z - \overline{z_0}}. \end{aligned} \tag{4.7.11}$$

Contour L is formed by sections AO and OA', extending from A to A' (Fig. 4.6), c_1 and c_2 are arbitrary constants, whose significance will become clear from the subsequent discussion. Functions $\Phi_0(z)$ and $\Omega_0(z)$ are analytic everywhere outside the slit $y = 0$, $x < a$. By utilizing (4.7.11) the boundary value problem (4.7.9) is reduced to

$$\begin{aligned} &\Phi_0(z) + \overline{\Phi_0(z)} + \Omega_0(z) = F(x) \quad \text{for } y = 0,\ -\infty < x < 0, \\ &\Phi_0(z) + \overline{\Phi_0(z)} + \Omega_0(z) = 2\tau + F(x) \quad \text{for } y = 0,\ 0 < x < a, \end{aligned} \tag{4.7.12}$$

where

$$F(x) = -\frac{1}{\pi i}\int_l \frac{g_1(t)\,dt}{t - x} - \frac{1}{2\pi i}\int_l \frac{g_2(t)\,dt}{t - x} - 2\,\mathrm{Re}\left\{\frac{2c_1 + c_2}{x - z_0}\right\}.$$

Since stresses in the ideally elastic–plastic body are bounded, the solutions for problem (4.7.12) must be sought in a class of totally bounded functions. We should note that due to conditions of symmetry with respect to the x axis, function $F(x)$ is real. Therefore, based on (4.7.12), $\operatorname{Im}\Omega_0 = 0$ on the entire real axis. Consequently, taking into account the condition at infinity we obtain $\Omega_0(z) = 0$.

The solution to the Dirichlet problem for function $\Phi_0(z)$ is written as

$$\Phi_0(z) = \frac{\sqrt{z-a}}{2\pi i}\int_{-\infty}^{a}\frac{F(x)\,dx}{\sqrt{x-a}\,(x-z)} + \tau\left(1 - \frac{1}{\pi i}\ln\frac{i\sqrt{a}-\sqrt{z-a}}{i\sqrt{a}+\sqrt{z-a}}\right). \quad (4.7.13)$$

Here $\sqrt{z-a}$ is considered positive when $y=0$, $x>a$, the root under the integral represents the value of function $\sqrt{z-a}$ on the upper face of the slit $y=0$, $x<a$. In accordance with the condition at infinity, the following relation must be satisfied:

$$-\frac{1}{\pi i}\int_{-\infty}^{a}\frac{F(x)\,dx}{\sqrt{x-a}} + \frac{4\tau\sqrt{a}}{\pi} = 1. \quad (4.7.14)$$

Constants c_1 and c_2 are obtained from the absence of concentrated forces at the end of the plastic line when $z = be^{\pm i\pi/4}$ and the displacement-vector jump in tracing around the same point is specified. The latter additional condition is needed due to differentiation of the displacement jump when the boundary condition (4.7.7) is transformed into (4.7.9). These conditions have the form

$$\left[\varphi(z) + z\overline{\varphi'(z)} + \overline{\psi(z)}\right]_\Lambda = 0,$$

$$\left[\varkappa\varphi(x) - z\overline{\varphi'(z)} - \overline{\psi(z)}\right]_\Lambda = 2f(b)e^{\pm i\pi/4}.$$

Here, Λ is the contour around points $z = be^{\pm i\pi/4}$ (Fig. 4.6). Following the execution of some elementary operations, we obtain

$$c_1 = \frac{f(b)e^{-i\pi/4}}{\pi i(1+\varkappa)},$$
$$c_2 = \frac{b\sqrt{2}}{2\pi}g_1(be^{-i\pi/4}) - \frac{\sqrt{2}f(b)}{\pi i(1+\varkappa)}. \quad (4.7.15)$$

Thus, formulas (4.7.10)–(4.7.15) give a complete solution for a plastic line (dislocation) of an arbitrary power.

3. Approximate solution to the boundary value problem for plane strain

From the physical considerations we specify that function $f(r)$ satisfies the condition

$$\lim f(r) = \lim f'(r) = 0 \quad \text{for } r \to b.$$

In this case the solution of (4.7.11), accounting for (4.7.13) and (4.7.15), will have the form

$$\Phi(z)=\frac{1}{2\pi i}\int_L\frac{g_1(t)\,dt}{t-z}+\frac{\sqrt{z-a}}{2\pi i}\int_{-\infty}^{a}\frac{F(x)\,dx}{\sqrt{x-a}\,(x-z)}$$

$$+\tau\left(1-\frac{1}{\pi i}\ln\frac{i\sqrt{a}-\sqrt{z-a}}{i\sqrt{a}+\sqrt{z-a}}\right), \tag{4.7.16}$$

$$\Omega(z)=\frac{1}{2\pi i}\int_L\frac{g_2(t)\,dt}{t-z},$$

where

$$F(x)=-\frac{1}{\pi i}\int_L\frac{g_1(t)\,dt}{t-x}-\frac{1}{2\pi i}\int_L\frac{g_2(t)\,dt}{t-x}.$$

By using the Poincaré–Bertrand rearrangement formula [10], the expression for function $\Phi(z)$ and relation (4.7.14) can be expressed in terms of single integration:

$$\Phi(z)=\frac{1}{2\pi i}\int_L\frac{g_1(t)\,dt}{t-z}+\frac{\sqrt{z-a}}{4\pi i}\left\{\int_L\frac{2g_1(t)+g_2(t)}{(t-z)\sqrt{t-a}}dt\right.$$

$$\left.-\frac{1}{\sqrt{z-a}}\int_L\frac{2g_1(t)+g_2(t)}{t-z}dt\right\}+\tau\left[1-\frac{1}{\pi i}\omega(z)\right], \tag{4.7.17}$$

$$\omega(z)=\ln\frac{i\sqrt{a}-\sqrt{z-a}}{i\sqrt{a}+\sqrt{z-a}}-\frac{1}{2\pi i}\left\{\int_L\frac{2g_1(t)+g_2(t)}{\sqrt{t-a}}dt\right\}+\frac{4\tau\sqrt{a}}{\pi}-1. \tag{4.7.18}$$

We shall later discuss the case $a=0$, which corresponds to plane strain.

The solution of the boundary value problem for the plastic lines of arbitrary power in the case of plane-strain state will have the form

$$\begin{aligned}\Phi(z)&=\frac{\sqrt{z}}{4\pi i}\int_L\frac{2g_1(t)+g_2(t)}{\sqrt{t}\,(t-z)}dt-\frac{1}{4\pi i}\int_L\frac{g_2(t)\,dt}{t-z},\\ \Omega(z)&=\frac{1}{2\pi i}\int_L\frac{g_2(t)\,dt}{t-z}.\end{aligned} \tag{4.7.19}$$

The solvability condition (4.7.14) of the boundary value problem taking into account (4.7.10) is

$$\int_0^b\frac{rf''(z)+2f'(r)}{\sqrt{r}}dr=-\frac{\pi(1+\varkappa)}{2(\sin(\pi/8)+\cos(\pi/8))}. \tag{4.7.20}$$

Function $f(r)$ will be sought in the form

$$f(r)=A(b-r)^2, \tag{4.7.21}$$

where A and b are to be determined. In this case, in addition to condition (4.7.20), we shall use the approximate relation

$$\int_0^b \tau_{r\theta}\,dr = \tau b. \tag{4.7.22}$$

By using (4.7.19), (4.7.10), and (4.7.21) and performing the necessary calculations, we obtain the following expressions for functions $\Phi(z)$ and $\Omega(z)$:

$$\begin{aligned}\Phi(z) = \frac{2A}{\pi(1+\varkappa)}\Bigg\{ & i\left(b - z\cos\frac{\pi}{4}\right)\ln\frac{z-\bar{z}_0}{z-z_0} - z\sin\frac{\pi}{4}\ln\frac{(z-\bar{z}_0)(z-z_0)}{z^2} \\ & -2i\left(b-3z\cos\frac{\pi}{4}\right)\ln\frac{\sqrt{z}+\sqrt{\bar{z}_0}}{\sqrt{z}+\sqrt{z_0}} \\ & +2b\ln\frac{\left(\sqrt{z}+\sqrt{\bar{z}_0}\right)\left(\sqrt{z}+\sqrt{z_0}\right)}{z} \\ & +6\sqrt{b}\sqrt{z} - 3b\Bigg\},\end{aligned} \tag{4.7.23}$$

$$\begin{aligned}\Omega(z) = -\frac{4A}{\pi(1+\varkappa)}\Bigg\{ & \left(b - z\cos\frac{\pi}{4}\right)\ln\frac{(z-z_0)(z-\bar{z}_0)}{z^2} \\ & + i\sqrt{2}\,z\ln\frac{z-\bar{z}_0}{z-z_0} - 3b\Bigg\}.\end{aligned}$$

Here, $z_0 = be^{-i\pi/4}$. Following integration, (4.7.20) can be written as

$$4Ab^{3/2} = \pi(1+\varkappa)/2(\sin\pi/8 + \cos\pi/8). \tag{4.7.24}$$

Relations (4.7.23), (4.7.6), and (4.7.8) determine $\Phi(z)$ and $\Psi(z)$. By satisfying condition (4.7.22), we shall have

$$\tau = 7.52Ab/\pi(1+\varkappa). \tag{4.7.25}$$

Finally, from (4.7.24) and (4.7.25) we obtain

$$b = 0.502\tau^{-2}, \qquad A = 0.81(1+\varkappa)\tau^3. \tag{4.7.26}$$

By using (4.7.6), (4.7.8), and (4.7.26), we find the crack tip opening

$$\delta = 0.308(1+\varkappa)\tau^{-1}. \tag{4.7.27}$$

After switching to dimensional units, the plastic zone size can be expressed as $b = 1.004K_I^2/\pi\sigma_s^2$. The relation (4.7.27) for a material with the Poisson ratio $\nu = 0.3$, in dimensional units, gives $\delta = 2.25K_I^2/\pi\sigma_s E$. We recall that the extent of the plastic zone and the crack tip opening for the plane stress condition amount to $d = \pi K_I^2/8\sigma_s^2$ and $\delta = K_I^2/E\sigma^s$, respectively.

Comparison of these results for plane strain and plane stress conditions shows that the plastic zone size b for plane strain amounts to 81 percent of the value for plane stress, and the crack tip opening δ amounts to 72 percent of the value for plane stress state. These values are close to the results obtained by Rice [86] using an entirely different method.

An approximate analysis by Rice [86] was based on a method, which bypasses a detailed solution of the boundary value problems. The approach consists in introducing a path independent line integral

$$I = \int_{\Gamma} \left(W\,dy - \vec{T}\frac{\partial \vec{u}}{\partial x} ds \right),$$

where $W = \int_0^{\varepsilon} \sigma_{ij}\, d\varepsilon_{ij}$ is the strain energy, Γ is a curve surrounding the tip of the slit (crack), $\vec{T}$ is the stress vector, determined with respect to the external normal along curve Γ, $T_i = \sigma_{ij} n_j$, and ds is an element of arc length along Γ.

§ 8. CRACK

Determination of stresses and strains near the tip of a crack with normal (mode I) opening, in its most typical state of plane strain, is very important. Let a body, under conditions of plane strain, contain an arbitrary crack with normal (mode I) opening. The material of the body is ideally elastic–plastic, with the Young modulus E and yield point under simple shear τ_s. The material satisfies the Guber–Mises plasticity condition. Let us consider the region around the crack tip, which is small compared with the characteristic linear dimension of the body, but large compared with the characteristic linear dimension of the plastic zone. The crack on plane xy is modeled as a semiinfinite slit along the real x axis, which is free from external loads. The problem is reduced to determination of strains and stresses in region D, which consists of an elastic D_1 and plastic D_2 parts. The boundary L between the elastic and plastic regions needs to be determined. We assume that the displacements, strain components, and stresses are continuous on the elastic–plastic boundary.

1

Considering that the relations of the deformation theory (1.2.12) with condition $T = \tau_s$ hold true, we can use the principle of minimum total energy, according to which the real displacements u, v ensure the minimum of functional

$$I(u,v) = \frac{E}{6(1-2\nu)} \iint \varepsilon^2\,dx\,dy + \frac{G}{2} \iint \Gamma^2\,dx\,dy + \tau_s \iint \left(\Gamma - \frac{\tau_s}{2G} \right) dx\,dy,$$

$$\Gamma = \frac{2}{3}\sqrt{3}\left(\varepsilon_x^2 + \varepsilon_y^2 - \varepsilon_x\varepsilon_y + 3\varepsilon_{xy}^2\right)^{1/2}; \qquad \varepsilon_{xy} = \frac{1}{2}\left(\frac{\partial u}{\partial y} + \frac{\partial v}{\partial x}\right), \quad (4.8.1)$$

$$G = \frac{E}{2(1+\nu)}; \qquad \varepsilon = \varepsilon_x + \varepsilon_y; \qquad \varepsilon_x = \frac{\partial u}{\partial x}, \qquad \varepsilon_y = \frac{\partial v}{\partial y}.$$

The boundary between the elastic and plastic regions is determined from condition $\Gamma = \Gamma_0 = \tau_s / G$.

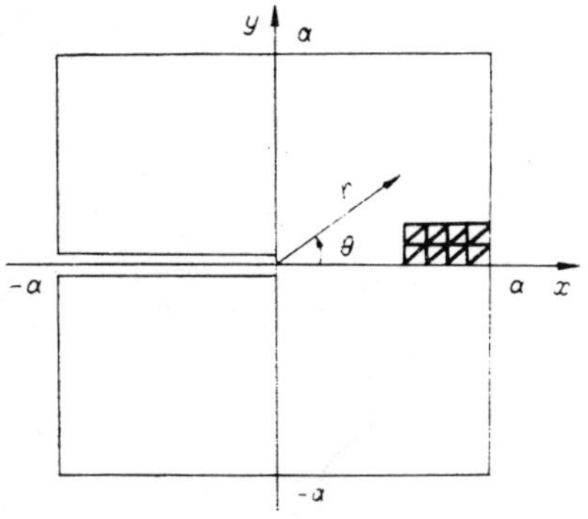

FIG. 4.7

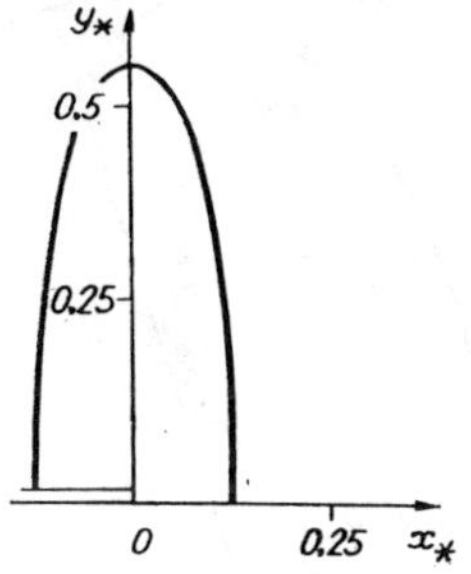

FIG. 4.8.

The elastic–plastic problem is solved numerically by the finite-element method, which involves a subdivision of the region into right-angled isosceles triangles, and by the method of local variations [72].

Region D in plane xy is represented by a square with a slit along the real axis side $2a$ and the center at the origin of the coordinates (Fig. 4.7). The dimensions of the square must, evidently, be such that it is possible to replace the infinite region by a finite one, and the components of displacements and stresses at the crack tip are hardly affected by the boundary conditions given on the sides of the square.

Within the bounds of each element it is assumed that displacements u and v vary linearly.

At the nodal points on the boundary of the square, the displacements are specified in the form adopted from the solution of the problem of an elastic surface with a semiinfinite slit ([8], Chapter 2) under tension at infinity:

$$u = \frac{K_\text{I}}{G}\sqrt{\frac{r}{2\pi}}\cos\frac{\theta}{2}\left(1 - 2\nu + \sin^2\frac{\theta}{2}\right),$$

$$v = \frac{K_\text{I}}{G}\sqrt{\frac{r}{2\pi}}\sin\frac{\theta}{2}\left(2 - 2\nu - \cos^2\frac{\theta}{2}\right).$$

A numerical solution of this elastic–plastic problem, for the Poisson ratio $\nu = 0.46$, showed that the approximating equation of contour L, which separates the elastic and plastic zones has the form [23]

$$\left(\frac{y}{0.55}\right)^2 + \left(\frac{x}{0.11}\right)^2 = \left(\frac{\pi K_\text{I}^2}{8\sigma_s^2}\right)^2. \tag{4.8.2}$$

The plastic zone is shown in Fig. 4.8 in dimensionless coordinates

$$x_* = \frac{8\sigma_s^2}{\pi K_\text{I}^2}x, \qquad y_* = \frac{8\sigma_s^2}{\pi K_\text{I}^2}y,$$

where K_I is the stress intensity factor.

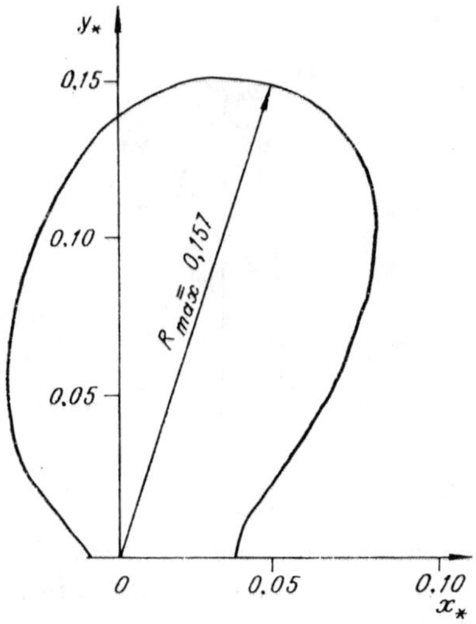

FIG. 4.9

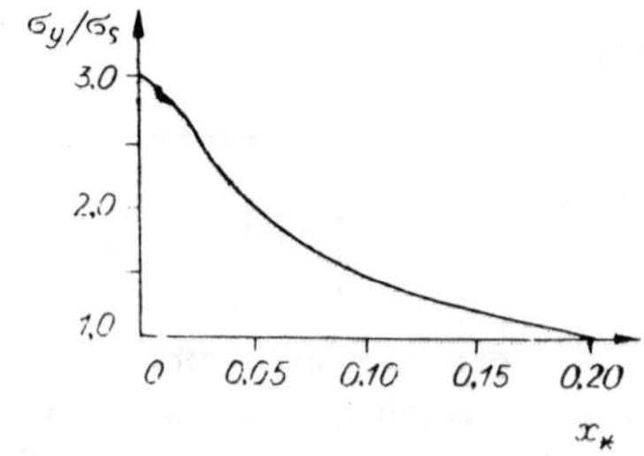

FIG. 4.10.

The region of plastic deformations is represented by the interior of an ellipse with the center at the crack tip. It is apparent that the plastic deformations in this case are concentrated along a relatively narrow layer, coming from the crack tip and perpendicular to its surface.

2

The elastic–plastic problem for the crack tip of normal (mode I) opening under conditions of plane strain in an ideal elastic–plastic material, satisfying the Guber–Mises plasticity condition, was studied by Rice and co-workers [83].

The finite element method with elements obtained by the intersection of concentric circles ($r = \text{const}$) and rays ($\theta = \text{const}$), was used for a numerical analysis of the problem. The Poisson ratio was taken equal to 0.3. The results for the elastic–plastic boundary and the stress on the extension of the crack are presented in Figs. 4.9 and 4.10, respectively.

The stress can be approximated by the relationship

$$\frac{\sigma_y}{\sigma_s} = \frac{1}{\sqrt{2\pi}\left(\sqrt{x_* + 0.1} - 0.18\right)},$$

$$x_* = \frac{\sigma_s^2}{K_{\mathrm{I}}^2} x, \qquad y_* = \frac{\sigma_s^2}{K_{\mathrm{I}}^2} y. \tag{4.8.3}$$

Contour L, separating the elastic and plastic regions, has a shape approximating an ellipse in dimensionless coordinates x_* and y_*, with the major axis length of 0.170 and the minor axis length of 0.100. The center is located at point $x_* \approx -0.02$, $y_* \approx 0.07$ and the major axis is inclined at an angle of 70° relative to axis x_*. Contour L intersects axis x_* at point $x_* \approx 0.040$, and the crack tip opening is equal to $2v_0 = 0.425K_{\mathrm{I}}^2/E\sigma_s$.

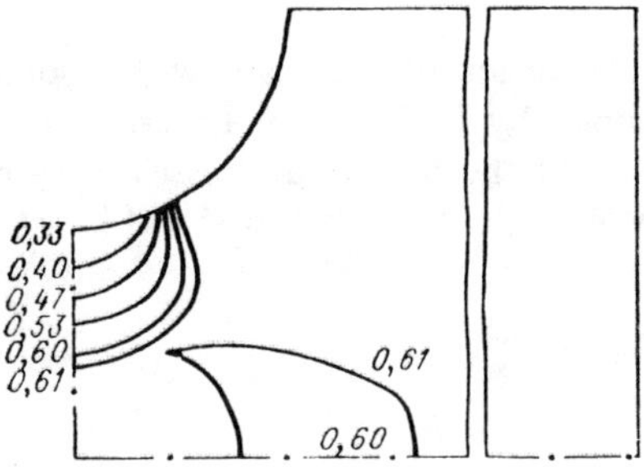

FIG. 4.11

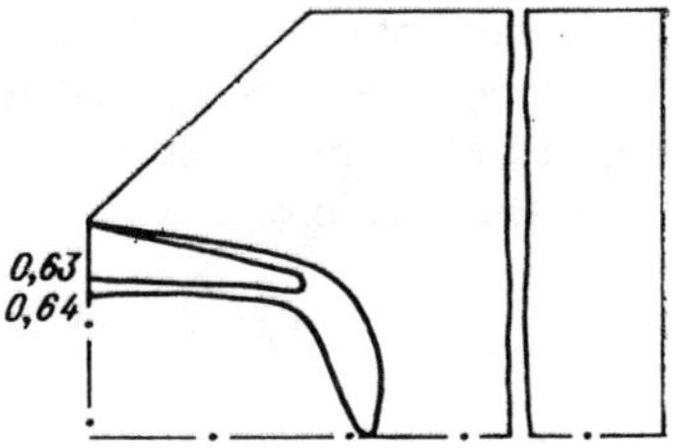

FIG. 4.12.

§ 9. SOME NUMERICAL RESULTS

1. Strip weakened by semicircular notches

The elastic–plastic problem for a strip with semicircular notches in tension was solved by Southwell and Allen using the relaxation method [88]. The material was considered ideal elastic–plastic having yield point τ_s in simple shear and satisfying the Guber–Mises plasticity condition. The calculations were conducted for a strip whose width is equal to four radii of the semicircular notch. The gradual development of plastic zones is illustrated in Fig. 4.11. The numbers at the elastic–plastic boundaries denote the mean stress in fractions of $2\tau_s$.

2. Strip weakened by angular notches

Calculations were conducted by the relaxation method for a strip with angular notches of 90° opening and a depth of one-quarter width of the strip [88]. The strip material was assumed ideal elastic–plastic with yield point τ_s under simple shear and obeying the Guber–Mises plasticity condition. The development of the plastic regions is illustrated in Fig. 4.12. The numbers at the elastic–plastic boundaries denote the mean stress in fractions of $2\tau_s$.

3. Plane weakened by an elliptical hole

Elastic–plastic boundaries for a plane, weakened by an ellipse with axes $2a$ and $2b$ are presented in Figs. 4.13 and 4.14 [84]. Stresses $\sigma_x^\infty = 2.5\tau_s$, $\sigma_y^\infty = 3\tau_s$ are applied at infinity.

The contour of the hole is free from loads. The material is ideal elastic–plastic with the yield point τ_s under simple shear.

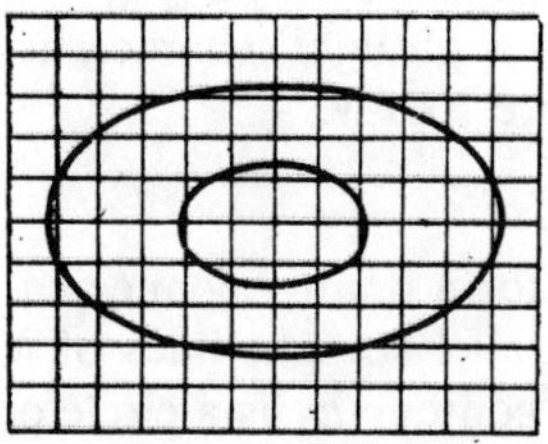

FIG. 4.13

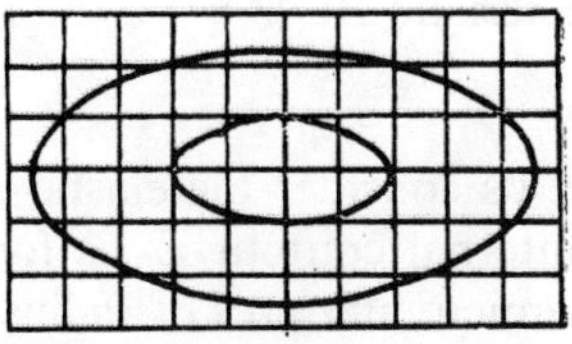

FIG. 4.14.

4. Plane with a square hole

A solution to the elastic–plastic problem for an infinite plane with a square hole was obtained by the method developed by P. I. Perlin. External forces $\sigma_n = p$, $\tau_{tn} = 0$ were applied to the contour of the hole and stresses $\sigma_x^\infty = \sigma_y^\infty$ applied at infinity. The solution to the problem is given below [59, 60].

In the elastic region,

$$\sigma_x + \sigma_y = 4\,\mathrm{Re}\,\varphi_*(\zeta),$$

$$\sigma_y - \sigma_x + 2i\tau_{xy} = \frac{\overline{\omega}(\zeta)}{\omega'(\zeta)}\varphi_*'(\zeta) + \psi_*(\zeta),$$

$$\omega(\zeta) = R\left(\zeta - \frac{1}{6\zeta^3}\right),$$

$$\varphi_*(\zeta) = a_0 + \frac{a_1}{\zeta^4} + \frac{a_2}{\zeta^8} + \frac{a_3}{\zeta^{12}}, \tag{4.9.1}$$

$$\psi_*(\zeta) = \frac{b_1}{\zeta^2} + \frac{b_2}{\zeta^6} + \frac{b_3}{\zeta^{10}},$$

$$a_0 = 1.09\tau_s + 0.5p, \qquad a_1 = -0.43\tau_s,$$

$$a_2 = -0.19\tau_s, \qquad a_3 = 0.03\tau_s,$$

$$b_1 = 1.21\tau_s, \qquad b_2 = -1.4\tau_s, \qquad b_3 = -0.44\tau_s.$$

The elastic–plastic boundary L passes through points $\zeta_1 = 1$, $\zeta_2 = 1.10e^{i\pi/4}$, $\zeta_3 = 1.15e^{-i\pi/9}$, and $\zeta_4 = 0.98e^{i\pi/36}$. The following stresses at infinity correspond to this contour:

$$\sigma_x^\infty = \sigma_y^\infty = 2.18\tau_s + p. \tag{4.9.2}$$

In the plastic region,

$$\begin{aligned} \sigma_x + \sigma_y &= 2\tau_s(1 + 2\varkappa) + 2p, \\ \sigma_y - \sigma_x + 2i\tau_{xy} &= -2\tau_s e^{2i\varphi}. \end{aligned} \tag{4.9.3}$$

Here, $\varkappa$ is a dimensionless variable, φ is the angle between the principle stress σ_1 and axis Ox.

Figure 4.15 shows values of $\varkappa$ and φ in typical regions within the grid of characteristics: $\varkappa = 0$, $\varphi = \pi/2$ in $\triangle ABC$, $\varkappa = 0$, $\varphi/2 = \pi/2 + \theta$, in the circular sector BCN (θ is the angle between a straight characteristic and characteristic BC), $\varkappa = \pi/4$, $\varphi = 3\pi/4$ in square $BDFN$.

5

Let us consider the elastic–plastic problem for a doubly connected region. The internal contour L_2 is formed by two parallel straight lines of length d and semicircular arcs of radius r. The external contour L_1 is a circle of radius R_1 (Fig. 4.16). The solution of this problem has been given in [52]. Ellipse

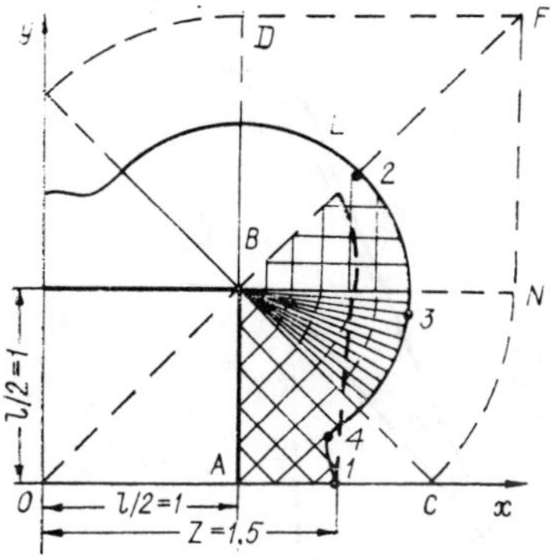

FIG. 4.15

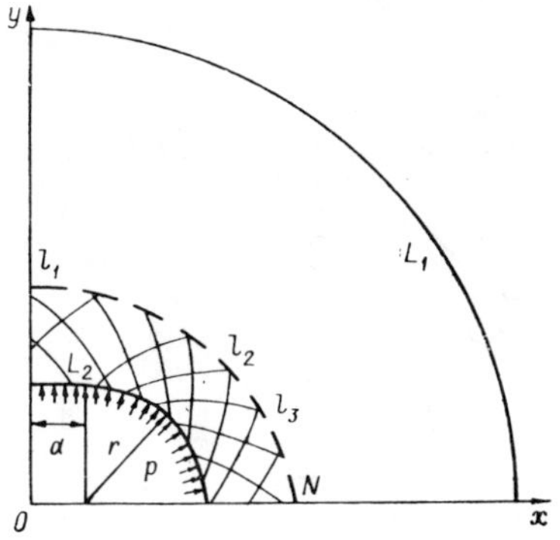

FIG. 4.16.

$\omega(\zeta) = c(\zeta + \gamma/\zeta)$ was adopted as the contour to which the analytical extension was conducted. The calculations were performed with the following values of the dimensionless parameters: $d/r = 0.8$, $R_1/r = 4$, $a/r = 2.2$, $c/r = 2$, and $\gamma = 0.1$, and with two intermediate points ($l_1 = 1.8ri$, $l_2 = 1.54r + i1.26r$), where a is the distance from the origin to the intersection point of the elastic–plastic boundary and axis Ox. Below, σ_s is the yield strength in tension.

The normal force on the inside contour turned out to be $p = 0.855\sigma_s$. Consider an infinite plane, weakened by an identical hole L_2, under the normal force $p = 0.855\sigma_s$ on L_2 and tensile forces at infinity. If the elastic–plastic boundary passes through points $1.8r$ and $2.2r$, the stresses at infinity are

$$\sigma_x^\infty = -0.065\sigma_s, \qquad \sigma_y^\infty = 0.165\sigma_s .$$

6. Problem of punch indentation in elastic–plastic body

a. Let us consider compression of a rectangular bar between two rigid punches. The upper punch exerts pressure on section $y = 1$, $|x| \le 0.5$, and the lower punch covers the entire lateral surface $y = 0$. Friction on the contact surfaces is neglected. A numerical solution of this problem using the method of local variations yields the following results [7] (the Prandtl–Reuss flow theory was used with the Guber–Mises plasticity condition).

At the onset of plastic deformations in the region adjoining the corner of the punch, $t_* = 1.197$. Figure 4.17 shows the elastic–plastic boundary with the numbers $n = 5, 10, 20, 25, 42$ for time intervals $t_n = t_* + n\Delta t$, $\Delta t = 0.02$.

With increasing n, plastic deformation spreads deeper into the body. The total pressure on the upper punch $P = P_*/l\tau_s$ is given in Fig. 4.18 as a function of t (l is a characteristic linear dimension of the body, τ_s is the yield point under simple shear). Curve 1 in Fig. 4.18 depicts the total pressure on the punch when the rectangular bar is compressed by two rigid punches, which traverse the entire lateral surface $y = 0$, $y = 1$. The dashed line represents the exact solution.

b. Let us present the results of a numerical solution to the indentation problem for two flat punches on a strip of finite width. The material of the

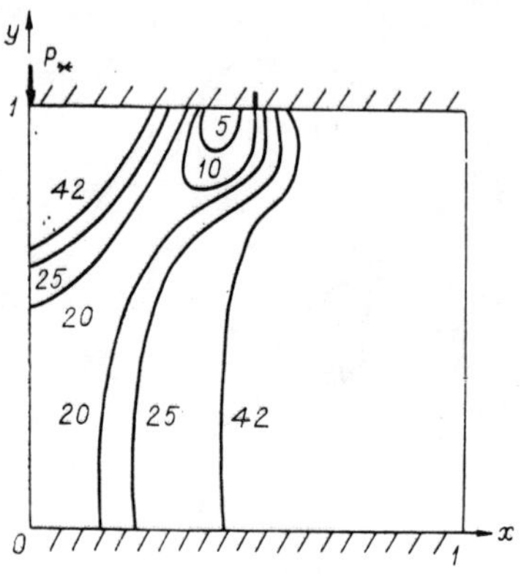

FIG. 4.17

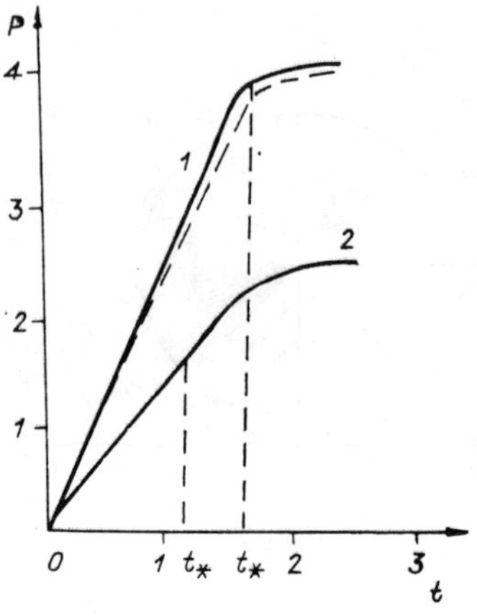

FIG. 4.18.

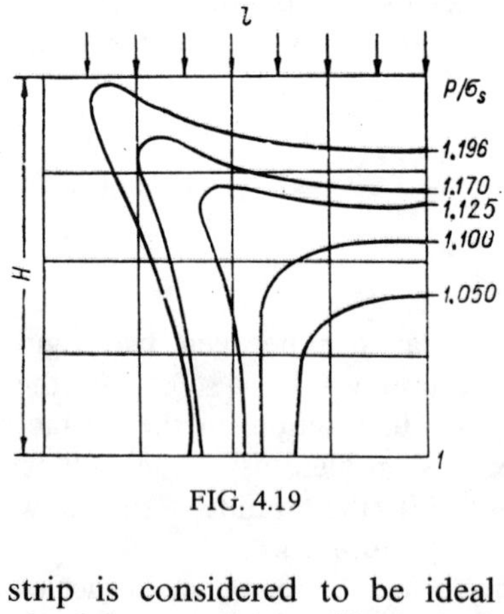

FIG. 4.19

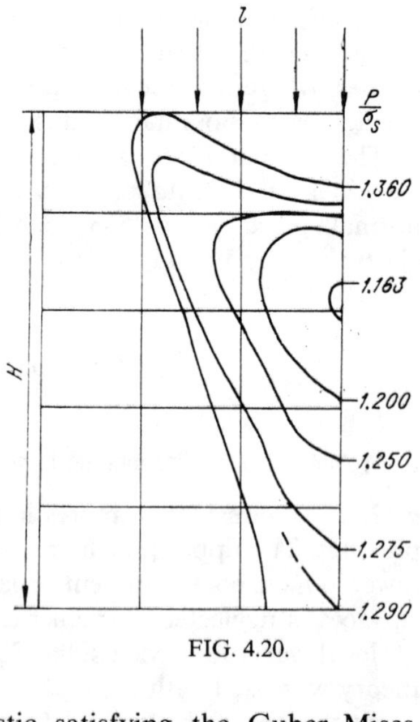

FIG. 4.20.

strip is considered to be ideal elastic–plastic satisfying the Guber–Mises plasticity condition. The calculations were conducted using the relaxation method [89].

Example 1. ***$H/l = 1.14$.*** ($2H$ is the strip thickness, $2l$ is the punch width.) Figure 4.19 shows a gradual growth of the plastic region. The numbers at the elastic–plastic boundaries denote the applied load on the punch divided by the shear yield point τ_s. Plastic deformations first appear when the applied load is $P = \tau_s$.

Example 2. ***$H/l = 2.5$.*** In this case, the region of plastic deformations appears at some distance away from the horizontal axis of strip symmetry. Figure 4.20 illustrates a gradual growth of the plastic region as a function of the applied load. Plastic flow first appears at $P/\tau_s = 1.163$ at a point in the range $0.358 < y/H < 0.395$. When the load reaches $P/\tau_s = 1.290$, the plastic zone spreads to the middle plane of the strip and touches it at the intersection point of the two axes of symmetry.

§ 10. CONTACT PROBLEM (NUMERICAL SOLUTION)

Let us consider the elastic–plastic problem of indentation of a rigid punch into a strip of finite dimensions. The material of the strip is elastic–plastic with work hardening; it satisfies the Guber–Mises plasticity condition and the Prandtl–Reuss flow theory, which accounts for work hardening.

Let us apply the extremum principle of the theory of plastic flow, according to which the real displacement velocities and strain rates realize the minimum of functional

$$\Phi = \tfrac{1}{2}\int_V \left(\dot{\vec{\sigma}}^T \dot{\vec{\varepsilon}}\right) dV - \int_S \left(\dot{\vec{U}}^T \dot{\vec{F}}\right) dS. \qquad (4.10.1)$$

Here, $\dot{\vec{\sigma}}$ is the stress rate vector, $\dot{\vec{\varepsilon}}$ is the strain rate vector, $\dot{\vec{U}}$ is the velocity vector, and $\dot{\vec{F}}$ is the vector of a given surface force rate; the superscript T denotes transposition.

For numerical solution of the contact problem we employ the finite-element method. The region V is broken up into a number of finite elements with two conditions being observed within each element. First, the displacement components vary with position such that the displacement of any point inside the element is determined by displacements of the nodal points. Second, continuity of displacements from one element to another is ensured with an arbitrary choice of nodal displacements. Thus, functional Φ can be approximated by the function of rates $\dot{\vec{u}}_k$ of the nodal points:

$$\Phi \cong \varphi\left(\dot{\vec{u}}_1, \dot{\vec{u}}_2, \ldots, \dot{\vec{u}}_n\right) = \sum_m^M \varphi^{(m)}\left(\dot{\vec{u}}_k^m\right), \qquad k = 1, \ldots, n, \qquad (4.10.2)$$

where $n = 1, 2, \ldots$ denotes nodes, $\dot{\vec{u}}_k^{(m)}$ are velocity vectors at the nodes of the mth element, and M is the number of finite elements.

The relation between the velocities inside the mth element and the velocities of the nodal points of this element is approximated by expression

$$\dot{U}^{(m)} = [N]^{(m)} \dot{\vec{u}}^{(m)}. \qquad (4.10.3)$$

Here, $[N]^{(m)}$ is the distribution matrix for the mth element. The strain rate vector $\dot{\varepsilon}^{(m)}$ of the mth element is expressed in terms of the nodal point velocities of the same element as

$$\dot{\vec{\varepsilon}}^{(m)} = [B]^{(m)} \dot{\vec{u}}^{(m)}. \qquad (4.10.4)$$

The stress and strain rates are linked by Hooke's law in the elastic region and by the Prandtl–Reuss generalized equation during loading in the plastic

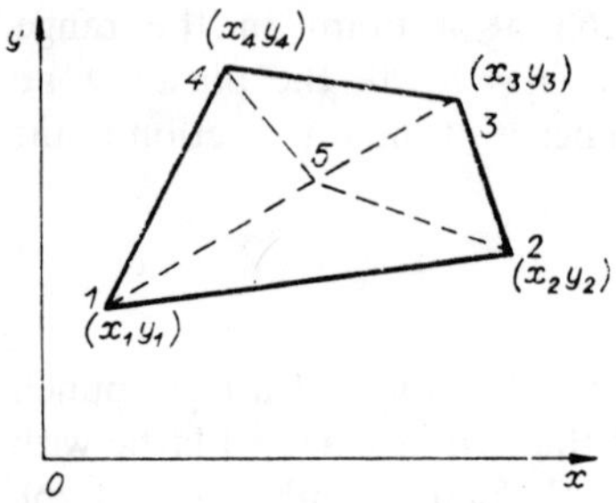

FIG. 4.21.

zone:

$$\dot{\sigma}_{ij} = 2G\left(\dot{\varepsilon}_{ij} + \frac{\nu}{1-2\nu}\varepsilon_{ii}\delta_{ij} - s_{ij}\frac{s_{kl}\cdot\dot{\varepsilon}_{kl}}{Q}\right), \tag{4.10.5}$$

where s_{ij} is the stress deviator, $Q = \frac{2}{3}\sigma_0[1 + H'/3]$, and H' corresponds to the slope of the effective curve of stress and plastic strain, expressed in terms of $\sigma_0 = H(\int d\varepsilon_0^p)$, $\sigma_0^2 = \frac{3}{2}s_{ij}s_{ij}$.

In the matrix form, for the mth element we have

$$\dot{\vec{\sigma}}^{(m)} = [D]^{(m)}\dot{\vec{\varepsilon}}^{(m)}. \tag{4.10.6}$$

Matrix $[D]^{(m)}$ is symmetrical.

By substituting relations (4.10.3), (4.10.4), and (4.10.6) into (4.10.1), we obtain

$$\varphi^{(m)} = \frac{1}{2}\int_V \left([D]^{(m)}[B]^{(m)}\dot{\vec{u}}^{(m)}\right)^T [B]^{(m)}\dot{\vec{u}}^{(m)}\,dV - \int_S \left([N]^{(m)}\dot{\vec{u}}^{(m)}\right)^T \dot{\vec{F}}^{(m)}\,dS, \tag{4.10.7}$$

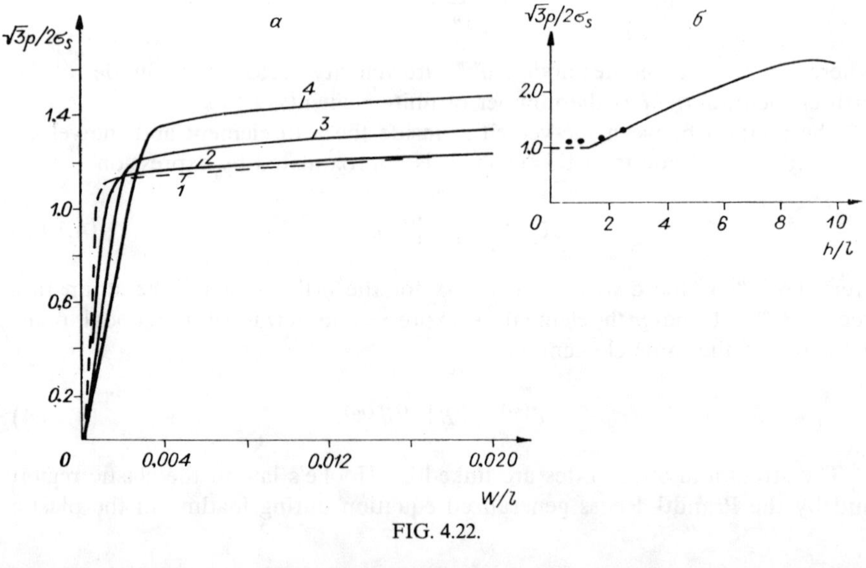

FIG. 4.22.

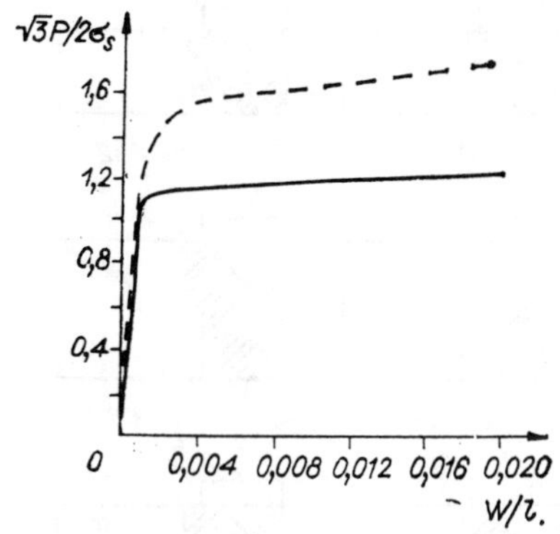

FIG. 4.23.

or, if the stiffness matrix and the force rate matrix are introduced in the element, relation (4.10.7) acquires the form

$$\varphi^m = \tfrac{1}{2}\dot{\vec{u}}^{(m)T}[K]^{(m)}\dot{\vec{u}}^{(m)} - \dot{\vec{u}}^{(m)T}\left[\dot{\vec{R}}\right]^{(m)}. \tag{4.10.8}$$

The stiffness matrix and the force rate matrix of the entire system, comprised of M finite elements, are obtained by summing the matrices of individual elements.

A body was broken up into quadrilateral elements. Each of the elements consisted of four triangular elements (Fig. 4.21). A linear velocity distribution was assumed within each triangular element.

Results obtained by the aforementioned numerical procedure [82] are presented below. A rigid punch of width $2l$ was indented in a strip of width $2L$ and height h. Computations were performed for $E = 1.0 \times 10^6$ psi, $\nu = 0.38$, and $H' = 2.10^4$ psi with an initial yield point $\sigma_s = 1.3 \times 10^4$ psi. These values describe the behavior of aluminum 1100-0.

Curves 1–4 in Fig. 4.22a illustrate the effect of punch displacement on the mean punch pressure at $h/l = 0.6, 1.0, 1.7, 2.5$ in the absence of friction. All computations were made for $L/l = 2.7$. The pressure sharply rises first linearly and then suddenly it tapers off. The bend in the curves corresponds to the yield point of the material. Comparison of these yield points with the values

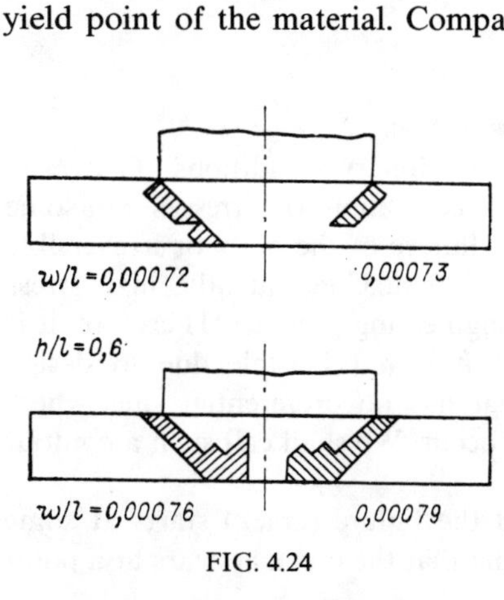

FIG. 4.24

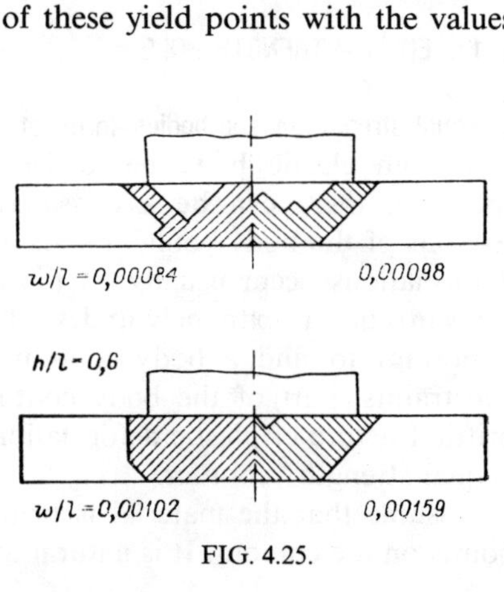

FIG. 4.25.

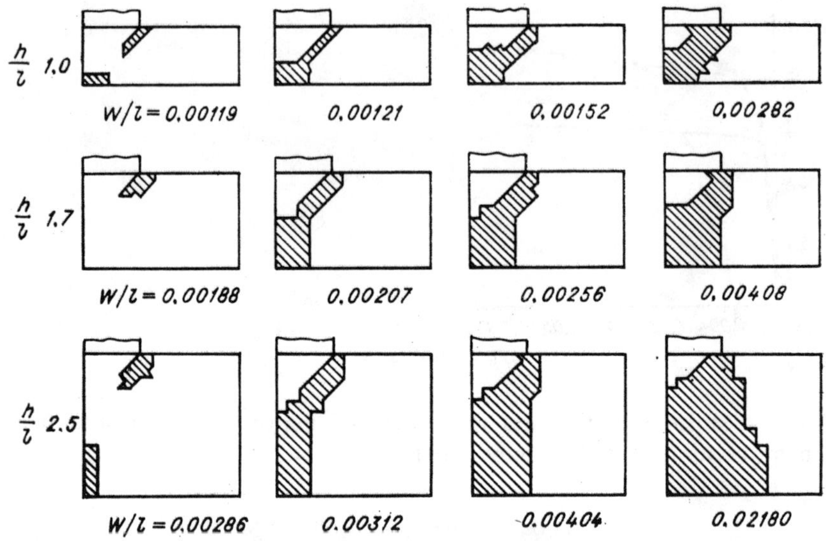

FIG. 4.26.

obtained from the theory of slip lines for an ideally rigid plastic body is presented in Fig. 4.22*b*.

The effect of friction on the contact surface is of particular interest. Computations were conducted at $h/l = 0.6, 2.5$ assuming complete adhesion on the contact surface. In case of $h/l = 2.5$, the curve obtained for contact with complete adhesion is identical to the smooth punch curve without friction. However, for $h/l = 0.6$ friction results in a significant increase of pressure compared with a smooth punch (Fig. 4.23). Gradual development of plastic deformation for $h/l = 0.6$ as a function of punch displacement is shown in Figs. 4.24 and 4.25 for smooth and rough punches. Figure 4.26 shows the plastic zones at $h/l = 1.0, 1.7, 2.7$ at various stages of indentation for a smooth punch. In all the cases illustrated in Fig. 4.26, the plastic flow appeared in two zones: at the corner of the punch and in the middle of the strip base.

§ 11. EQUAL-STRENGTH HOLE

1. Equal-strength elastic bodies–formulation of the problem

Let an elastic body be loaded under stationary conditions. Generally speaking, there can be a considerable concentration of stresses on some sections of the body boundary. As a result, failure of the body or irreversible deformations occur usually on the boundary first. Consequently, high stress concentration is extremely undesirable in engineering practice. Therefore, it is important to find a body contour (or, if it is not feasible due to design constraints, part of the body contour) that has no preferential sites where brittle fracture or plastic deformation may occur. We shall call such a contour "equal-strength" contour.

Assume that the material is brittle and the failure (crack) starts at some points on the contour. It is natural to assume that the crack appears at a point of the boundary when the stress σ_t at this point reaches a critical value equal to σ_* (t and n are tangential and normal directions to the body contour). Generally speaking, the value of σ_* depends on the dimensions of the initial defect on the boundary of the body and on the shape of the boundary. As a first approximation this effect can be neglected and σ_* assumed constant along contour L (or its section). Thus, an additional boundary condition must be satisfied on the "equal-strength" contour of the brittle body L,

$$\sigma_t = \sigma_* = \text{const on } L. \tag{4.11.1}$$

Let us assume now that the body is elastic–plastic and plastic deformation first appears on the body contour. Assume also that the plasticity condition of the material has the form

$$f(\sigma_n, \sigma_t, \tau_{tn}) = 0, \tag{4.11.2}$$

where f is a given function. We shall require that at the outset the plastic region encloses the entire contour instantly prior to penetration inside the body. Evidently, such a body will be considered stronger as regards the uniform distribution of stress concentration along the contour.

The loads given on the boundary $\sigma_n = p$, $\tau_{tn} = \tau$ are considered constant; then it follows from condition (4.11.2) that stress σ_t is constant everywhere on the unknown body boundary. Thus, we again arrive at the boundary condition (4.11.1). In particular, by assuming that the material of the body follows the Prandtl diagram in the plastic region, we obtain condition

$$|\sigma_y - \sigma_x + 2i\tau_{xy}| = 2k, \tag{4.11.3}$$

where k is a constant of the material.

Since the left-hand side of (4.11.3) is invariant with respect to the rotation of the coordinate axes at a given point of the body, we find, from (4.11.3) and by accounting for the loads given on the boundary, that the following condition is true at all points of the unknown contour L:

$$\sigma_t = p \pm 2\sqrt{k^2 - \tau^2} = \sigma_* = \text{const.} \tag{4.11.4}$$

The plus or minus sign in front of the radical is chosen from physical considerations.

We can arrive at a similar boundary condition if we simply require the absence of stress concentration, i.e., require that stress σ_t is constant. However, in this case, constant σ_* in the boundary condition (4.11.1) is not given. Instead it is determined in the process of solution.

2

Let us assume that a cylindrical body with a cross section F, which is a doubly connected region with smooth boundary $\Gamma_1 \cup \Gamma_2$ (Fig. 4.27a), is in plane strain conditions under a self-balanced system of external normal and shear forces applied to Γ_1 and Γ_2. We denote the region bounded by Γ_j by F_j, $j = 1, 2$, and assume that the origin of xy coordinates belongs to region F_1. The stress state in the body is characterized by tensor $\sigma_x^0, \sigma_y^0, \tau_{xy}^0$ and is assumed to be known. Let us denote the Kolosov–Muskhelishvili potentials of this stress

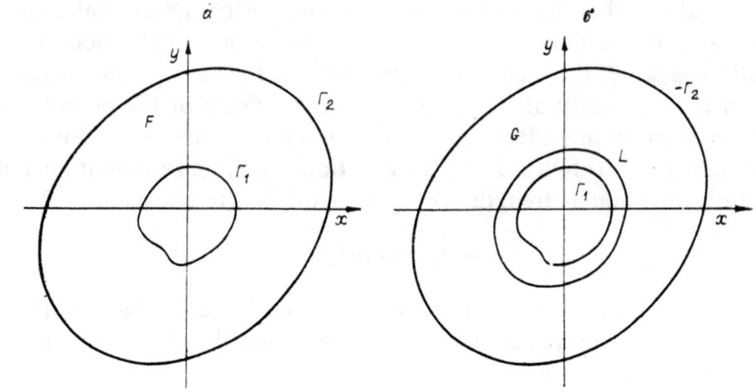

FIG. 4.27.

state by $\Phi_0(z)$, $\Psi_0(z)$:

$$\sigma_x^0 + \sigma_y^0 = 4\,\mathrm{Re}\,\Phi_0(z), \qquad z = x + iy,$$

$$\sigma_y^0 - \sigma_x^0 + 2i\tau_{xy}^0 = 2\left[\bar{z}\frac{d\Phi_0(z)}{dz} + \Psi_0(z)\right]. \tag{4.11.5}$$

A cylindrical hole with a cross section D^+ introduced in the body (Fig. 4.27*b*) represents a singly connected region with boundary L, where $F_2 \supset D^+ \supset F_1$. We shall seek such a contour L that the following condition is satisfied on L:

$$\sigma_n = p(s), \qquad \tau_{nt} = r(s), \qquad \sigma_t = q(s). \tag{4.11.6}$$

Here, n is the normal, t is the tangent to L, σ_n, σ_{i}, τ_{nt} are the components of the stress tensor in system n, t with orientation similar to system x, y and $p(s), q(s), r(s)$ are continuous functions of arc length s of contour L. It is assumed that the external system of stresses $p(s), r(s)$, $0 \le s \le S$, where S is the length of L, is self-balancing. In particular, we assume that

$$q(s) = \sigma^* = \text{const} \tag{4.11.7}$$

or that $q(s)$ is determined from equation (4.11.2),

$$f(p(s), q(s), r(s)) = 0. \tag{4.11.8}$$

Let us denote by G the doubly connected region, bounded by L and Γ_2 (Fig. 4.27*b*). The stress state in the body with a cross section G, characterized by tensor $\sigma_x^1, \sigma_y^1, \tau_{xy}^1$, can be presented as

$$\sigma_x^1 = \sigma_x^0 + \sigma_x, \qquad \sigma_y^1 = \sigma_y^0 + \sigma_y, \qquad \tau_{xy}^1 = \tau_{xy}^0 + \tau_{xy}. \tag{4.11.9}$$

Here, $\sigma_x, \sigma_y, \tau_{xy}$ are additional stresses caused by a new hole. By virtue of (4.11.9) we shall have the following representation for the Kolosov–

Muskhelishvili potentials:

$$\Phi_1(z) = \Phi_0(z) + \Phi(z), \qquad \Psi_1(z) = \Psi_0(z) + \Psi(z), \qquad z \in L + G + \Gamma_2. \tag{4.11.10}$$

Assuming that the distance from any point of contour L to Γ_2 is much greater than the diameter of region D^+, we shall determine the additional stresses approximately by replacing the doubly connected region G by an infinite region D^- bounded by L. Thus, it is required to find contour $L \subset F$ bounding the infinite region D^-. It is also required to find the analytical functions $\Phi(z)$ and $\Psi(z)$ in D^- such that the following conditions are satisfied on L

$$\begin{gathered} \operatorname{Re}\Phi(z) = \tfrac{1}{4}(p(s) + q(s)) - \operatorname{Re}\Phi_0(z), \\ \bar{z}\frac{d\Phi}{dz} + \Psi(z) = -m(s)\frac{d\bar{z}}{dz} - \bar{z}\frac{d\Phi_0}{dz} - \Psi_0(z), \\ m(s) \equiv \tfrac{1}{2}(q(s) - p(s) + 2ir(s)), \end{gathered} \tag{4.11.11}$$

and when $|z| \to \infty$,

$$\Phi(z) = O(z^{-2}), \qquad \Psi(z) = O(z^{-2}). \tag{4.11.12}$$

Here, $\Phi_0(z)$, $\Psi(z)$ are given functions, which are defined in region F. In particular, region F_1 can degenerate into a point, which can be a removable singular point for functions $\Phi_0(z)$, $\Psi_0(z)$. Instead of parameter s, other geometrical parameters can be used, for example, $r = (x^2 + y^2)^{1/2}$, $\varphi = \arctan(y/x)$.

We shall consider only the simplest case of problem (4.11.11) and (4.11.12) assuming that

$$\begin{gathered} p(s) = p_0, \qquad q(s) = q_0, \qquad r(s) = 0, \\ \Phi_0(z) = a + a_1 z, \qquad \Psi_0(z) = b + b_1 z, \end{gathered} \tag{4.11.13}$$

where p_0, q_0, a_1 are real constants and a, b, b_1 are complex constants. The fact that constant a_1 is real does not diminish generality, since it can always be achieved by rotating the initial coordinates x, y.

Let us map, one-to-one and conformally, region $K^- + C$: $|\zeta| \geq 1$ of the complex variable ζ onto region $D^- + L$ using function

$$z = \omega(\zeta) = c\zeta + c_0 + c_1\zeta^{-1} + c_2\zeta^{-2} + \cdots, \qquad c > 0. \tag{4.11.14}$$

Let us assume

$$\Phi(\omega(\zeta)) \equiv \varphi(\zeta), \qquad \Psi(\omega(\zeta)) \equiv \psi(\zeta). \tag{4.11.15}$$

For $|\zeta| \to \infty$, based on (4.11.12) and (4.11.14), we have

$$\varphi(\zeta) = O(\zeta^{-2}), \qquad \psi(\zeta) = O(\zeta^{-2}). \tag{4.11.16}$$

For points on the contour $t \in C$, the following relations are satisfied:

$$t\bar{t} = 1, \qquad dz = \omega'(t)\,dt,$$
$$\overline{dz} = \overline{\omega'(t)}\,\overline{dt} = -\overline{\omega'(t)t^2}\,dt. \tag{4.11.17}$$

Hence, the boundary conditions of (4.11.11) on C acquire the form

$$\operatorname{Re}\varphi(t) = \tfrac{1}{4}(p_0 + q_0) - \operatorname{Re}(a + a_1\omega(t)), \tag{4.11.18}$$

$$\overline{\omega(t)}\frac{\varphi'(t)}{\omega'(t)} + \psi(t) = m_0\frac{\overline{\omega'(t)t^2}}{\omega'(t)} - a_1\overline{\omega(t)} - b - b_1\omega(t), \tag{4.11.19}$$

$$m_0 = \tfrac{1}{2}(q_0 - p_0).$$

From condition (4.11.18) we find ($\zeta \in K^- + C$)

$$\varphi(\zeta) = -\{a_1[\omega(\zeta) - c\zeta - c_0] + a_1c_1/\zeta\}. \tag{4.11.20}$$

In this case the following relation must be satisfied as a result of (4.11.16):

$$\tfrac{1}{2}(p_0 + q_0) - (a + \bar{a} + a_1c_0 + a_1\bar{c}_0) = 0,$$
$$a_1(c_1 + c) = 0. \tag{4.11.21}$$

By eliminating function $\varphi(\zeta)$ from the boundary condition (4.11.19) using (4.11.20), we obtain on C $(a_1 \neq 0)$,

$$\overline{\omega(t)}a_1c(1 + t^{-2}) - m_0\overline{\omega'(t)t^2} = -(\psi(t) + b + b_1\omega(t))\omega'(t). \tag{4.11.22}$$

Let us introduce function

$$\Omega(\zeta) = \begin{cases} \omega(\zeta)a_1c(1 + \zeta^2) - m_0\omega'(\zeta)\zeta^2, & |\zeta| \geq 1, \\ -[\bar{\psi}(1/\zeta) + \bar{b} + \bar{b}_1\bar{\omega}(1/\zeta)]\bar{\omega}'(1/\zeta), & |\zeta| \leq 1. \end{cases} \tag{4.11.23}$$

From (4.11.22), the following equality is satisfied on C:

$$\Omega^-(t) = \Omega^+(t). \tag{4.11.24}$$

The singular points of function $\Omega(\zeta)$ in the expanded plane ζ will be points $\zeta = 0$ and $\zeta = \infty$. At point $\zeta = 0$, function $\Omega(\zeta)$ has a pole with a main term $-\bar{b}_1c^2\zeta^{-1}$; at point $\zeta = \infty$, function $\Omega(\zeta)$ also has a pole with the main term

$$a_1c[c\zeta^3 + c_0\zeta^2 + (c_1 + c)\zeta + c_0 + c_2] - m_0(c\zeta^2 - c_1).$$

On the basis of Liouville's generalized theorem for all $\zeta \in K^- + C + K^+$,

$$\Omega(\zeta) = \delta + a_1c[c\zeta^3 + c_0\zeta^2 + (c_1 + c)\zeta + c_0 + c_2] - m_0(c\zeta^2 - c_1) - c^2\bar{b}_1\zeta^{-1}. \tag{4.11.25}$$

Now, from (4.11.23) for $\zeta \in K^- + C$,

$$\psi(\zeta) = -\bar{\Omega}(\zeta^{-1})/\omega'(\zeta) - b - b_1\omega(\zeta). \tag{4.11.26}$$

From condition (4.11.16),

$$\bar{\delta} + a_1 c(\bar{c}_0 + \bar{c}_2) + m_0\bar{c}_1 + c(b + b_1 c_0) = 0, \qquad a_1(\bar{c}_1 + c) = 0. \tag{4.11.27}$$

Function $\omega(\zeta)$ is determined from the differential equation ($|\zeta| \geq 1$)

$$-m_0\zeta^2\frac{d\omega(\zeta)}{d\zeta} + a_1 c(1+\zeta^2)\omega(\zeta) = \delta + a_1 c\left[c\zeta^3 + c_0\zeta^2 + (c_1 + c)\zeta + c_0 + c_2\right]$$
$$-m_0\left(c\zeta^2 - c_1\right) - \frac{\bar{b}_1 c^2}{\zeta}. \tag{4.11.28}$$

The solution of equation (4.11.28) can be sought in the form of (4.11.14) or in the form

$$\omega(\zeta) = c\zeta \exp\left(u_1/\zeta + u_2/\zeta^2 + u_3/\zeta^3 + \cdots\right).$$

This leads to a system of nonlinear equations with respect to $c, c_0, c_1, \ldots$ or $c, u_1, u_2, u_3, \ldots$.

In particular, if $a_1 = 0$, from (4.11.21) and (4.11.18) we have $\omega(\zeta) = c\zeta + c_1/\zeta + c_0 + c_2/\zeta^2$. If $a \neq 0$, from (4.11.21) we have

$$c_1 + c = 0. \tag{4.11.29}$$

The area of the finite region, bounded by L, is nonnegative; therefore,

$$\pi\left(c^2 - \sum_{n=1}^{\infty} n\,|c_n^2|\right) \geq 0.$$

Taking into account condition (4.11.29), we shall have $c_2 = c_3 = c_4 = \cdots = 0$, $c_1 = -c$. This means that L is a straight-line section.

If $m(s) = \text{const}$, the approach discussed above is also applicable when $\Phi_0(z)$ and $\Psi(z)$ have the form

$$\Phi_0(z) = \sum_{k=-n}^{m} a_k z^k; \qquad \Psi_0(z) = \sum_{k=-n_1}^{m_1} b_k z^k.$$

Detailed investigation for $n = n_1 = 0$ by a somewhat different method is given by Ostrosablin [48].

3

The indicated equal-strength condition (4.11.1) can prove sometimes structurally inadmissible. Therefore, we shall solve the problem for a less stringent condition, namely, we assume that the following boundary condition is implemented on the unknown hole contour L:

$$\sigma_t = \sigma + \gamma \sin^2\alpha. \tag{4.11.30}$$

Here, $\sigma = \text{const}$, $\gamma = \text{const}$, and α is the angle between axes x and n. Hence, it is required to find contour L and the Kolosov–Muskhelishvili potentials $\Phi(z)$ and $\Psi(z)$ according to condition (4.11.30) and conditions

$$\sigma_n = p, \qquad \tau_{tn} = 0 \quad \text{on } L, \tag{4.11.31}$$

when $|z| \to \infty$,

$$\Phi(z) = \tfrac{1}{4}(\sigma_y^\infty + \sigma_x^\infty) + O(z^{-2}),$$

$$\Psi(z) = \tfrac{1}{2}(\sigma_y^\infty - \sigma_x^\infty) + O(z^{-2}).$$

Outside the unit circle of the parametric plane we have the boundary value problem

$$4\,\mathrm{Re}\,\varphi(\zeta) = p + \sigma - \frac{\gamma\left[\zeta\omega'(\zeta) - \overline{\zeta\omega'(\zeta)}\right]^2}{4\,|\omega'(\zeta)|^2} \quad \text{for } |\zeta| = 1,$$

$$2\frac{\zeta^2\omega'(\zeta)}{\overline{\omega'(\zeta)}}\left[\frac{\varphi'(\zeta)}{\omega'(\zeta)}\overline{\omega(\zeta)} + \psi(\zeta)\right] = \sigma - p - \frac{\gamma\left[\zeta\omega'(\zeta) - \overline{\zeta\omega'(\zeta)}\right]^2}{4\,|\omega'(\zeta)|^2}, \tag{4.11.32}$$

when $|\zeta| \to \infty$,

$$\omega(\zeta) = O(\zeta), \qquad \varphi(\zeta) = \tfrac{1}{4}(\sigma_x^\infty + \sigma_y^\infty) + O(\zeta^{-2}),$$

$$\psi(\zeta) = \tfrac{1}{2}(\sigma_y^\infty - \sigma_x^\infty) + O(\zeta^{-2}).$$

For the solution of the nonlinear boundary value problem (4.11.32) we adopt the method of functional equations. We write the boundary value problem (4.11.32) in the form of a system of two functional equations, which hold true in the entire plane ζ:

$$2\varphi(\zeta) + 2\overline{\varphi}\left(\frac{1}{\zeta}\right) = p + \sigma - \frac{\gamma}{4\omega'(\zeta)\overline{\omega}'(1/\zeta)}\left[\zeta\omega'(\zeta) - \frac{1}{\zeta}\overline{\omega}'\left(\frac{1}{\zeta}\right)\right]^2,$$

$$\begin{aligned} 2\zeta^2\omega'(\zeta)&\left[\frac{\varphi'(\zeta)}{\omega'(\zeta)}\overline{\omega}\left(\frac{1}{\zeta}\right) + \psi(\zeta)\right] \\ &= \overline{\omega}'\left(\frac{1}{\zeta}\right)\left\{\sigma_* - p - \frac{\gamma}{4\omega'(\zeta)\overline{\omega}'(1/\zeta)} \times\left[\zeta\omega'(\zeta) - \frac{1}{\zeta}\overline{\omega}'\left(\frac{1}{\zeta}\right)\right]^2\right\}. \end{aligned} \tag{4.11.33}$$

We seek function $\omega(\zeta)$ as the polynomial

$$\omega(\zeta) = c\zeta + c_1\zeta^{-1} + c_3\zeta^{-3} + \cdots + c_{2n+1}\zeta^{-2n-1}. \tag{4.11.34}$$

The constants $c_1, c_3, \ldots, c_{2n+1}$ from the symmetry conditions are real. We determine n by comparing the orders of the values in the right- and left-hand parts of the second functional equation (4.11.33) when $\zeta \to \infty$. We then obtain $n = 0$. Thus, $\omega(\zeta)$ is equal to

$$\omega(\zeta) = c\zeta + c_1/\zeta. \tag{4.11.35}$$

From the first functional equation (4.11.33), using (4.11.35), $\varphi(\zeta)$ can be found as

$$\varphi(\zeta) = \frac{1}{4}\left(\sigma_x^\infty + \sigma_y^\infty\right) - \frac{\gamma\left(c^2 - c_1^2\right)}{8c\left(c\zeta^2 - c_1\right)}. \tag{4.11.36}$$

Here, the following condition should be satisfied:

$$\frac{c_1}{c} = \frac{2}{\gamma}\left(\sigma_x^\infty + \sigma_y^\infty - p - \sigma\right) - 1. \tag{4.11.37}$$

Function $\psi(\zeta)$ is determined by the second equation in (4.11.33):

$$\psi(\zeta) = \frac{\bar{\omega}'(1/\xi)}{2\zeta^2\omega'(\zeta)}\left\{\sigma_* - p - \frac{\gamma}{4\omega'(\zeta)\bar{\omega}'(1/\zeta)}\left[\zeta\omega'(\zeta) - \frac{1}{\zeta}\bar{\omega}'\left(\frac{1}{\zeta}\right)\right]^2\right\}$$
$$- \frac{\bar{\omega}(1/\zeta)}{\omega'(\zeta)}\varphi'(\zeta).$$

From the condition at infinity it follows that

$$\gamma = \frac{\left(\sigma_x^\infty + \sigma_y^\infty - p - \sigma\right)\left(3p - \sigma - \sigma_x^\infty - \sigma_y^\infty\right)}{\sigma_y^\infty - \sigma_x^\infty + p - \sigma}.$$

4

Now we shall seek an equal-strength contour for a plane weakened by two identical holes. Let the centers of the holes be located on the x axis and at a distance $2l$ from each other. The origin of the coordinates is chosen at a point equidistant from the centers of the holes. Constant external forces are applied to the holes

$$\sigma_n = p, \qquad \tau_{tn} = 0$$

and a uniform stress state exists at infinity:

$$\sigma_x = \sigma_x^\infty, \qquad \sigma_y = \sigma_y^\infty, \qquad \tau_{xy} = 0.$$

As an equal-strength condition we shall use the condition of uniform distribution of stresses

$$\sigma_t = \sigma_*.$$

On the basis of the boundary conditions and relations (4.1.9), we write the boundary problem for determination of functions $\Phi(z)$ and $\Psi(z)$ and an unknown contour L:

$$\begin{gathered} 4\,\mathrm{Re}\,\Phi(z) = \sigma_* + p - \sigma_x^\infty - \sigma_y^\infty, \\ 2[(\bar{z} - z)\Phi'(z) + \psi_*(z)] = (\sigma_* - p)e^{-2i\alpha} - \sigma_y^\infty + \sigma_x^\infty \quad \text{on } L. \end{gathered} \tag{4.11.38}$$

Let us go over to the parametric plane ζ using transformation

$$z - l = R\left(\zeta + \sum_{k=1}^{\infty} c_k \zeta^{-k}\right) = R\omega(\zeta).$$

Here, R and c_k are unknown parameters, characterizing the size and shape of the required holes. Let us represent functions $\Phi(z)$ and $\psi_*(z)$ as

$$\begin{gathered} \Phi(z) = \varphi_1(\zeta) + \varphi_2(\zeta) = \sum_{k=2}^{\infty} a_k \zeta^{-k} + A_0 + A_1\omega(\zeta) + A_2[\omega(\zeta)]^2, \\ \psi_*(z) = \psi_1(\zeta) + \psi_2(\zeta) = \sum_{k=2}^{\infty} b_k \zeta^{-k} + B_0 + B_1\omega(\zeta) + B_2[\omega(\zeta)]^2. \end{gathered} \tag{4.11.39}$$

We have limited ourselves to the terms containing the perturbation $\varepsilon = R/2l$ raised to a power ≤ 4. Here

$$\begin{gathered} A_0 = (\varepsilon^2 + 2c_1\varepsilon^4)a_2 - \varepsilon^3 a_3 + \varepsilon^4 a_4, \\ A_1 = -2\varepsilon^3 a_2 + 3\varepsilon^4 a_3, \qquad A_2 = 3\varepsilon^4 a_2. \end{gathered} \tag{4.11.40}$$

Constants B_0, B_1, B_2 are determined from equations (4.11.40) if a_k is substituted by b_k.

By solving the Dirichlet problem (4.11.38) and satisfying the conditions at infinity, we obtain

$$\varphi_1(\zeta) = \tfrac{1}{4}(\sigma_* + p - \sigma_x^\infty - \sigma_y^\infty) = 0.$$

The second boundary condition from (4.11.38) becomes

$$(\sigma_* - p)\overline{\omega'(\zeta)} = 2\zeta^2\omega'(\zeta)\left\{\tfrac{1}{2}\left(\sigma_y^\infty - \sigma_x^\infty\right) + \psi_1(\zeta) + B_0 + B_1\omega(\zeta) + B_2[\omega(\zeta)]^2\right\}. \quad (4.11.41)$$

To solve the nonlinear boundary value problem (4.11.41) we use the method of functional equations. Let us consider a functional equation, which holds true in the entire plane ζ,

$$(\sigma_* - p)\overline{\omega}'(1/\zeta) = 2\zeta^2\omega'(\xi)\left\{\tfrac{1}{2}\left(\sigma_y^\infty - \sigma_x^\infty\right) + \psi_1(\zeta) + B_0 + B_1\omega(\zeta) + B_2[\omega(\zeta)]^2\right\}. \quad (4.11.42)$$

By substituting the expression for $\omega(\zeta)$ in the functional equation (4.11.42), and expanding all the functions into series in the vicinity of an infinitely distant point, it is easy to see that $k = 3$ so that

$$\omega(\zeta) = \zeta + c_1/\zeta + c_2/\zeta^2 + c_3/\zeta^3. \quad (4.11.43)$$

Function $\psi_1(\zeta)$ is determined from the functional equation (4.11.42),

$$\psi_1(\zeta) = (\sigma_* - p)\frac{\overline{\omega}'(1/\zeta)}{2\zeta^2\omega'(\zeta)} - \frac{1}{2}\left(\sigma_y^\infty - \sigma_x^\infty\right) - B_0 - B_1\omega(\zeta) - B_2[\omega(\zeta)]^2. \quad (4.11.44)$$

By satisfying the conditions at infinity we obtain, for determination of constants c_k, b_2, b_3, b_4, a system of nonlinear algebraic equations,

$$b_2 - c_1\left[\tfrac{1}{2}\left(\sigma_y^\infty - \sigma_x^\infty\right) + B_0\right] - c_2B_1\left(c_1^2 + c_3\right)B_2 = \sigma_* - p,$$

$$b_3 - 2c_2\left[\tfrac{1}{2}\left(\sigma_y^\infty - \sigma_x^\infty\right) + B_0\right] - \left(c_1^2 + 2c_3\right)B_1 - 4c_1c_2B_2 = 0,$$

$$b_4 - c_1b_2 - 3c_3\left[\tfrac{1}{2}\left(\sigma_y^\infty - \sigma_x^\infty\right) + B_0\right] - 3c_1c_2B_1 - \left(c_1^3 + 3c_2 + 6c_1c_2\right)B_2 = 0, \quad (4.11.45)$$

$$\tfrac{1}{2}\left(\sigma_y^\infty - \sigma_x^\infty\right) + B_0 + c_1B_2 + \bar{c}_1(\sigma_* - p) = 0,$$

$$B_1 + 2\bar{c}_2(\sigma_* - p) = 0, \qquad B_2 + 3\bar{c}_3(\sigma_* - p) = 0, \qquad c_k = 0, \qquad k \geq 4.$$

The algebraic system of equations is readily solved by the perturbation method:

$$c_1 = c_{10} + c_{12}\varepsilon^2 + c_{14}\varepsilon^4, \qquad c_2 = -c_{12}\varepsilon^3,$$

$$c_3 = c_{12}\varepsilon^4, \qquad c_{14} = c_{10}c_{12} + 7c_{10}c_{12},$$

$$c_{10} = \frac{1}{2}\frac{(\sigma_y^\infty - \sigma_x^\infty)}{\sigma_* - p}, \qquad c_{12} = c_{10}^2 - 1,$$

$$b_2 = b_{20} + b_{22}\varepsilon^2 + b_{24}\varepsilon^4, \qquad b_4 = b_{40} + b_{42}\varepsilon^2 + b_{44}\varepsilon^4,$$

$$b_3 = -2\left[c_{12}\tfrac{1}{2}(\sigma_y^\infty - \sigma_x^\infty) + c_{10}^2 b_{20}\right]\varepsilon^3, \tag{4.11.46}$$

$$b_{20} = (\sigma_* - p) + \tfrac{1}{2}c_{10}(\sigma_y^\infty - \sigma_x^\infty), \qquad b_{22} = c_{10}b_{20} + \tfrac{1}{2}c_{12}(\sigma_y^\infty - \sigma_x^\infty),$$

$$b_{24} = (c_{14}/2)(\sigma_y^\infty - \sigma_x^\infty) + c_{10}(b_{22} + 6c_{10}b_{20}) + c_{12}b_{20},$$

$$b_{40} = c_{10}b_{20}, \qquad b_{42} = c_{12}b_{20} + c_{10}b_{22},$$

$$b_{44} = c_{10}(b_{24} + 3c_{10}^2 b_{20}) + c_{12}b_{22} + c_{14}b_{20} + \frac{3c_{12}}{2}(\sigma_y^\infty - \sigma_x^\infty).$$

The final solution of the problem [71], when the "equal-strength" condition is of the form (4.11.30), becomes

$$\varphi_1(\zeta) = \frac{1}{2}(p + \sigma_* - \sigma_x^\infty - \sigma_y^\infty) + \frac{\gamma}{4}c_1 + \frac{\gamma}{4} - \frac{\gamma(1 - c_1^2)}{8(\zeta^2 - c_1)} - 2a_2\varepsilon^2,$$

$$p + \sigma_* - \sigma_x^\infty - \sigma_y^\infty + \frac{\gamma}{2}c_1 + \frac{\gamma}{2} - 4a_2\varepsilon^2 = 0,$$

$$\psi_1(\zeta) = \frac{\overline{\omega}'(\zeta^{-1})}{2\zeta^2\omega'(\zeta)}\left\{\sigma_* - \rho - \frac{\gamma[\zeta\omega'(\zeta) - (1/\zeta)\overline{\omega}'(\zeta^{-1})]^2}{4\omega'(\zeta)\overline{\omega}'(\zeta^{-1})}\right\}$$

$$- \frac{\overline{\omega}'(\zeta^{-1}) - \omega(\zeta)}{\omega'(\zeta)}\varphi'(\zeta) - \frac{1}{2}(\sigma_y^\infty - \sigma_x^\infty) - b_2\varepsilon^2, \tag{4.11.47}$$

$$a_2 = \frac{\gamma(1 - c_1^2)}{8}, \qquad \omega(\zeta) = \zeta + \frac{c_1}{\zeta},$$

$$c_1\left[\sigma_* - p + \frac{\gamma(1 + c_1)^2}{4c_1}\right] + (\sigma_y^\infty - \sigma_x^\infty) + 2b_2\varepsilon^2 = 0,$$

$$b_2 = \frac{1}{2}(\sigma_* - p) - \frac{\gamma(1 + c_1)^2}{4c_1} + \frac{\gamma(1 - c_1)^2(1 + c_1)}{4}.$$

We have limited ourselves to the terms containing ε raised to a power less than or equal to 2.

§ 12. EQUAL-STRENGTH OPENING IN A ROCK MASS

The problem of finding an equal-strength opening in a heavy rock mass is of significant interest in mining practice. Let us discuss this problem with the assumption that the rock is a homogeneous and isotropic body and the opening is located sufficiently far from the ground.

1

It is required to build an equal strength tunnel in a heavy elastic half space $y < H$. The shape of the tunnel is cylindrical and its axis is parallel to the surface of the space. A constant normal pressure $-p$ ($p > 0$) is applied to the yet unknown surface of the tunnel, and the tangential load is equal to zero. Additionally, at a distance from the hole the entire elastic half space is deformed by stresses

$$\sigma_x = \sigma_x^\infty, \qquad \sigma_y = \sigma_y^\infty, \qquad \tau_{xy} = 0. \tag{4.12.1}$$

The problem is considered to be a plane problem. Evidently, the Cartesian coordinates x and y can be selected to coincide with the axes of symmetry of the unknown opening contour L. The surface of the half space $y = H$ is subjected to a constant normal load $\sigma_y = \sigma_y^\infty$ and the tangential load is equal to zero.

Let us write three boundary conditions on the unknown contour of the opening as

$$\begin{gathered} \sigma_n = -p, \qquad \tau_{tn} = 0, \qquad \sigma_t = \sigma_* + \gamma y, \\ \sigma_* = \text{const}, \qquad \gamma = \text{const} \end{gathered} \tag{4.12.2}$$

(t and n are directions of a tangent and normal to the contour). The last condition of (4.12.2) is the equal-strength condition for the opening. In the general case, when $\gamma = 0$ there is no solution (see §11.2). This means that no opening with nonzero surface area, which did not experience stress concentration, exists. It is natural, in this case, to allow only those stress concentrations, which are determined by formulas (4.12.2) when $\gamma \neq 0$. We assume that the hole is situated sufficiently far from the surface of the half space. In this case, the boundary conditions on the hole contour will be satisfied exactly, whereas the conditions on the boundary of the half space can only be satisfied approximately, asymptotically. Comparison with the solutions of problems for a hole near the boundary of the body shows that it is practically sufficient to satisfy condition $H > 2L_0$, where H is the depth and L_0 is a characteristic linear dimension of the hole.

For stresses σ_x, σ_y, and τ_{xy}, it is easy to obtain the following fundamental representations:

$$\begin{gathered} \sigma_x + \sigma_y = 2\left[\Phi(z) + \overline{\Phi(z)}\right] + \frac{\rho g}{2i(1-\nu)}(z - \bar{z} - 2Hi), \\ \sigma_y - \sigma_x + 2i\tau_{xy} = 2[\bar{z}\Phi'(z) + \Psi(z)] + \frac{\rho g(1-2\nu)}{2i(1-\nu)}(z - \bar{z} - 2Hi). \end{gathered} \tag{4.12.3}$$

Here, ρ is the density of the body, g is the acceleration of gravity $\Phi(z)$ and $\Psi(z)$ are Kolosov–Muskhelishvili potentials [when $\Phi(z)=0$ and $\Psi(z)=0$, formulas (4.12.3) give the stresses in a heavy half space $y<H$ with a free surface].

On the basis of (4.12.1)–(4.12.3), we formulate the boundary problem for determination of the functions $\Phi(z)$ and contour

$$2\left[\Phi(z)+\overline{\Phi(z)}\right]+\frac{\rho g}{2i(1-\nu)}(z-\bar{z}-2Hi)=\sigma_{*}-p+\gamma y, \tag{4.12.4}$$

$$2[\bar{z}\Phi'(z)+\Psi(z)]+\frac{\rho g(1-2\nu)}{2i(1-\nu)}(z-\bar{z}-2Hi)=e^{-2i\alpha}(p+\sigma_{*}+\gamma y).$$

When $|z|\to\infty$,

$$\Phi(z)=\tfrac{1}{4}\left(\sigma_x^{\infty}+\sigma_y^{\infty}\right)+O(z^{-1}),$$

$$\Psi(z)=\tfrac{1}{2}\left(\sigma_y^{\infty}-\sigma_x^{\infty}\right)+O(z^{-1}).$$

Here, α is the angle between axes n and x.

Let us go over to the parametric plane ζ with the help of transformation $z=\omega(\zeta)$,

$$\omega(\zeta)=c\zeta+O(\zeta^{-1})\quad \text{when } |\zeta|\to\infty. \tag{4.12.5}$$

The analytical function $\omega(\zeta)$ gives conformal mapping of the outside of the unit circle in plane ζ onto the outside of contour L in plane z; the function must be determined from the solution of the problem. Constant c can be considered positive without any loss of generality.

Based on equality

$$e^{2i\alpha}=\frac{\zeta^2\omega'(\zeta)}{\rho_1^2\overline{\omega'(\zeta)}},\qquad \zeta=\rho_1 e^{i\theta}, \tag{4.12.6}$$

and conditions (4.12.4), for determination of the three analytical functions $\varphi(\zeta)=\Phi[\omega(\zeta)]$, $\psi(\zeta)=\Psi[\omega(\zeta)]$, and $\omega(\zeta)$ we obtain the nonlinear boundary value problem

$$\begin{gathered}2\left[\varphi(\zeta)+\overline{\varphi(\zeta)}\right]+\frac{\rho g}{2i(1-\nu)}\left[\omega(\zeta)-\overline{\omega(\zeta)}-2Hi\right]\\ =\sigma_{*}-p+\frac{\gamma}{2i}\left[\omega(\zeta)-\overline{\omega(\zeta)}\right],\end{gathered} \tag{4.12.7}$$

$$2\left[\frac{\overline{\omega(\zeta)}}{\omega(\zeta)}\varphi'(\zeta)+\psi(\zeta)\right]+\frac{\rho g(1-2\nu)}{2i(1-\nu)}\left[\omega(\zeta)-\overline{\omega(\zeta)}-2Hi\right]$$

$$=\frac{\overline{\omega'(\zeta)}}{\zeta^2\omega'(\zeta)}\left\{p+\sigma_{*}+\frac{\gamma}{2i}\left[\omega(\zeta)-\overline{\omega(\zeta)}\right]\right\}\quad \text{when } |\zeta|=1.$$

When $|\zeta| \to \infty$,

$$\varphi(\zeta) = \frac{1}{4}(\sigma_x^\infty + \sigma_y^\infty) + O(\zeta^{-1}), \qquad \omega(\zeta) = c\zeta + O(\zeta^{-1}),$$

$$\psi(\zeta) = \frac{1}{2}(\sigma_y^\infty - \sigma_x^\infty) + O(\zeta^{-1}).$$

Solution of a boundary value problem. Function $F^-(\zeta)$,

$$F^-(\zeta) = 2\varphi(\zeta) + \frac{\rho g}{2i(1-\nu)}\omega(\zeta) - \frac{\gamma}{2i}\omega(\zeta), \tag{4.12.8}$$

is analytic outside the unit circle with the exception of an infinitely distant point, where it has a first order pole. Function $F^+(\zeta)$,

$$F^+(\zeta) = -2\bar{\varphi}\left(\frac{1}{\zeta}\right) + \frac{\rho g}{2i(1-\nu)}\left[\bar{\omega}\left(\frac{1}{\zeta}\right) + 2Hi\right] + \sigma_* - p - \frac{\gamma}{2i}\bar{\omega}\left(\frac{1}{\zeta}\right), \tag{4.12.9}$$

is analytic inside the unit circle with the exception of point $\zeta = 0$, at which it has a first order pole. Based on the first boundary condition (4.12.7), functions $F^+(\zeta)$ and $F^-(\zeta)$ are analytical extensions of each other through the unit circle

$$F^+(\zeta) = F^-(\zeta) \quad \text{for } |\zeta| = 1. \tag{4.12.10}$$

According to the Liouville theorem, the analytical function is

$$F(\zeta) = b_0\zeta + b_1/\zeta + b_2, \tag{4.12.11}$$

where b_0, b_1, b_2 are constants which have not been determined yet. By expanding function $F^-(\zeta)$ in the vicinity of an infinitely distant point and function $F^+(\zeta)$ in the vicinity of zero, and comparing with (4.12.11), we obtain

$$b_0 = b_1 = \frac{c}{2i}\left[\frac{\rho g}{(1-\nu)} - \gamma\right], \qquad b_2 = \frac{1}{2}(\sigma_x^\infty + \sigma_y^\infty). \tag{4.12.12}$$

The following condition has to be satisfied:

$$\sigma_* = p + \sigma_x^\infty + \sigma_y^\infty - \rho g H/(1-\nu). \tag{4.12.13}$$

In a problem with a given σ_*, condition (4.12.13) is the solvability condition. On the basis of formulas (4.12.8), (4.12.11), and (4.12.12) we obtain

$$\varphi(\zeta) = \frac{1}{4}(\sigma_x^\infty + \sigma_y^\infty) + \frac{1}{4i}\left(\frac{\rho g}{1-\nu} - \gamma\right)\left[c\left(\zeta + \frac{1}{\zeta}\right) - \omega(\zeta)\right]. \tag{4.12.14}$$

Using relation (4.12.14), the second boundary condition of (4.12.7) can be written as

$$\frac{1}{2i}\left(\frac{\rho g}{1-\nu}-\gamma\right)\frac{\overline{\omega(\zeta)}}{\omega'(\zeta)}\left[c\left(1-\frac{1}{\zeta^2}\right)-\omega'(\zeta)\right]+2\psi(\zeta)$$

$$+\frac{\rho g(1-2\nu)}{2i(1-\nu)}\left[\omega(\zeta)-\overline{\omega(\zeta)}-2Hi\right]$$

$$=\frac{\overline{\omega'(\zeta)}}{\zeta^2\omega'(\zeta)}\left\{p+\sigma_*+\frac{\gamma}{2i}\left[\omega(\zeta)-\overline{\omega(\zeta)}\right]\right\}. \qquad (4.12.15)$$

Let us consider the functional equation

$$\frac{1}{2i}\left(\frac{\rho g}{1-\nu}-\gamma\right)\frac{\overline{\omega}(1/\zeta)}{\omega'(\zeta)}\left[c\left(1-\frac{1}{\zeta^2}\right)-\omega'(\zeta)\right]+2\psi(\zeta)$$

$$+\frac{\rho g(1-2\nu)}{2i(1-\nu)}\left[\omega(\zeta)-\overline{\omega}\left(\frac{1}{\zeta}\right)-2Hi\right]$$

$$=\frac{\omega'(1/\zeta)}{\zeta^2\omega(\zeta)}\left\{p+\sigma_*+\frac{\gamma}{2i}\left[\omega(\zeta)-\overline{\omega}\left(\frac{1}{\zeta}\right)\right]\right\}. \qquad (4.12.16)$$

We seek the solution of the functional equation (4.12.16) in the form

$$\omega(\zeta)=c\zeta+\overline{P}_\nu(1/\zeta). \qquad (4.12.17)$$

Here, P_ν is a polynomial of order ν with, as yet, undetermined coefficients. By substituting (4.12.17) in the functional equation (4.12.16) and expanding all functions into series in the vicinity of an infinitely distant point, it is readily seen that $\nu=1$ so that, when function $\omega(\zeta)$ has the form of (4.12.17), it must necessarily be

$$\omega(\zeta)=c\zeta+c_1/\zeta. \qquad (4.12.18)$$

From the symmetry condition it follows that constants c and c_1 are real. From the functional equation (4.12.16), using formula (4.12.18), we obtain the function

$$\psi(\zeta)=\frac{c_1\zeta^2-c}{2(c\zeta^2-c_1)}\left[p+\sigma+\frac{\gamma}{2i}(c-c_1)\left(\zeta+\frac{1}{\zeta}\right)\right]$$

$$-\frac{\rho g(1-2\nu)}{4i(1-\nu)}\left[(c-c_1)\left(\zeta-\frac{1}{\zeta}\right)-2Hi\right]$$

$$-\frac{1}{4i}\left(\frac{\rho g}{1-\nu}-\gamma\right)\frac{(c_1-c)(c_1\zeta^2+c)}{\zeta(c\zeta^2-c_1)}. \qquad (4.12.19)$$

When $|\zeta| \to \infty$, function $\psi(\zeta)$, as can be seen, behaves as follows:

$$\psi(\zeta) = \frac{c - c_1}{4i}\left(\gamma\frac{c_1}{c} - \rho g\frac{1-2\nu}{1-\nu}\right)\zeta + \frac{\rho g H(1-2\nu)}{2(1-\nu)}$$

$$+ \frac{c_1}{2c}(p + \sigma_*) + O(\zeta^{-1}). \tag{4.12.20}$$

By satisfying the conditions at infinity (4.12.7), we obtain

$$\gamma\frac{c_1}{c} - \rho g\frac{1-2\nu}{1-\nu} = 0,$$

$$\frac{\rho g H(1-2\nu)}{2(1-\nu)} + \frac{c_1}{2c}(p + \sigma_*) = \frac{1}{2}(\sigma_y^\infty - \sigma_x^\infty). \tag{4.12.21}$$

Therefore,

$$\gamma = \frac{\rho g(1-2\nu)\left[2p + \sigma_x^\infty + \sigma_y^\infty - \rho g H/(1-\nu)\right]}{(1-\nu)(\sigma_y^\infty - \sigma_x^\infty) - \rho g H(1-2\nu)}, \qquad c_1 = c\frac{1-2\nu}{1-\nu}\cdot\frac{\rho g}{\gamma}. \tag{4.12.22}$$

In order for function $\omega(\zeta)$ (4.12.18) to be analytic everywhere outside the unit circle, it is both necessary and sufficient that the following condition is satisfied:

$$|c_1/c| < 1. \tag{4.12.23}$$

Formulas (4.12.14), (4.12.18), (4.12.19), (4.12.22), and (4.12.13) provide a complete solution of the stated problem. The contours of the hole turn out to be a family of similar ellipses.

The formulas giving the solution when $\sigma_x^\infty = \sigma_y^\infty = p = 0$, i.e., when the walls of the hole and the boundary of the half space are free from loads, acquire a particularly simple form:

$$\varphi(\zeta) = 0, \qquad \omega(\zeta) = c\left(\zeta + \frac{1-2\nu}{\zeta}\right),$$

$$\psi(\zeta) = \frac{\rho g}{(1-\nu)}\left\{\frac{1-2\nu}{4i}\left[2Hi - 2\nu c\left(\zeta - \frac{1}{\zeta}\right)\right]\right.$$

$$\left. - \frac{(1-2\nu)\zeta^2 - 1}{2(\zeta^2 - 1 + 2\nu)}\cdot\left[H + i\nu c\left(\zeta - \frac{1}{\zeta}\right)\right]\right\}, \tag{4.12.24}$$

$$\gamma = \frac{\rho g}{1-\nu}, \qquad \sigma_* = -\frac{\rho g H}{1-\nu}, \qquad c \ll H.$$

2

Let a heavy elastic half space $y < H$ be weakened by a periodic system of cylindrical tunnels parallel to the surface of the half space. Let us consider the problem of determining the shape of the tunnels, which ensure maximum strength [35, 38].

The centers of the holes are located on the x axis at a distance l from each other. We call hole L, with the center at the origin, the principal hole. The boundary condition (4.12.2) exists on an unknown contour of the opening. In addition, all the elastic half space at a distance from the hole is deformed by the stresses (4.12.1), and the surface of the half space $y = H$ is subjected to a constant normal load $\sigma_y = \sigma_y^\infty$, while the shear load is equal to zero. The stresses are determined by equations

$$\sigma_x + \sigma_y = \sigma_x^\infty + \sigma_y^\infty + 4\,\mathrm{Re}\,\Phi(z) + \frac{\rho g}{2i(1-\nu)}(z - \bar{z} - 2Hi),$$

$$\sigma_y - \sigma_x + 2i\tau_{xy} = \sigma_y^\infty - \sigma_x^\infty + 2[(\bar{z} - z)\Phi'(z) + \Psi_*(z)] \tag{4.12.25}$$

$$+ \frac{\rho g(1-2\nu)}{2i(1-\nu)}(z - \bar{z} - 2Hi).$$

We go over to the parametric plane ζ using transformation

$$z - nl = R\left(\zeta + \sum_{k=1}^{\infty} c_k \zeta^{-k}\right) = R\omega(\zeta). \tag{4.12.26}$$

By retaining only those terms which contain the perturbation $\varepsilon = R/l$ raised to a power less than or equal to 2, we present functions $\Phi(z)$ and $\Psi_*(z)$ as

$$\Phi(z) = \varphi(\zeta) + \varphi_1(\zeta) = \sum_{k=1}^{\infty} a_k \zeta^{-k} + A_0 + A_1\omega(\zeta),$$

$$\Psi_*(z) = \psi(\zeta) + \psi_1(\zeta) = \sum_{k=1}^{\infty} b_k \zeta^{-k} + B_0 + B_1\omega(\zeta). \tag{4.12.27}$$

Here

$$A_0 = \frac{\pi^2}{3} a_2 \varepsilon^2, \qquad A_1 = -\frac{\pi^2}{3} a_1 \varepsilon^2,$$

$$B_0 = \frac{\pi^2}{3} b_2 \varepsilon^2, \qquad B_1 = -\frac{\pi^2}{3} b_1 \varepsilon^2. \tag{4.12.28}$$

The boundary value problem on the parametric plane ζ for determination of the analytical functions $\varphi(\zeta)$, $\psi(\zeta)$, and $\omega(\zeta)$ can be written as

$$2\left[\varphi(\zeta)+\overline{\varphi(\zeta)}\right]+\frac{\rho g}{2i(1-\nu)}\left[\omega(\zeta)-\overline{\omega(\zeta)}-2Hi\right]$$

$$=\sigma_*-p+\frac{\gamma}{2i}\left[\omega(\zeta)-\overline{\omega(\zeta)}\right]-\left(\sigma_x^\infty+\sigma_y^\infty\right)$$

$$-4A_0-2A_1\left[\omega(\zeta)+\overline{\omega(\zeta)}\right] \quad \text{when } |\zeta|=1, \qquad (4.12.29)$$

$$2\left[\frac{\omega(\zeta)-\overline{\omega(\zeta)}}{\omega'(\zeta)}\varphi'(\zeta)+\psi(\zeta)\right]+\frac{\rho g(1-2\nu)}{2i(1-\nu)}\left[\omega(\zeta)-\overline{\omega(\zeta)}-2Hi\right]$$

$$=-\left(\sigma_y^\infty-\sigma_x^\infty\right)-2B_0-2B_1\omega(\zeta)-2A_1\left[\overline{\omega(\zeta)}-\omega(\zeta)\right]+\frac{\overline{\omega'(\zeta)}}{\zeta^2\omega'(\zeta)}$$

$$\times\left\{p+\sigma_*+\frac{\gamma}{2i}\left[\omega(\zeta)-\bar{\omega}(\zeta)\right]\right\} \quad \text{when } |\zeta|=1. \qquad (4.12.30)$$

Solution of a boundary value problem. Function $F^-(\zeta)$, which is

$$F^-(\zeta)=2\varphi(\zeta)+\left[\frac{\rho g}{2i(1-\nu)}-\frac{\gamma}{2i}+2A_1\right]\omega(\zeta), \qquad (4.12.31)$$

is analytic outside the unit circle with the exception of an infinitely distant point, where it has a pole with the main term

$$\left[\frac{1}{2i}\frac{\rho g}{(1-\nu)}-\frac{\gamma}{2i}+2A_1\right]\zeta, \qquad (4.12.32)$$

while function $F^+(\zeta)$, which is equal to

$$F^+(\zeta)=-2\bar{\varphi}\left(\frac{1}{\zeta}\right)+\frac{\rho g}{2i(1-\nu)}\left[\bar{\omega}\left(\frac{1}{\zeta}\right)+2Hi\right]+\sigma_*-p$$

$$-\frac{\gamma}{2i}\bar{\omega}\left(\frac{1}{\zeta}\right)-\left(\sigma_x^\infty+\sigma_y^\infty\right)-4A_0-2A_1\bar{\omega}\left(\frac{1}{\zeta}\right), \qquad (4.12.33)$$

is analytic inside the unit circle with the exception of point $\zeta=0$, at which it has a pole with the main term

$$\left[\frac{1}{2i}\left(\frac{\rho g}{1-\nu}-\gamma\right)-2A_1\right]\frac{1}{\zeta}. \qquad (4.12.34)$$

On the basis of the boundary condition (4.12.29), functions $F^+(\zeta)$ and $F^-(\zeta)$ are analytical extensions of each other through the unit circle:

$$F^+(\zeta)=F^-(\zeta) \quad \text{for } |\zeta|=1. \qquad (4.12.35)$$

In accordance with the Liouville theorem, the analytical function $F(\zeta)$ is

$$F(\zeta)=\left[\frac{1}{2i}\left(\frac{\rho g}{1-\nu}-\gamma\right)-2A_1\right]\frac{1}{\zeta}+\left[\frac{1}{2i}\left(\frac{\rho g}{1-\nu}-\gamma\right)+2A_1\right]\zeta, \quad (4.12.36)$$

and the following condition is satisfied:

$$\sigma_*-p-\sigma_x^\infty-\sigma_y^\infty+\rho gH/(1-\nu)-4A_0=0. \quad (4.12.37)$$

On the basis of formulas (4.12.31) and (4.12.36) we obtain

$$\varphi(\zeta)=\frac{1}{4i}\left(\frac{\rho g}{1-\nu}-\gamma\right)\left[\left(\zeta+\frac{1}{\zeta}\right)-\omega(\zeta)\right]+A_1\left[\zeta-\frac{1}{\zeta}-\omega(\zeta)\right]. \quad (4.12.38)$$

By accounting for the conditions at infinity, we obtain

$$a_1=\frac{1}{4i}\left(\frac{\rho g}{1-\nu}-\gamma\right)(1-c_1)-A_1(1+c_1), \qquad a_2=0. \quad (4.12.39)$$

Using relation (4.12.38), the boundary condition (4.12.30) is written in the form

$$\frac{\overline{\omega(\zeta)}-\omega(\zeta)}{\omega'(\zeta)}\left\{\frac{1}{2i}\left(\frac{\rho g}{1-\nu}-\gamma\right)\left[1-\frac{1}{\zeta^2}-\omega'(\zeta)\right]+2A_1\left(1+\frac{1}{\zeta^2}\right)\right\}+2\psi(\zeta)$$
$$+\frac{\rho g(1-2\nu)}{2i(1-\nu)}\left[\omega(\zeta)-\overline{\omega(\zeta)}-2Hi\right]$$
$$=-(\sigma_y^\infty-\sigma_x^\infty)-2B_0-2B_1\omega(\zeta)$$
$$+\frac{\overline{\omega'(\zeta)}}{\zeta^2\omega'(\zeta)}\left\{p+\sigma_*+\frac{\gamma}{2i}\left[\omega(\zeta)-\overline{\omega(\zeta)}\right]\right\} \quad \text{for } |\zeta|=1. \quad (4.12.40)$$

Let us consider the functional equation

$$\frac{\bar{\omega}(1/\zeta)-\omega(\zeta)}{\omega'(\zeta)}\left\{\frac{1}{2i}\left(\frac{\rho g}{1-\nu}-\gamma\right)\left[1-\frac{1}{\zeta^2}-\omega'(\zeta)\right]+2A_1\left[1+\frac{1}{\zeta^2}\right]\right\}$$
$$+2\psi(\zeta)+\frac{\rho g(1-2\nu)}{2i(1-\nu)}\left[\omega(\zeta)-\bar{\omega}\left(\frac{1}{\zeta}\right)-2Hi\right]$$
$$=-(\sigma_y^\infty-\sigma_x^\infty)-2B_0-2B_1\omega(\zeta)$$
$$+\frac{\bar{\omega}'(1/\zeta)}{\zeta^2\omega'(\zeta)}\left\{p+\sigma_*+\frac{\gamma}{2i}\left[\omega(\zeta)-\bar{\omega}\left(\frac{1}{\zeta}\right)\right]\right\}. \quad (4.12.41)$$

By substituting expression (4.12.26) for $\omega(\zeta)$ in the functional equation (4.12.41) and expanding all functions into series in the vicinity of an infinitely distant point, it is easy to notice that $k=1$ so that, if $\omega(\zeta)$ has the form of

(4.12.26), it must then be

$$\omega(\zeta) = \zeta + c_1/\zeta. \tag{4.12.42}$$

From the symmetry conditions it follows that constant c_1 is real. From the functional equation (4.12.41), using (4.12.42), we find $\psi(\zeta)$.

By satisfying the conditions at infinity, we obtain

$$2B_1 + \frac{\gamma}{2i}c_1(1-c_1) + 2A_1(c_1-1) + \frac{\rho g(1-2\nu)}{2i(1-\nu)}(1-c_1) = 0,$$

$$\frac{\rho g H(1-2\nu)}{1-\nu} - c_1(p+\sigma_*) = \sigma_y^\infty - \sigma_x^\infty + 2B_0,$$

$$\begin{aligned} 2b_1 = &-2B_1c_1 - \frac{\gamma}{2i}c_1(c_1-1) + \frac{\gamma}{2i}(1-c_1^2)(1-c_1) \\ &-(c_1-1)\left[\frac{1}{2i}\left(\frac{\rho g}{1-\nu}-\gamma\right)(c_1-1)+2A_1\right] - 2A_1(1+c_1^2) \\ &-\frac{\rho g(1-2\nu)}{2i(1-\nu)}(c_1-1), \end{aligned} \tag{4.12.43}$$

$$2b_2 = (p+\sigma_*)(1-c_1^2).$$

The system of nonlinear algebraic equations (4.12.43) as well as equations (4.12.37) and (4.12.39) are used to determine constants σ_*, γ, c_1, a_1, b_1, a_2, and b_2. The algebraic system of equations is easily solved by the perturbation method:

$$c_1 = c_{10} + c_{12}\varepsilon^2, \qquad c_{10} = -\frac{\rho g}{\gamma}\frac{1-2\nu}{1-\nu},$$

$$\gamma = \frac{\rho g(1-2\nu)\left[2p + \sigma_x^\infty + \sigma_y^\infty - \rho g H/(1-\nu)\right]}{(1-\nu)(\sigma_y^\infty - \sigma_x^\infty) - \rho g H(1-2\nu)},$$

$$c_{12} = \frac{\pi^2}{3}(c_{10}^2 - 1), \qquad a_1 = a_{10} + a_{12}\varepsilon^2, \qquad a_2 = 0,$$

$$a_{10} = \frac{1}{4i}\left(\frac{\rho g}{1-\nu} - \gamma\right)(1-c_{10}), \qquad a_{12} = \frac{\pi^2}{3}\frac{1}{2i}\left(\frac{\rho g}{1-\nu} - \gamma\right)(1-c_{10}^2),$$

$$\begin{aligned} b_1 = b_{10} + b_{12}\varepsilon^2, \qquad b_{10} = &\frac{\rho g(1-2\nu)(1-c_{10})}{4i(1-\nu)} - \frac{1}{4i}\left(\frac{\rho g}{1-\nu}\right)(c_{10}-1)^2 \\ &-\frac{\gamma}{4i}(2c_{10}^2 - 1 - c_{10}^3), \end{aligned} \tag{4.12.44}$$

$$b_{12} = \frac{\pi^2}{3} b_{10} c_{10} - \frac{\gamma}{4i} c_{12}\left(2 - 3c_{10}^2\right) - \frac{\rho g(1-2\nu)}{4i(1-\nu)} c_{12}$$

$$+ \frac{\pi^2}{3} a_{10}\left(c_{10} + c_{10}^2\right) + \frac{1}{2i}\left(\frac{\rho g}{1-\nu} - \gamma\right) c_{12}\left(c_{10} - 1\right),$$

$$\sigma_* = p + \sigma_x^\infty + \sigma_y^\infty - \frac{\rho g H}{1-\nu}, \qquad b_2 = b_{20} + b_{22}\varepsilon^2,$$

$$b_{20} = \frac{1}{2}(p + \sigma_*)\left(1 - c_{10}^2\right), \qquad b_{22} = \frac{\pi^2}{3} c_{10}\left(1 - c_{10}^2\right)(p + \sigma_*).$$

As a first approximation, the contours of the hole, which ensure maximum strength, turn out to be a family of similar ellipses (4.12.42).

If in expansion (4.12.27) we account for the terms containing ε raised to higher powers, then the lines will be near elliptic rather than elliptic. In particular, it can be shown that if we retain terms containing ε raised to a power less than or equal to 4, then the contours of the hole will be

$$\omega(\zeta) = \zeta + c_1/\zeta + c_3/\zeta^3. \tag{4.12.45}$$

Let us present the formulas that provide a solution when $\sigma_x^\infty = \sigma_y^\infty = p = 0$, i.e., when the walls of the hole and the boundary of the half space are free from loads:

$$\varphi(\zeta) = 0, \qquad \omega(\zeta) = \zeta + \frac{c_1}{\zeta},$$

$$\psi(\zeta) = \frac{\rho g}{1-\nu}\left\{\frac{1-2\nu}{4i}\left[2Hi - (1-c_1)\left(\zeta - \frac{1}{\zeta}\right)\right] - \frac{1 - c_1\zeta^2}{2(\zeta^2 - c_1)}\right.$$

$$\left.\times\left[H - \frac{1}{2i}(1-c_1)\left(\zeta - \frac{1}{\zeta}\right)\right]\right\} \frac{\pi^2}{3} b_2\varepsilon^2 + \frac{\pi^2}{3} b_1\varepsilon^2\left(\zeta + \frac{c_1}{\zeta}\right),$$

$$\gamma = \frac{\rho g}{1-\nu}, \qquad \sigma_* = -\frac{\rho g H}{1-\nu}, \qquad c_1 = c_{10} + c_{12}\varepsilon^2,$$

$$c_{10} = 1 - 2\nu, \qquad c_{12} = \frac{4\pi^2}{3}\nu(\nu - 1), \qquad a_1 = 0, \qquad a_2 = 0, \tag{4.12.46}$$

$$b_1 = b_{10} + b_{12}\varepsilon^2, \qquad b_{10} = \frac{\rho g(1-2\nu)}{i(1-\nu)}\nu,$$

$$b_{12}=\frac{\pi^2}{3}\frac{\rho g \nu}{i(1-\nu)}(5-18\nu+26\nu^2-12\nu^3),$$

$$b_2=b_{20}+b_{22}\varepsilon^2, \qquad b_{22}=\frac{4}{3}\pi^2\rho gH(1-\nu)\nu,$$

$$b_{20}=\frac{2\rho gH}{\nu}.$$

3

Let us consider the problem of determining the contour of an equal-strength opening in a heavy elastic half space $y < H$, weakened by two identical tunnels of cylindrical shape. The axis of each cylinder is parallel to the surface of the half space. The centers of the holes are located on the x axis and spaced at a distance $2l$ from each other. The origin is chosen at a point equidistant from the centers of the holes. The boundary conditions are the same as in the previous section. The solution of this problem is determined by formulas (4.12.27), (4.12.38), (4.12.39), and (4.12.41)–(4.12.43) so that

$$z-l=R\omega(\zeta), \qquad A_0=-a_1\varepsilon+a_2\varepsilon^2, \qquad A_1=a_1\varepsilon^2,$$

$$B_0=-b_1\varepsilon+b_2\varepsilon^2, \qquad B_1=b_1\varepsilon^2, \qquad \varepsilon=R/2l.$$

Due to the fact that the expressions for the constants σ_*, γ, c_1, a_1, b_1, and b_2 are cumbersome, they are not presented here.

It can be shown that if the terms containing ε raised to a power less than or equal to 4 are accounted for, then the contours of the hole will be

$$\omega(\zeta)=\zeta+c_1/\zeta+c_2/\zeta^2+c_3/\zeta^3.$$

5. PLANE STRESS

§ 1. FUNDAMENTAL EQUATIONS. BRIEF REVIEW

The plane stress state approximately describes stresses in a thin plate loaded only in its plane. Since the plate thickness is small compared with two other dimensions, and since the base of the plate is free from loads, the components of the stress tensor $\sigma_z, \tau_{xz}, \tau_{yz}$ are small in comparison with stresses $\sigma_x, \sigma_y, \tau_{xy}$, which vary little across thickness. The stresses $\sigma_x, \sigma_y, \tau_{xy}$ can, hereafter, be replaced by their mean values and stresses $\sigma_z, \tau_{xz}, \tau_{yz}$ are assumed to be zero. The plate thickness in this case is not important; therefore, it will be assumed equal to unity.

The stress components, in the absence of body forces, satisfy two differential equilibrium equations:

$$\frac{\partial \sigma_x}{\partial x} + \frac{\partial \tau_{xy}}{\partial y} = 0,$$
$$\frac{\partial \tau_{xy}}{\partial x} + \frac{d\sigma_y}{\partial y} = 0. \tag{5.1.1}$$

The following relations exist in the elastic region:
Hooke's law:

$$\varepsilon_x = E^{-1}(\sigma_x - \nu\sigma_y),$$
$$\varepsilon_y = E^{-1}(\sigma_y - \nu\sigma_x), \tag{5.1.2}$$
$$\varepsilon_{xy} = \tau_{xy}/(2G).$$

Strain compatibility, which can be expressed in terms of stresses:

$$\Delta(\sigma_x + \sigma_y) = 0. \tag{5.1.3}$$

Relation between displacements and strains:

$$\varepsilon_x = \frac{\partial u}{\partial x}, \qquad \varepsilon_y = \frac{\partial v}{\partial y}, \qquad \varepsilon_{xy} = \frac{1}{2}\left(\frac{\partial u}{\partial y} + \frac{\partial v}{\partial x}\right). \tag{5.1.4}$$

In the plastic region, the following relations apply:
Plasticity condition:

$$f(\sigma_x, \sigma_y, \tau_{xy}) - k = 0. \tag{5.1.5}$$

Associated law of plastic flow:

$$d\varepsilon_x^p = \lambda \frac{\partial f}{\partial \sigma_x}, \qquad d\varepsilon_y^p = \lambda \frac{\partial f}{\partial \sigma_y}, \qquad d\varepsilon_{xy}^p = \frac{\lambda}{2} \frac{\partial f}{\partial \tau_{xy}}. \tag{5.1.6}$$

Kinematic relations:

$$\begin{aligned} d\varepsilon_x &= d\varepsilon_x^e + d\varepsilon_x^p, \\ d\varepsilon_y &= d\varepsilon_y^e + d\varepsilon_y^p, \\ d\varepsilon_{xy} &= d\varepsilon_{xy}^e + d\varepsilon_{xy}^p. \end{aligned} \tag{5.1.7}$$

The increments of elastic strains in the plastic region are related to the stresses by Hooke's law.

As noted in Chapter 4, the plane problem of the theory of ideal plasticity is statically determinate if the boundary conditions are given in stresses.

In 1963, G. P. Cherepanov presented an exact solution for the problem of stress distribution in the vicinity of a circular hole under plane stress conditions. Constant normal forces were applied to the contour and the stress distribution at infinity was uniform [23, 24]. It was assumed that the material satisfied the Tresca–Saint Venant plasticity condition. The solution was obtained by the method of functional equations (see Chapter 4, § 2).

The elastic–plastic problem in this case, generally speaking, includes an entire range of types of equations. In the region next to the contour of the body, the governing system of equations is of hyperbolic type (two families of the characteristics are logarithmic spirals). Next follows the region where the governing equations are of parabolic type (the only family of characteristics is formed by radial straight lines). In the external elastic region the governing equations are of elliptical type (biharmonic equations). The preceding problem was studied earlier by Sokolov [19], Zaslavsky [9], Faerberg [21], Perlin [17], and Ivlev [19 of Chapter 6].

Khoma [22] discussed the elastic–plastic problem for a thin plate with an infinite row of identical circular holes. Panferov [16], using the method of elastic solution by A. A. Ilyushin, discussed the problem with an elliptical hole.

The method developed by A. A. Ilyushin, in its finite difference interpretation, was successfully used by Korotkikh [11, 12] for the numerical solution of nonlinear elastic problems. Numerical methods had not been extensively used to solve elastic–plastic problems under plane stress conditions. Apart from papers by Perlin [17], Solodilov [20], and Birger [1, 2], very few publications utilizing numerical methods can be quoted.

Southwell and Allen discussed a strip with two symmetrical semicircular and angular notches [30]. A numerical solution of some elastic–plastic problems is available [13] for thin plates with rectangular slits. An elastic–plastic analysis of an infinite plate with circular holes was given in [10]. The plate was loaded in uniaxial tension and its material was characterized by power-law hardening.

In the case of plane stress state (in plates), there is a set of elastic–plastic solutions that is of a particularly great interest for fracture mechanics [27]. Plastic strains, in many plane stress problems, are concentrated along lines ("neck"); the solution of such problems is found by the methods of the theory of elasticity. Dugdale [29] was the first to present a solution of that type for one slit in a plate under tension applied at infinity in the direction perpendicular to the line of the slit. Dugdale also confirmed this solution experimentally. Almost simultaneously, similar solutions were thoroughly studied by M. Y. Leonov with his co-workers, P. M. Vitvitsky and C. Y. Yarema [3–7, 15].

The assumption of plastic strain concentration along a section on the extension of a crack was named "Dugdale's hypothesis." The latter was also discussed by Goodier and Field [28]. In this formulation, L. A. Galin and G. P. Cherepanov obtained a solution of the contact elastic–plastic problem under conditions of plane stress for both rigid punch and contact of two elastic–plastic bodies [8].

§ 2. PLATE WITH A CIRCULAR HOLE

1

An infinite plate in plane stress state has a circular hole of radius R with the center at the origin. Constant normal load $\sigma_r = p$ is applied to the contour of the hole and the shear load is equal to zero, $\tau_{r\theta} = 0$. At an infinitely distant point, there exists a uniform stress distribution

$$\sigma_x = \sigma_x^\infty, \qquad \sigma_y = \sigma_y^\infty, \qquad \tau_{xy} = 0. \tag{5.2.1}$$

We assume that the plastic region encloses the entire circular hole under the given system of external forces. We shall use the plasticity condition of Tresca–Saint Venant and assume that the inequality $\sigma_\theta \geq \sigma_r > 0$ exists in the plastic region. Hence, the characteristics in the plastic region will be radial straight lines, and the stresses will be defined as:

$$\sigma_r = \sigma_s + (p - \sigma_s) R/r, \qquad \sigma_\theta = \sigma_s, \qquad \tau_{r\theta} = 0 \tag{5.2.2}$$

(σ_s is the yield point under tension).

To satisfy inequality $\sigma_\theta \geq \sigma_r > 0$, load p must, evidently, satisfy condition $0 \leq p \leq \sigma_s$. The Kolosov–Muskhelishvili representations hold true in the elastic region

$$\begin{aligned} \sigma_r + \sigma_\theta &= 4\,\mathrm{Re}\,\Phi(z), \\ \sigma_\theta - \sigma_r + 2i\tau_{r\theta} &= 2e^{2i\theta}[\bar{z}\Phi'(z) + \Psi(z)]. \end{aligned} \tag{5.2.3}$$

By using (5.2.2) and (5.2.3), the boundary conditions on an unknown contour L separating the elastic and plastic regions are presented as

$$\begin{aligned} 4\,\mathrm{Re}\,\Phi(z) &= 2\sigma_s + \frac{R(p - \sigma_s)}{r} \quad \text{on } L, \\ \bar{z}\Phi'(z) + \Psi(z) &= \frac{R(\sigma_s - p)}{2r} e^{-2i\theta}. \end{aligned} \tag{5.2.4}$$

When $|z| \to \infty$,

$$\Phi(z) = \tfrac{1}{4}(\sigma_x^\infty + \sigma_y^\infty) + O(z^{-2})$$

$$\Psi(z) = \tfrac{1}{2}(\sigma_y^\infty - \sigma_x^\infty) + O(z^{-2}).$$

Let us go over to the parametric plane of the complex variable ζ using transformation $z = \omega(\zeta)$. The analytical function $\omega(\zeta)$ conformally maps the outside of a unit circle in plane ζ onto the outside of an unknown contour L in plane z with correspondence of infinitely distant points $\omega(\infty) = \infty$. This function must be determined in the process of problem solving. Assuming

$$\varphi(\zeta) = \Phi[\omega(\zeta)], \qquad \psi(\zeta) = \Psi[\omega(\zeta)],$$

we obtain, in accordance with (5.2.4), the following boundary value problem on plane ζ for determination of the three analytical functions $\varphi(\zeta)$, $\psi(\zeta)$, and $\omega(\zeta)$:

$$\begin{aligned} &4\,\mathrm{Re}\,\varphi(\zeta) = 2\sigma_s + (p - \sigma_s)\frac{R}{|\omega(\zeta)|}, \\ &\frac{\overline{\omega(\zeta)}}{\omega'(\zeta)}\varphi'(\zeta) + \psi(\zeta) = \frac{R(\sigma_s - p)\overline{\omega(\zeta)}}{2\omega(\zeta)\,|\omega(\zeta)|} \quad \text{when } |\zeta| = 1. \end{aligned} \tag{5.2.5}$$

On the basis of (5.2.4), functions $\varphi(\zeta)$ and $\psi(\zeta)$ at infinity behave as follows:

$$\begin{aligned} &\varphi(\zeta) = \tfrac{1}{4}(\sigma_x^\infty + \sigma_y^\infty) + O(\zeta^{-2}), \\ &\psi(\zeta) = \tfrac{1}{2}(\sigma_y^\infty - \sigma_x^\infty) + O(\zeta^{-2}), \\ &\omega(\zeta) = O(\zeta). \end{aligned} \tag{5.2.6}$$

Let us consider a functional equation

$$\frac{\varphi'(\zeta)}{\omega'(\zeta)}\overline{\omega}\left(\frac{1}{\zeta}\right) + \Psi(\zeta) = \frac{R(\sigma_s - p)\overline{\omega}(1/\zeta)}{2\omega(\zeta)\sqrt{\omega(\zeta)\overline{\omega}(1/\zeta)}}, \qquad |\zeta| > 1. \tag{5.2.7}$$

We seek a solution of the functional equation in the form (see Chapter 4, § 2)

$$\omega(\zeta) = c\zeta + \overline{P}_\nu(1/\zeta), \tag{5.2.8}$$

where $\overline{P}_\nu$ is a polynomial of order ν with, as yet undetermined coefficients. Substituting (5.2.8) in (5.2.7) and expanding all functions into series in the vicinity of an infinitely distant point, it can be readily seen that $\nu = 3$. This means that if function $\omega(\zeta)$ has the form of (5.2.8), then it must be equal to

$$\omega(\zeta) = c\zeta + c_1/\zeta + c_2/\zeta^2 + c_3/\zeta^3. \tag{5.2.9}$$

From the symmetry condition, constant c_2 is equal to zero, and the remaining constants are real. We shall show that they can be selected in such a way that the right-hand part of the functional equation (5.2.7) is analytic outside the unit circle. Function $\omega(\zeta)$, evidently, has four zeros inside the unit circle $|\zeta| < 1$, and all four zeros of function $\bar{\omega}(1/\zeta)$ are located outside the unit circle. In order to ensure that the right-hand side of the functional equation (5.2.7) is analytic outside the unit circle, it is necessary and sufficient to require that the zeros of function $\bar{\omega}(1/\zeta)$ coincide in pairs. Since function

$$\bar{\omega}(1/\zeta) = 1/\zeta(c + c_1\zeta^2 + c_3\zeta^4), \tag{5.2.10}$$

then for the coincidence in pairs it is sufficient to require that the discriminant of the biquadratic equation be equal to zero:

$$c_1^2 = 4cc_3. \tag{5.2.11}$$

Using condition (5.2.11), functions $\omega(\zeta)$ and $\omega(\zeta^{-1})$ can be written as

$$\begin{aligned} \omega(\zeta) &= \frac{c}{\zeta^3}\left(\zeta^2 + \frac{c_1}{2c}\right)^2, \\ \bar{\omega}\left(\frac{1}{\zeta}\right) &= \frac{c_3}{\zeta}\left(\zeta^2 + \frac{c_1}{2c_3}\right)^2. \end{aligned} \tag{5.2.12}$$

From the functional equation (5.2.7),

$$\psi(\zeta) = \frac{R(\sigma_s - p)\sqrt{c_3}\,(\zeta^2 + c_1/2c_3)\zeta^4}{2c^{3/2}(\zeta^2 + c_1/2c)^3} - \frac{\varphi'(\zeta)}{\omega'(\zeta)}\bar{\omega}\left(\frac{1}{\zeta}\right). \tag{5.2.13}$$

Let us find function $\varphi(\zeta)$. The first boundary condition of (5.2.5), using formula (5.2.12), is convenient to express as

$$\begin{aligned} 4\,\mathrm{Re}\,\varphi(\zeta) &= 2\sigma_s + \frac{R(p - \sigma_s)\zeta^2}{\sqrt{cc_3}\,(\zeta^2 + c_1/2c_3)(\zeta^2 + c_1/2c)} \\ &= 2\sigma_s + \frac{2R(p - \sigma_s)\sqrt{cc_3}}{c_1(c_3 - c)}\left[\frac{\zeta^2}{\zeta^2 + c_1/2c_3} - \frac{\zeta^2}{\zeta^2 + c_1/2c}\right], \quad |\zeta| = 1, \end{aligned} \tag{5.2.14}$$

Function $F^+(\zeta)$, which is equal to

$$F^+(\zeta) = -2\bar{\varphi}\left(\frac{1}{\zeta}\right) + \frac{2R\sqrt{cc_3}\,(p - \sigma_s)}{c_1(c_3 - c)}\,\frac{\zeta^2}{\zeta^2 + c_1/2c_3}, \tag{5.2.15}$$

is analytic everywhere inside the unit circle $|\zeta| < 1$, and function $F^-(\zeta)$, which is equal to

$$F^-(\zeta) = 2\varphi(\zeta) - 2\sigma_s + \frac{R(p-\sigma_s)\sqrt{cc_3}}{c_1(c_3-c)} \frac{\zeta^2}{\zeta^2 + c_1/2c}, \tag{5.2.16}$$

is analytic everywhere outside the unit circle $|\zeta| > 1$.

The boundary condition (5.2.14) can be presented as

$$F^+(\zeta) = F^-(\zeta) \quad \text{for } |\zeta| = 1. \tag{5.2.17}$$

Consequently, functions $F^+(\zeta)$ and $F^-(\zeta)$ are analytical extensions of each other through the unit circle. According to the Liouville theorem, both functions are identically equal to the same constant. Consequently, from the condition at infinity (5.2.6), it is easy to obtain $\varphi(\zeta)$:

$$\varphi(\zeta) = \sigma_s - \frac{\sigma_y^\infty + \sigma_x^\infty}{4} - \frac{R(p-\sigma_s)\sqrt{cc_3}}{c_1(c_3-c)} \cdot \frac{\zeta^2}{(\zeta^2 + c_1/2c)}, \tag{5.2.18}$$

where the following relation must exist between the three yet unknown constants

$$\sigma_x^\infty + \sigma_y^\infty - 2\sigma_s + \frac{2R(p-\sigma_s)\sqrt{cc_3}}{c_1(c_3-c)} = 0. \tag{5.2.19}$$

From formula (5.2.18) we obtain

$$\varphi'(\zeta) = -\sqrt{\frac{c_3}{c}} \frac{R(p-\sigma_s)}{(c_3-c)\zeta^3} + O(\zeta^{-4}), \quad \text{when } |\zeta| \to \infty. \tag{5.2.20}$$

By using the conditions at infinity (5.2.6) and (5.2.20), we obtain from formula (5.2.13) one more relationship between c, c_1, and c_3:

$$\sigma_y^\infty - \sigma_x^\infty = \frac{R(\sigma_s - p)\sqrt{c_3}}{c\sqrt{c}} \cdot \frac{c+c_3}{c-c_3}. \tag{5.2.21}$$

Thus, functions $\varphi(\zeta)$, $\psi(\zeta)$, and $\omega(\zeta)$ are determined by formulas (5.2.12), (5.2.13), and (5.2.18), and the constants are found from the solution of the three equations (5.2.11), (5.2.19), and (5.2.21). It is easy to check that all boundary conditions and the conditions at infinity are satisfied in this case.

The solution of equations (5.2.11), (5.2.19), and (5.2.21) can be presented as

$$c = \frac{4R}{\alpha(a^2-4)}, \qquad c_1 = \frac{4aR}{\alpha(a^2-4)},$$
$$c_3 = \frac{a^2R}{\alpha(a^2-4)}, \qquad \alpha = \frac{\sigma_x^\infty + \sigma_y^\infty - 2\sigma}{\sigma_s - p}, \tag{5.2.22}$$

where a is a real root of the cubic equation

$$a^3 + 4a + \beta = 0, \qquad \beta = 8(\sigma_y^\infty - \sigma_x^\infty)/(\sigma_s - p). \tag{5.2.23}$$

Since the discriminant $D = -4^4 - 27\beta^2$ of the cubic equation (5.2.23) is always negative, equation (5.2.23) always has one real and two conjugate complex roots.

Finally, the initial functions $\omega(\zeta)$, $\varphi(\zeta)$, and $\psi(\zeta)$ according to formulas (5.2.12), (5.2.13), (5.2.18), and (5.2.22) acquire the form

$$\omega(\zeta) = \frac{R(2\zeta^2 + a)^2}{\alpha(a^2 - 4)\zeta^3},$$

$$\varphi(\zeta) = \sigma_s - \frac{1}{4}(\sigma_x^\infty + \sigma_y^\infty) - \frac{\alpha(p - \sigma_s)\zeta^2}{2\zeta^2 + a}, \tag{5.2.24}$$

$$\psi(\zeta) = \alpha(p - \sigma_s)\frac{a\zeta^4(\zeta^2 + 1)[2(a^2 + 4)\zeta^2 - a(4 - 3a^2)]}{2(2\zeta^2 + a)^3(2\zeta^2 - 3a)}.$$

2

Let us investigate the obtained solution (5.2.24). The equation of contour L separating the elastic and plastic regions can be presented in the parametric form

$$x^*(t) = 4(1 + a)\cos t + a^2\cos 3t,$$

$$y^*(t) = 4(1 - a)\sin t - a^2\sin 3t, \qquad 0 \le t \le 2\pi, \tag{5.2.25}$$

$$[Rx^*(t) = \alpha(a^2 - 4)x(t), \qquad Ry^*(t) = \alpha(a^2 - 4)y(t)].$$

The parameters a and α must satisfy certain inequalities following from the condition of complete enclosure of the circular hole by the plastic zone and from the univalence condition for function $\omega(\zeta)$. These inequalities determine the domain of the existence for solution (5.2.24). Here, we assume

$$\sigma_y^\infty \ge \sigma_x^\infty \ge 0.$$

Additionally, the inequality $\sigma_s \ge \sigma_y^\infty$ must evidently be satisfied; hence, $\alpha < 0$, $\beta < 0$, $a > 0$. In order for the plastic zone to enclose the entire hole, inequality $iR < \omega(i)$ must take place; therefore, the parameters must satisfy one more inequality,

$$a < 2(1 + \alpha)/(1 - \alpha). \tag{5.2.26}$$

By grouping all inequalities together, we obtain

$$0 > \alpha > -1, \qquad 2(1 + \alpha)/(1 - \alpha) > a > 0, \qquad 2 > a > 0. \tag{5.2.27}$$

The derivative of expression (5.2.24) for $\omega(\zeta)$ will be

$$\omega'(\zeta)=\frac{R(2\zeta^2+a)(2\zeta^2-3a)}{\alpha(a^2-4)\zeta^4}. \tag{5.2.28}$$

For the mapping produced by the analytical function $\omega(\zeta)$ to be conformal it is necessary for the first derivative to be different from zero everywhere outside the unit circle. Otherwise, a nonuniqueness loop, which has no physical meaning, forms on contour L separating the elastic and plastic regions. For the univalence condition to be met, parameter a, in accordance with (5.2.28), must satisfy inequality

$$0\le a\le \tfrac{2}{3}. \tag{5.2.29}$$

The final domain of existence for solution (5.2.26) is determined by inequalities

$$2(1+\alpha)/(1-\alpha)\ge a\ge 0, \qquad 0\le a\le \tfrac{2}{3}, \qquad -1<\alpha<-\tfrac{1}{2}. \tag{5.2.30}$$

As can be seen, solution (5.2.24) exists only until a cusp is encountered on the contour separating the elastic and plastic regions.

3

A solution of the boundary value problem (5.2.24) was sought in a class of functions, bounded everywhere in the elastic region, including the unknown boundary. Another solution, which is unbounded in some points of the elastic region (i.e., essentially bounded), can be constructed for this problem. Indeed, let the solution of the functional equation (5.2.7) be expressed as before by formulas (5.2.12) and (5.2.13). For definiteness, assume $\sigma_y^\infty\ge\sigma_x^\infty\ge 0$. We seek a solution of the Dirichlet boundary value problem (5.2.5) for function $\varphi(\zeta)$ in the class of functions having poles at points $\zeta=\pm1$. Since there is no concentrated force at the corresponding points of the unknown contour in the physical plane z, the following condition must be satisfied at points $\zeta=\pm1$:

$$\omega'(\pm1)=0. \tag{5.2.31}$$

Condition (5.2.31) means that the elastic–plastic boundary, which is unknown beforehand, always has cusps at the points corresponding to $\zeta=\pm1$. In this case, stresses in the elastic region near the cusp have a singularity of type $1/\sqrt{s}$. The mathematical assumption about stress unboundedness near the cusp is physically explained by the presence of local statically indeterminate plastic zones near the cusp. Therefore, the second solution is approximate.

From condition (5.2.31), taking into account formula (5.2.12), we obtain

$$2c=3c_1. \tag{5.2.32}$$

It is easy to show that the solution to the Dirichlet problem (5.2.14) in the indicated class of functions, which does not satisfy the condition at infinity yet, has the form

$$\varphi(\zeta) = A\frac{\zeta^2+1}{\zeta^2-1} - \frac{3R(p-\sigma_s)}{16c}\cdot\frac{1}{\zeta^2+1/3} + \frac{1}{2}\sigma_s + \frac{9R(p-\sigma_s)}{32c}, \quad (5.2.33)$$

where A is a real constant. In this case, relations (5.2.11) and (5.2.32) were used.

For determination of constants c and A, the two conditions at infinity (5.2.6) are used. Using relation (5.2.6), as well as formulas (5.2.13), (5.2.11), and (5.2.33), we obtain

$$c = \frac{6R(\sigma_s - p)}{4\sigma_s + 7\sigma_y^\infty - 11\sigma_x^\infty},$$
$$A = \frac{37}{64}\sigma_y^\infty - \frac{17}{64}\sigma_x^\infty - \frac{5}{16}\sigma_s. \quad (5.2.34)$$

Functions $\omega(\zeta)$ and $\psi(\zeta)$ can be written as

$$\omega(\zeta) = \frac{c}{\zeta^3}\left(\zeta^2 + \frac{1}{3}\right)^2,$$
$$\psi(\zeta) = \frac{R(\sigma_s - p)\zeta^4(\zeta^2+3)}{6c(\zeta^2+1/3)^3} - \frac{\varphi'(\zeta)}{\omega'(\zeta)}\overline{\omega}\left(\frac{1}{\zeta}\right). \quad (5.2.35)$$

Let us write the equation of the elastic–plastic boundary in parametric form

$$x(t) = 1/3c(5\cos t + 1/3\cos 3t),$$
$$y(t) = 1/3c(\sin t - 1/3\sin t), \quad (5.2.36)$$
$$0 \le t \le 2\pi.$$

As can be seen, the elastic–plastic boundary for the loading parameters remains similar to the contour of solution (5.2.24) for the critical value of parameter a, which is equal to $\frac{2}{3}$. When this value is exceeded, solution (5.2.24) loses its physical meaning. Function $\omega(\zeta)$ is univalent for any values of the loading parameters outside of the unit circle.

The solution (5.2.33)–(5.2.35) exists for any values of the loading parameters that result in a complete enclosure of the circular hole,

$$4c \ge 9R. \quad (5.2.37)$$

4

Let us consider the existence and uniqueness of the solution of the initial elastic–plastic problem. Let us first find the regions of existence for solutions (5.2.24) and (5.2.33)–(5.2.35). These solutions depend on three loading parameters, i.e., $p^* = p/\sigma_s$, $\sigma_1 = \sigma_x^\infty/\sigma_s$, and $\sigma_2 = \sigma_y^\infty/\sigma_s$, which form a three-dimensional space of the loading parameters. For the sake of simplicity and clarity, we assume $p^* = 0$, so that the region where σ_1 and σ_2 vary is a triangle with

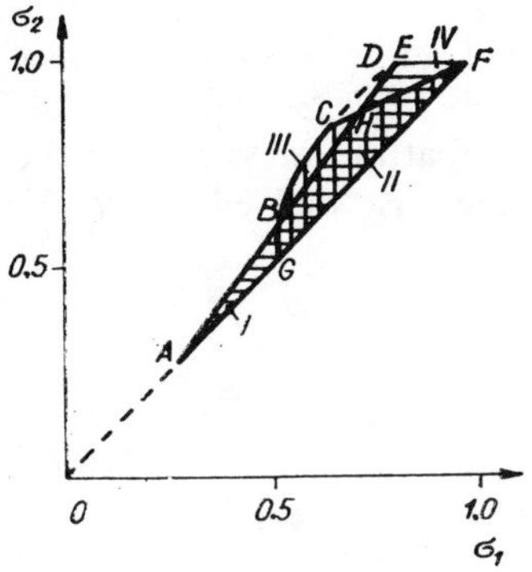

FIG. 5.1.

vertices (0, 0), (0, 1), and (1, 1) in plane $\sigma_1\sigma_2$ (as a consequence of inequality $\sigma_s \geq \sigma_y^\infty \geq \sigma_x^\infty \geq 0$).

The regions of existence of the solutions are depicted in Fig. 5.1. The horizontal lines cover the region of existence for solution (5.2.33)–(5.2.35), and the vertical lines cover the region of existence for solution (5.2.24). The straight line *ABHE* gives the loading parameters at which the elastic–plastic boundary of solution (5.2.33)–(5.2.35) is tangent to the circular holes; its equation on the basis of (5.2.37) has the form

$$21\sigma_2 - 33\sigma_1 + 4 = 0. \tag{5.2.38}$$

The straight line *AGF* is defined by $\sigma_2 - \sigma_1 = 0$. Curve *GBCD* gives the loading parameters at which the elastic–plastic boundary in accordance with solution (5.2.24) is tangent to the circular hole. The equation of the curve based on (5.2.30), (5.2.22), and (5.2.23) looks like

$$\left(\frac{\sigma_1 + \sigma_2 - 1}{3 - \sigma_1 - \sigma_2}\right) + \frac{\sigma_1 + \sigma_2 - 1}{3 - \sigma_1 - \sigma_2} = \frac{\sigma_2 - \sigma_1}{2 - \sigma_1 - \sigma_2}. \tag{5.2.39}$$

The straight line *CHF* gives the loading parameters at which a cusp appears on the elastic–plastic boundary in solution (5.2.24); its equation on the basis of (5.2.30), (5.2.22), and (5.2.23) is written as

$$37\sigma_2 - 17\sigma_1 - 20 = 0. \tag{5.2.40}$$

As Fig. 5.1 shows, the solution in regions I, III, and IV is unique, while in region II there are two solutions. It can be shown that when the uniqueness of the solution is assumed a priori in the class of essentially bounded functions, univalent solutions, other than (5.2.24) and (5.2.33)–(5.2.35), of the initial boundary problem (5.2.4) do not exist anymore. Indeed, the uniqueness of solution (5.2.24) in the class of totally bounded functions follows from the uniqueness of the Dirichlet problem (5.2.5) for function $\varphi(\zeta)$. The uniqueness of solution (5.2.33)–(5.2.35) in the class of unbounded functions is a consequence of the fact that $\omega'(\zeta)$ can vanish only at points $\zeta = \pm 1$ of the unit circle (when $\sigma_y^\infty \geq \sigma_x^\infty > 0$).

5

In the present section it has been assumed that load p satisfies inequality $0 \leq p \leq \sigma_s$. Let now load p vary within the bounds $0 \geq p \geq -\sigma_s$. From the Tresca–Saint Venant plasticity condition, it follows that the stress state in the plastic region is described by formulas that were first derived by V. V. Sokolovsky:

When $R \leq r \leq R\exp(-p/\sigma_s)$,

$$\begin{aligned} &\sigma_r = p + \sigma_s \ln(r/R), \qquad \tau_{r\theta} = 0, \\ &\sigma_\theta = p + \sigma_s + \sigma_s \ln(r/R), \qquad \sigma_\theta - \sigma_r = \sigma_s; \end{aligned} \tag{5.2.41}$$

when $r > R\exp(-p/\sigma_s)$,

$$\sigma_r = \sigma_s + R(p - \sigma_s)/r, \qquad \sigma_\theta = \sigma_s, \qquad \tau_{r\theta} = 0.$$

Obviously, all previously obtained solutions of the elastic–plastic problem will be true only when the elastic–plastic boundary completely encloses a circle of radius $R\exp(-p/\sigma_s)$. For this purpose, it is sufficient to formally replace p by zero and R by $R\exp(-p/\sigma_s)$ everywhere in the solutions. It is interesting to note that the elastic–plastic problem includes a range of equations. In region $R \leq r \leq R\exp(-p/\sigma_s)$, the governing equations are of hyperbolic type (two families of characteristics are logarithmic spirals). In the region between the circle of radius $R\exp(-p/\sigma_s)$ and the elastic–plastic boundary, the governing equation is of parabolic type (one family of characteristics formed by radial straight lines). In the elastic region, the governing equations are of elliptical type.

6

For determination of displacements in the plastic region, we use the associated rule of plastic flow (5.1.7). In this case, we have

$$d\varepsilon_r^p = 0, \qquad d\varepsilon_\theta^p = \lambda, \qquad d\varepsilon_{r\theta}^p = 0, \tag{5.2.42}$$

and since the strains are small, it is possible to go over to the velocity components v_r and v_θ:

$$\begin{aligned} &\dot{\varepsilon}_r = \frac{\partial v_r}{\partial r}, \qquad \dot{\varepsilon}_\theta = \frac{1}{r}\frac{\partial v_\theta}{\partial \theta} + \frac{v_r}{r}, \\ &\dot{\gamma}_{r\theta} = \frac{1}{r}\frac{\partial v_r}{\partial \theta} + \frac{\partial v_\theta}{\partial r} - \frac{v_\theta}{r} \end{aligned} \tag{5.2.43}$$

(the dot denotes differentiation with respect to time).

For velocities v_r and v_θ in the plastic region, we obtain equations

$$\frac{\partial v_r}{\partial r} = 0, \qquad \frac{\partial v_r}{\partial \theta} + r\frac{\partial v_\theta}{\partial r} - v_\theta = 0. \tag{5.2.44}$$

By integrating the first equation we obtain

$$v_r = -f(\theta). \tag{5.2.45}$$

Accounting for (5.2.45), the second equation is written as

$$-f'(\theta) + r\frac{\partial v_\theta}{\partial r} - v_\theta = 0, \tag{5.2.46}$$

Assume $\rho = \ln r$. Then equation (5.2.46) becomes

$$\frac{\partial v_\theta}{\partial \rho} - v_\theta = f'(\theta). \tag{5.2.47}$$

Integration of the last equation gives

$$v_\theta = -f'(\theta) + C(\theta)e^{\rho} = -f'(\theta) + C(\theta)r. \tag{5.2.48}$$

Formulas (5.2.45) and (5.2.48) determine the velocities v_r and v_θ. On the above elastic–plastic boundary, v_r and v_θ must be equal to the corresponding elastic velocities of the boundary points known from the solution of the elastic–plastic problem.

The displacements in the elastic region are equal to ($\rho_1 = |\zeta|$)

$$\begin{aligned} 2G(u_r + iu_\theta) &= \frac{\bar{\zeta}\overline{\omega'(\zeta)}}{\rho_1 |(\omega'(\zeta))|}\left\{ -R\frac{\sigma_s - p}{(4-a^2)\zeta}\left[\varkappa(2\zeta^2 + 3a) - \frac{\bar{\zeta}^2(2\zeta^2+a)^2}{\zeta^2(2\bar{\zeta}^2+a)}\right]\right. \\ &\quad -(\varkappa - 1)\frac{R(4\sigma_s - \sigma_x^\infty - \sigma_y^\infty)(2\zeta^2 + a^2)}{4\alpha(4-a^2)\zeta^3} - \frac{aR(\sigma_r - \beta)}{2(4-a^2)} \\ &\quad \times\left[(a^2+4)\frac{\bar{\zeta}}{2} - \frac{(2-a)(4-a^2)\bar{\zeta}}{2(2\bar{\zeta}^2+a)}\right. \\ &\quad \left.\left. - \sqrt{2a}\,(2-a)\arctan\sqrt{\frac{2}{a}}\,\bar{\zeta}\right]\right\}. \end{aligned} \tag{5.2.49}$$

To determine functions $f(\theta)$ and $C(\theta)$, expression (5.2.49) for the elastic displacements with $\bar{\zeta} = 1/\zeta$ corresponding to L (a, α, p, σ_x^∞, and σ_y^∞ are functions of time) must be differentiated with respect to time, and the elastic velocities v_r^e and v_θ^e must be equated to the velocities v_r and v_θ.

We omit excessively cumbersome expressions involved and write only the resulting equation:

$$f(\theta) = -v_r^e, \qquad C(\theta) = \frac{v_\theta^e(\theta) - v_r'^e(\theta)}{r_0(\theta)}, \tag{5.2.50}$$

where $r = r_\theta(\theta)$ is the equation of contour L.

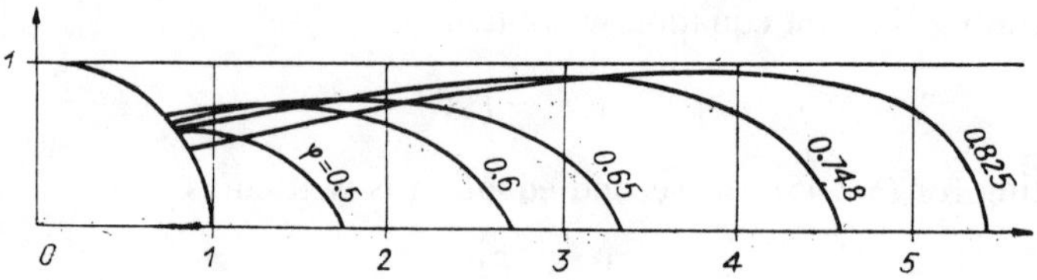

FIG. 5.2.

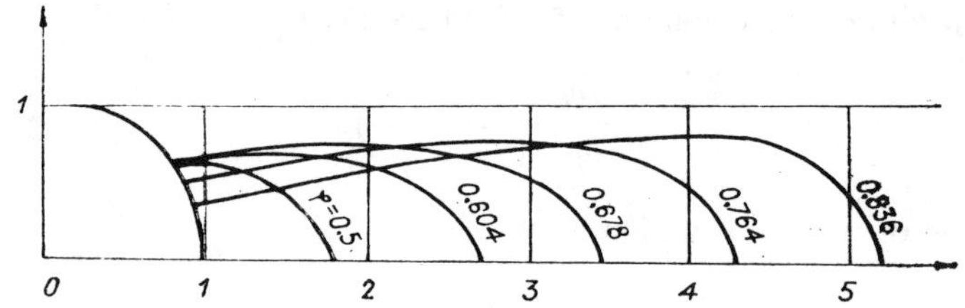

FIG. 5.3.

7

When a plate is subjected to tension along one of the axes, the region of plastic deformation might not completely enclose the circular hole. As indicated previously, there is no exact solution of the elastic–plastic problem with a partial enclosure of the circular hole by the plastic zone. Therefore, we shall present an approximate solution based on the theory of elastic–plastic bending of a curved beam [21].

We shall now consider a thin plate of thickness h with a round hole. The contour of the hole is free from external forces and the plate is stretched along one of the coordinates by forces $\sigma_y^\infty = \varphi\sigma_s h$. Here, φ is a dimensionless parameter $0 \leq \varphi \leq 1$, E is Young's modulus, and E_p is the hardening modulus.

When $\lambda = 0.33$, plastic deformation appears. Figures 5.2 and 5.3 show advance of the elastic–plastic boundary as a function of the external load (at $k = E_p/E = 0.0167$) in a plate made from an aluminum alloy (D-16 TC) for $n = b/2R = 7$ and $n = 6$, respectively. Here, n is the ratio of the plate width b to the diameter of the circular hole $2R$. The stress–strain diagram of the plate approximated by two straight lines OA and AB is shown in Fig. 5.4 ($k = 0.0167$). As Figs. 5.2 and 5.3 show, when load σ_y^∞ increased from $0.33\sigma_s h$ to

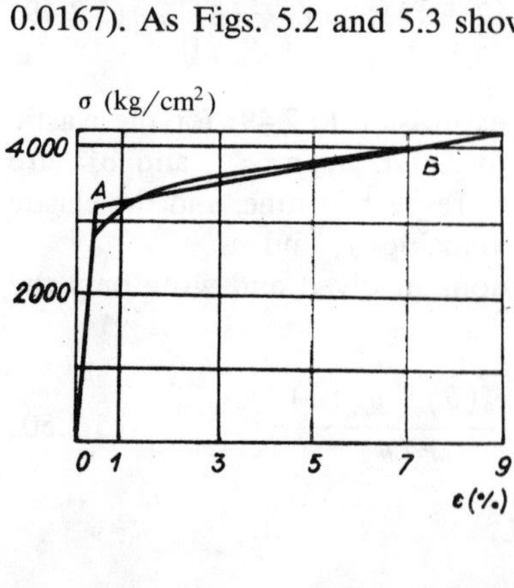

FIG. 5.4.

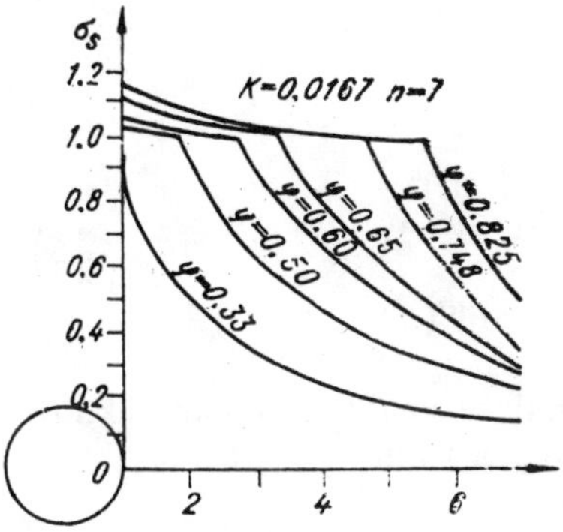

FIG. 5.5

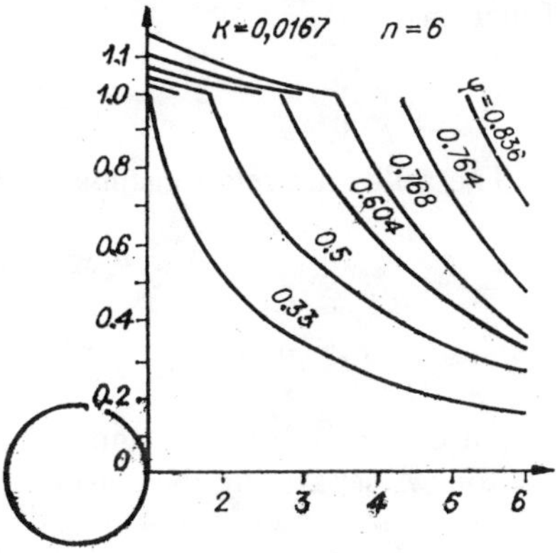

FIG. 5.6.

$0.65\sigma_s h$, plastic deformations spread quickly along the contour of the circular hole.

However, when the external force is further increased ($\sigma_y^\infty > 0.65\sigma_s h$) the region of plastic deformations grows only away from the hole, but shrinks near the hole. Depicted in Figs. 5.5 and 5.6 are the normal stress profiles along the width of the weakened section. As the external load increases, the stress profile levels off resulting in a decrease of stress concentration. This method was later used by Rybakina [18] for the case of large plastic deformations; the method was based on the true hardening law.

§ 3. PLATE WEAKENED BY AN INFINITE ROW OF IDENTICAL CIRCULAR HOLES

Let us consider a thin plate under plane stress with an infinite number of identical circular holes. The centers of the holes are located along the x axis with a spacing l. The radius of the holes is equal to R and normal force p is applied to the contour of the holes. Uniform stress distribution exists at infinity:

$$\sigma_x = \sigma_x^\infty, \qquad \sigma_y = \sigma_y^\infty, \qquad \tau_{xy} = 0. \tag{5.3.1}$$

Under the given forces, the plastic zones totally enclose the holes without joining or overlapping each other. It is required to determine the elastic–plastic boundaries as well as the stress state of the plate.

The hole at the origin is called here the principal hole and the elastic–plastic boundary near this hole is denoted by L. Due to the geometrical and force symmetry, the shape of the elastic–plastic boundary near each hole will be the same as near the principal hole. Since the stresses in the plastic region are determined only by the shape of the hole contour and the boundary load and are independent from the stress state in the elastic region, they will be exactly the same at each plastic region as in the case of one hole (5.2.2).

The stress state in the elastic region is determined by two analytical functions, $\Phi(z)$ and $\Psi(z)$. Function $\Phi(z)$, directly related to the stresses, will be periodic in the direction of the x axis. We introduce here a periodic function

$$\Psi_*(z) = \Psi(z) + z\Phi'(z).$$

The following relationships hold true:

$$\begin{aligned} \sigma_x + \sigma_y &= \sigma_x^\infty + \sigma_y^\infty + 4\operatorname{Re}\Phi(z), \\ \sigma_y - \sigma_x + 2i\tau_{xy} &= \sigma_y^\infty - \sigma_x^\infty + 2[(\bar{z} - z)\Phi'(z) + \Psi_*(z)]. \end{aligned} \tag{5.3.2}$$

On the basis of (5.2.2) and (5.3.2), the boundary conditions on an unknown contour L separating the elastic and plastic regions will be written as

$$\begin{aligned} 4\operatorname{Re}\Phi(z) &= 2\sigma_s + \frac{R(p - \sigma_s)}{r} - (\sigma_x^\infty + \sigma_y^\infty), \\ (\bar{z} - z)\Phi'(z) + \Psi_*(z) &= \frac{R(\sigma_s - p)}{2r} e^{-2i\theta} - \frac{1}{2}(\sigma_y^\infty - \sigma_x^\infty). \end{aligned} \tag{5.3.3}$$

The analytical functions $\Phi(z)$ and $\Psi_*(z)$ can be presented as

$$\begin{aligned} \Phi(z) &= \sum_{k=1}^{\infty} \frac{a_k}{[\zeta(z)]^k} + \sum_{n=-\infty}^{\infty}{}' \sum_{k=1}^{\infty} \frac{a_k}{[\zeta(z - nl)]^k}, \\ \Psi_*(z) &= \sum_{k=1}^{\infty} \frac{b_k}{[\zeta(z)]^k} + \sum_{n=-\infty}^{\infty}{}' \sum_{k=1}^{\infty} \frac{b_k}{[\zeta(z - nl)]^k}. \end{aligned} \tag{5.3.4}$$

The prime sign (′) denotes the absence of terms with the number $n = 0$ in the sums, and the variable ζ is linked with z by the relation

$$z - nl = \omega(\zeta) = c\zeta + \bar{P}_\nu(1/\zeta), \tag{5.3.5}$$

where $P_\nu(\zeta)$ is a polynomial of the νth degree with yet unknown coefficients.

The first sums $\varphi(\zeta)$ and $\psi(\zeta)$ are analytic outside the contour L, and the second sums $\varphi_1(\zeta)$ and $\psi_1(\zeta)$ are analytic inside L, where they are dependent on the perturbation $\varepsilon = R/l$.[1] By expanding in a power series of this parameter and keeping the terms containing, for example, ε^2, we obtain

$$\begin{aligned} \varphi_1(\zeta) &= \frac{\pi^2}{3} a_2 c^2 \varepsilon^2 - \frac{\pi^2}{3} a_1 c^2 \varepsilon^2 \omega(\zeta), \\ \psi_1(\zeta) &= \frac{\pi^2}{3} b_2 c^2 \varepsilon^2 - \frac{\pi^2}{3} b_1 c^2 \varepsilon^2 \omega(\zeta). \end{aligned} \tag{5.3.6}$$

[1]A more appropriate perturbation is $\varepsilon_1 = d/l$, where d is the largest distance of the points in the plastic region from the origin. Transition to this parameter is made by replacing $\varepsilon = \lambda\varepsilon_1$ and $d = \lambda R$. The value λ is determined by the largest value of the loads admissible for the given method following the construction of a formal analytical solution.

From (5.3.3), (5.3.4), and (5.3.6), we obtain a boundary value problem on a plane for determination of the analytical functions $\varphi(\zeta)$, $\psi(\zeta)$, and $\omega(\zeta)$ in the form:

when $|\zeta| = 1$,

$$4\,\mathrm{Re}\,\varphi(\zeta) = 2\sigma_s + \frac{R(p-\sigma_s)}{|\omega(\zeta)|} - (\sigma_x^\infty + \sigma_y^\infty) + \frac{4}{3}\pi^2 a_1 c^2 \varepsilon^2 \left[\omega(\zeta) + \overline{\omega(\zeta)}\right] - \frac{4}{3}\pi^2 a_2 c^2 \varepsilon^2, \tag{5.3.7}$$

when $|\zeta| = 1$,

$$\frac{\overline{\omega(\zeta)} - \omega(\zeta)}{\omega'(\zeta)}\varphi'(\zeta) + \psi(\zeta) = \frac{R(\sigma_s - p)\overline{\omega(\zeta)}}{2\omega(\zeta)\,|\omega(\zeta)|} - \frac{1}{2}(\sigma_y^\infty - \sigma_x^\infty) + \frac{\pi^2}{3} a_1 c^2 \varepsilon^2 \left[\overline{\omega(\zeta)} - \omega(\zeta)\right] + \frac{\pi^2}{3} b_1 c^2 \varepsilon^2 \omega(\zeta) - \frac{\pi^2}{3} b_2 c^2 \varepsilon^2.$$

The nonlinear boundary value problem (5.3.7) can be solved by writing it as a system of two functional equations, which hold true in the entire plane ζ:

$$2\varphi(\zeta) + 2\overline{\varphi}\left(\frac{1}{\zeta}\right) = 2\sigma_s + \frac{R(p-\sigma_s)}{\sqrt{\omega(\zeta)\overline{\omega}(1/\zeta)}} - (\sigma_x^\infty + \sigma_y^\infty) + \frac{4}{3}\pi^2 a_1 c^2 \varepsilon^2 [\omega(\zeta) + \overline{\omega}(1/\zeta)] - \frac{4}{3}\pi^2 a_2 c^2 \varepsilon^2,$$

$$\frac{\overline{\omega}(1/\zeta) - \omega(\zeta)}{\omega'(\zeta)}\varphi'(\zeta) + \psi(\zeta) = \frac{R(\sigma_s - p)\overline{\omega}(1/\zeta)}{2\omega(\zeta)\sqrt{\omega(\zeta)\overline{\omega}(1/\zeta)}} - \frac{1}{2}(\sigma_y^\infty - \sigma_x^\infty) + \frac{\pi^2}{3} a_1 c^2 \varepsilon^2 \left[\overline{\omega}\left(\frac{1}{\zeta}\right) - \omega(\zeta)\right] + \frac{\pi^2}{3} b_1 c^2 \varepsilon^2 \omega(\zeta) - \frac{\pi^2}{3} b_2 c^2 \varepsilon^2. \tag{5.3.8}$$

We shall seek function $\omega(\zeta)$ in the form

$$\omega(\zeta) = c\zeta + c_1/\zeta + c_2/\zeta^2 + c_3/\zeta^3. \tag{5.3.9}$$

From the symmetry, constant c_2 is equal to zero, and the other constants are real. It was shown in § 2 that it is necessary and sufficient to satisfy the equality

$$c_1^2 = 4cc_3 \tag{5.3.10}$$

for the right-hand side of the functional equation to become analytic outside the unit circle.

Accounting for condition (5.3.10), we can write functions $\omega(\zeta)$ and $\overline{\omega}(1/\zeta)$ as

$$\omega(\zeta) = \frac{c}{\zeta^3}\left(\zeta^2 + \frac{c_1}{2c}\right)^2, \qquad \omega\left(\frac{1}{\zeta}\right) = \frac{c_3}{\zeta}\left(\zeta^2 + \frac{c_1}{2c_3}\right)^2. \tag{5.3.11}$$

Let us find function $\varphi(\zeta)$. We can write the first functional equation as

$$\begin{aligned}
&2\varphi(\zeta) + 2\overline{\varphi}(1/\zeta)\\
&\quad = 2\sigma_s + \frac{2R(p-\sigma_s)\sqrt{cc_3}}{c_1(c_3-c)}\left[\frac{\zeta^2}{\zeta^2 + c_1/2c_3} - \frac{\zeta^2}{\zeta^2 + c_1/2c_3}\right]\\
&\quad -\left(\sigma_x^\infty + \sigma_y^\infty\right) + \frac{4}{3}\pi^2 a_1 \varepsilon^2 c^2\left[c\zeta + \frac{c_1}{\zeta} + \frac{c_3}{\zeta^3} + \frac{c}{\zeta} + c_1\zeta + c_3\zeta^3\right]\\
&\quad -\frac{4}{3}\pi^2 a_2 c^2 \varepsilon^2.
\end{aligned} \tag{5.3.12}$$

Formulas (5.3.11) have been taken into account here.

Function $F^+(\zeta)$, which is

$$\begin{aligned}
F^+(\zeta) &= -2\overline{\varphi}\left(\frac{1}{\zeta}\right) + \frac{2R(p-\sigma_s)\sqrt{cc_3}}{c_1(c_3-c)}\cdot\frac{\zeta^2}{\zeta^2 + c_1/2c_3}\\
&\quad + \frac{4}{3}\pi^2 a_1 c^2 \varepsilon^2\left(c\zeta + c_1\zeta + c_3\zeta^3\right),
\end{aligned} \tag{5.3.13}$$

is analytic everywhere inside the unit circle $|\zeta| < 1$, and function $F^-(\zeta)$, which is

$$\begin{aligned}
F^-(\zeta) &= 2\varphi(\zeta) - 2\sigma_s + \frac{2R(p-\sigma_s)\sqrt{cc_3}}{c_1(c_3-c)}\cdot\frac{\zeta^2}{\zeta^2 + c_1/2c} + \left(\sigma_x^\infty + \sigma_y^\infty\right)\\
&\quad - \frac{4}{3}\pi^2 a_1 c^2 \varepsilon^2\left(\frac{c_1}{\zeta} + \frac{c}{\zeta} + \frac{c_3}{\zeta^3}\right) + \frac{4}{3}\pi^2 a_2 c^2 \varepsilon^2,
\end{aligned} \tag{5.3.14}$$

is analytic everywhere outside the unit circle $|\zeta| > 1$. On the basis of the first boundary condition (5.3.12), functions $F^+(\zeta)$ and $F^-(\zeta)$ are analytical extensions of each other through the unit circle

$$F^+(\zeta) = F^-(\zeta) \quad \text{for } |\zeta| = 1.$$

According to the Liouville theorem, the unit analytical function $F(\zeta)$ is equal to zero. Hence, for function $\varphi(\zeta)$ we obtain

$$\varphi(\zeta) = \sigma_s - \frac{R(p-\sigma_s)\sqrt{cc_3}}{c_1(c_3-c)}\frac{\zeta^2}{\zeta^2 + c_1/2c} - \frac{1}{2}(\sigma_x^\infty + \sigma_y^\infty)$$

$$+\frac{2}{3}\pi^2 a_1 c^2 \varepsilon^2\left(\frac{c_1+c}{\zeta} + \frac{c_3}{\zeta^3}\right) - \frac{2\pi^2}{3}a_2 c^2\varepsilon^2. \tag{5.3.15}$$

By taking into consideration the conditions at infinity, we have

$$2\sigma_s - (\sigma_x^\infty + \sigma_y^\infty) - \frac{2R(p-\sigma_s)\sqrt{cc_3}}{c_1(c_3-c)} - \frac{4}{3}\pi^2 a_2 c^2\varepsilon^2 = 0, \tag{5.3.16}$$

$$\frac{4}{3}\pi^2 a_1 c^2 \varepsilon^2 (c + c_1) = a_1, \tag{5.3.17}$$

$$a_2 = \frac{R(p-\sigma_s)}{2(c_3-c)}\sqrt{\frac{c_3}{c}}. \tag{5.3.18}$$

From relation (5.3.17), it follows that

$$a_1 = 0. \tag{5.3.19}$$

From (5.3.15), we find

$$\varphi'(\zeta) = -\sqrt{\frac{c_3}{c_0}}\frac{R(p-\sigma_s)}{(c_3-c)\zeta^3} + \frac{c_1\sqrt{c_3}R(p-\sigma_s)}{c\sqrt{c}(c_3-c)\zeta^5} + O(\zeta^{-6}) \quad \text{when } |\zeta| \to \infty.$$

From the second functional equation we define the following function:

$$\psi(\zeta) = \frac{R(\sigma_s - p)\sqrt{c_3}(\zeta^2 + c_1/2c_3)\zeta^4}{2c^{3/2}(\zeta^2 + c_1/2c)^3} - \frac{\varphi'(\zeta)}{\omega'(\zeta)}\left[\overline{\omega}\left(\frac{1}{\zeta}\right) - \omega(\zeta)\right]$$

$$-\frac{1}{2}(\sigma_y^\infty - \sigma_x^\infty) + \frac{\pi^3}{3}b_1 c^2\varepsilon^2\omega(\zeta) - \frac{\pi^2}{3}b_2 c^2 \varepsilon^2. \tag{5.3.20}$$

Accounting for the condition at infinity, we obtain three more relationships:

$$\frac{R(\sigma_s - p)\sqrt{c_3}}{c^{3/2}}\frac{c+c_3}{c-c_3} - (\sigma_y^\infty - \sigma_x^\infty) - \frac{2}{3}\pi^2 b_2 c^2\varepsilon^2 = 0, \tag{5.3.21}$$

$$b_1 = 0, \tag{5.3.22}$$

$$\frac{R(\sigma_s - p)\sqrt{c_3}(c+c_3)}{c^{3/2}c_1} + \frac{R(\sigma_s - p)\sqrt{c_3}}{c^{3/2}(c_3-c)}\cdot\left(c - c_1 + \frac{c_1c_3}{c}\right) = b_2. \tag{5.3.23}$$

Functions $\varphi(\zeta)$, $\psi(\zeta)$, and $\omega(\zeta)$ are, therefore, determined from formulas (5.3.11), (5.3.15), and (5.3.20). Constants c, c_1, c_3, a_2, and b_2 in the preceding formulas, are found from the system of five equations (5.3.10), (5.3.16), (5.3.18), (5.3.21), and (5.3.23). It is readily seen that all the boundary conditions and the conditions at infinity in this case are satisfied.

If the terms containing ε raised to the fourth or higher power are included in expansion (5.3.4), the separation contour will not be represented by lines (5.3.11) but by lines close to (5.3.11). Evidently, these terms are small when ε is small and could not significantly alter the obtained boundaries.

Earlier an attempt was made [9] to obtain a solution of the problem (for $\sigma_r = p = 0$ on the contour of the hole) using an approximation of the stress function in the plastic region by a biharmonic function. It should be noted that a solution of the problem was sought in a class of functions (stresses) bounded everywhere in the elastic region including an unknown elastic–plastic boundary. On the basis of Chapter 5, § 2, it is expected that there exist other solutions of this problem in a class of unbounded functions. In this case, the elastic–plastic boundary must have a cusp at the corresponding point of unbounded functions.

§ 4. SOLUTION WITH PLASTIC DISCONTINUITY LINES

1

Let us consider a thin plate with a through crack of length $2l$ under tensile stresses normal to the crack. The material of the plate is assumed to be ideal elastic–plastic satisfying the Tresca–Saint Venant plasticity condition. The sides of the crack are free from loads. The origin of the Cartesian coordinates xy is taken at the center of the crack and the x axis is directed along the crack.

From experience, it is well known that there is a general tendency for plastic regions to form as narrow slip bands during their early stages of development, occupying an insignificant volume of the body compared with its elastic part. Such development of plastic deformations is characteristic of materials having clearly defined yield plateau such as mild steels, which display delayed yield. The plastic regions near the crack tips will be represented by sections of length d on the extension of the crack along its line [4, 29]. The thickness of the zone is equal to zero. In thin plates it can physically take the form of a slip plane oriented at an angle of 45° to the plane of the plate.

Thus, due to the localization of plastic deformations, the elastic–plastic problem can be reduced to the following boundary value problem of the plane theory of elasticity:

$$\begin{aligned}
\tau_{xy} &= 0 \quad \text{when } y = 0, \\
\sigma_y &= 0 \quad \text{when } y = 0,\ |x| < l, \\
\sigma_y &= \sigma_s \quad \text{when } y = 0,\ l < |x| < l + d, \\
v &= 0 \quad \text{when } y = 0,\ |x| > l + d, \\
\sigma_y &= \sigma_y^\infty \quad \text{when } z \to \infty.
\end{aligned} \tag{5.4.1}$$

With the help of the Kolosov–Muskhelishvili representations and the auxiliary analytical function $\Omega(z) = z\Phi'(z) + \Psi(z)$, the boundary problem can be written as:

when $y = 0,\ -l < x < l,$

$$\Phi(z) + \overline{\Phi(z)} + \Omega(z) = 0;$$

when $y = 0,\ l < |x| < l + d,$

$$\Phi(z) + \overline{\Phi(z)} + \Omega(z) = \sigma_s; \tag{5.4.2}$$

when $z \to \infty,$

$$\Phi(z) = \sigma_y^\infty/4, \qquad \Omega(z) = \sigma_y^\infty/2.$$

The solution of the boundary value problem (5.4.2) looks like

$$\Phi(z) = \frac{\sigma_y^\infty}{4} - \frac{\sigma_s}{2\pi i}\ln\frac{l - z}{l + z}$$

$$- \frac{\sigma_s}{2\pi i}\ln\frac{\sqrt{d(2l + d)\left((l + d)^2 - z^2\right)} + lz + (l + d)^2}{\sqrt{d(2l + d)\left((l + d)^2 - z^2\right)} - lz + (l + d)^2} \tag{5.4.3}$$

$$\Omega(z) = \frac{\sigma_y^\infty}{2}, \quad \frac{d}{l} = \sec\left(\frac{\pi\sigma_y^\infty}{2\sigma_s}\right) - 1.$$

Here, the root $\sqrt{(l + d^2) - z^2}$ is considered positive on the upper edge of the slit $(-l - d, l + d)$ along the x axis and

$$\ln F = \ln|F| + i\arg F, \qquad -\pi < \arg F < \pi.$$

The stresses and displacements in the plate are determined from equations

$$\sigma_x + \sigma_y = 4\,\mathrm{Re}\,\Phi(z), \qquad \sigma_y - \sigma_x + 2i\tau_{xy} = \sigma_y^\infty - 4iy\Phi'(z),$$

$$2G\left(\frac{\partial u}{\partial x} + i\frac{\partial v}{\partial x}\right) = \frac{3 - \nu}{1 + \nu}\Phi(z) - \overline{\Phi(z)} - \frac{1}{2}\sigma_y^\infty - 2iy\overline{\Phi'(z)}. \tag{5.4.4}$$

The obtained solution (5.4.3) of the elastic–plastic problem is true when the material outside the section $l \le |x| \le l + d$ of the x axis is in the elastic state. However, when the maximum shear stresses on the planes, which are perpendicular to the plane of the plate, reach the yield limit $\sigma_s/2$, new slip lines appear (see Chapter IV, § 7). The load at which lateral plasticity bands appear, is found from

$$\tau_{\max} = \tfrac{1}{2}|\sigma_y - \sigma_x + 2i\tau_{xy}| = \tfrac{1}{2}\sigma_s.$$

Finally we obtain

$$\sigma_y^\infty = \sigma_s\sqrt{1 - 2/\pi} = 0.603\sigma_s . \tag{5.4.5}$$

Thus, the solution of the elastic–plastic problem in this formulation (Dugdale's formulation) is significantly simplified. It is reduced to discontinuous solutions within the bounds of the theory of elasticity.

The preceding arguments can be similarly used in the general case of an arbitrary number of cracks, located along a straight line in an infinite plate, if only normal loads are applied to the crack faces. The plastic regions in the solution of the corresponding elastic–plastic problem (for the Tresca–Saint Venant condition) can also be represented by sections on the extension of the crack. The solution is constructed by the Muskhelishvili method: the linear dimensions of the zones are determined from the solvability conditions for the boundary value problem in a class of bounded functions (stresses). It is pertinent, however, to make certain that condition $|\sigma_1 - \sigma_2| < \sigma_s$ is also satisfied for the principal stresses in the elastic region. This condition is violated for some values of the loading parameters, and consequently, secondary plastic regions appear in the vicinity of crack tips. Slip in these regions occurs along planes normal to the plane of the plate.

Vitvitsky [6] discussed the elastic–plastic problem for a plate with two equal cracks and two semiinfinite slits. We shall present the main results of his analysis [6].

2. A plate with two equal cracks (Fig. 5.7)

$$\begin{aligned}\Phi(z) &= \Omega(z) - \frac{1}{2}\sigma_y^\infty \\ &= -\frac{\sigma_s}{\pi i}\ln\frac{\left[\sqrt{(z^2-c^2)(d^2-a^2)} + \sqrt{(d^2-z^2)(a^2-c^2)}\right]}{\left[\sqrt{(z^2-c^2)(d^2-b^2)} + \sqrt{(d^2-z^2)(b^2-c^2)}\right]} \\ &\quad \times \frac{\sqrt{b^2-z^2}}{\sqrt{z^2-a^2}} - \frac{\sigma_y^\infty}{4},\end{aligned} \tag{5.4.6}$$

$$\Omega(z) = \bar{\Phi}(z) + z\bar{\Phi}'(z) + \bar{\Psi}(z).$$

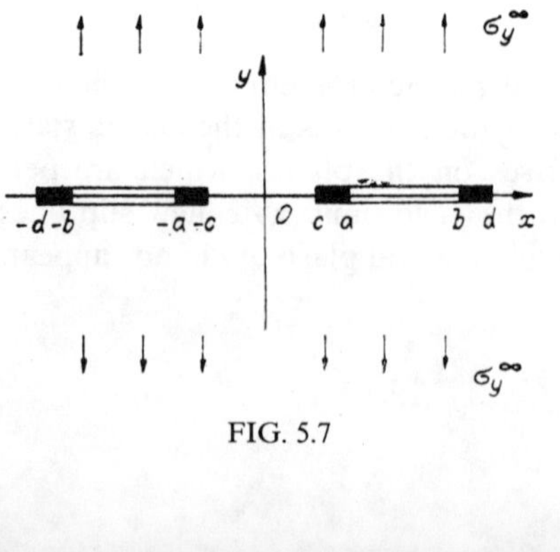

FIG. 5.7

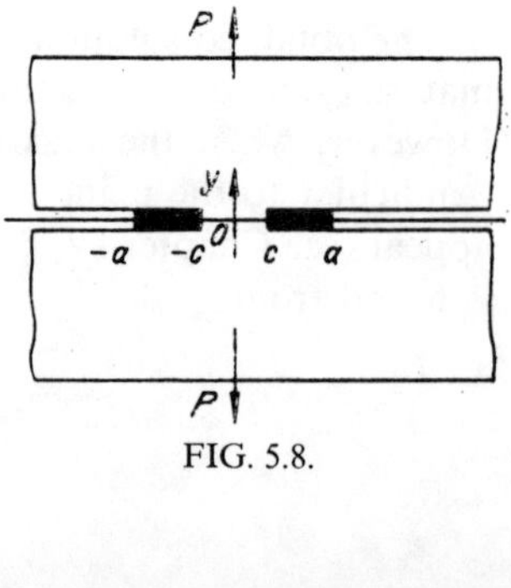

FIG. 5.8.

The length of plasticity bands is determined by equations

$$(d^2-c^2)\cos^2\frac{\pi\sigma_y^\infty}{2\sigma_s}=a^2+b^2-2c^2-2\sqrt{(a^2-c^2)(b^2-c^2)}\,\sin\frac{\pi\sigma_y^\infty}{2\sigma_s},$$

$$\sqrt{(d^2-a^2)(a^2-c^2)}\left[\mathbf{K}(k)+\frac{a^2}{d^2-a^2}\Pi\left(-\frac{d^2-c^2}{d^2-a^2},k\right)\right]$$

$$=\sqrt{(d^2-b^2)(b^2-c^2)}\left[\mathbf{K}(k)+\frac{b^2}{d^2-b^2}\Pi\left(-\frac{d^2-c^2}{d^2-b^2},k\right)\right], \quad (5.4.7)$$

where $\mathbf{K}$, Π denote complete elliptic integrals of the first and third kinds, respectively, and $k=(d^2-c^2)^{1/2}/d$.

If the internal plasticity bands join, we shall have

$$\Phi(z)=\Omega(z)-\frac{1}{2}\sigma_y^\infty=-\frac{1}{4}\sigma_y^\infty-\frac{\sigma_s}{\pi i}\ln\frac{\left[z\sqrt{d^2-b^2}+a\sqrt{d^2-z^2}\right]\sqrt{b^2-z^2}}{\left[z\sqrt{d^2-b^2}+b\sqrt{d^2-z^2}\right]\sqrt{z^2-a^2}}. \quad (5.4.8)$$

From the first of equations (5.4.7), we find

$$d=\sec\frac{\pi\sigma_y^\infty}{2\sigma_s}\sqrt{a^2+b^2-2ab\sin\frac{\pi\sigma_y^\infty}{2\sigma_s}}. \quad (5.4.9)$$

The second of equations (5.4.7) allows determination of the load σ_y^∞, at which the internal slip bands join:

$$\frac{b}{a}\sec\frac{\pi\sigma_y^\infty}{2\sigma_s}\sqrt{\frac{a^2}{b^2}+1-2\frac{a}{b}\sin\frac{\pi\sigma_y^\infty}{2\sigma_s}}$$

$$=\cosh\left[\frac{b}{a}\cosh^{-1}\left(\sec\frac{\pi\sigma_y^\infty}{2\sigma_s}\sqrt{\frac{a^2}{b^2}+1-2\frac{a}{b}\sin\frac{\pi\sigma_y^\infty}{2\sigma_s}}\right)\right]. \quad (5.4.10)$$

3. A plate with two semiinfinite slits (Fig. 5.8)

$$\Phi(z)=\Omega(z)=-\frac{\sigma_s}{2\pi i}\ln\frac{\sqrt{z^2-c^2}-\sqrt{a^2-c^2}}{\sqrt{z^2-c^2}+\sqrt{a^2-c^2}}. \quad (5.4.11)$$

The length of the plasticity bands is determined from the relationship

$$c=a\sqrt{1-p^2/4a^2\sigma_s^2}. \quad (5.4.12)$$

4. A plate with a periodic system of rectilinear through slits of length 2 located along the real ■ axis.

Let the centers of the slits be located at points $x = \pm 2nl$, where n is an integer. The faces of the slits are free from load, and the plate is subjected at infinity to a tensile stress $\sigma_y = \sigma_y^\infty$, $\sigma_x = 0$, and $\tau_{xy} = 0$ when $z \to \infty$.

By using transformation

$$\zeta = \sin(\pi z/2L), \tag{5.4.13}$$

we go over from plane z to a parametric plane of the complex variable ζ. Thus, the exterior of the periodic system of slits in the physical plane z transforms into an infinitely sheeted Riemann surface ζ with a slit

$$[(-\sin(\pi l/2L)), \sin(\pi l/2L)]. \tag{5.4.14}$$

By utilizing the Kolosov–Muskhelishvili relations (4.1.9) and the results obtained for a single crack in an unbounded plane (5.4.3), as well as transformation (5.4.13), the solution of the problem can be written in the form [7]:

$$\Phi(z) = \frac{\sigma_y^\infty}{4} - \frac{\sigma_s}{2\pi i}$$

$$\times \ln\left\{\frac{(\sin\alpha - \zeta)\left(\sqrt{\sin^2\alpha_1 - \sin^2\alpha}\sqrt{\sin^2\alpha_1 - \zeta^2}\right) + \zeta\sin\alpha + \sin^2\alpha_1}{(\sin\alpha + \zeta)\left(\sqrt{\sin^2\alpha_1 - \sin^2\alpha}\sqrt{\sin^2\alpha_1 - \zeta^2}\right) - \zeta\sin\alpha + \sin^2\alpha_1}\right\}. \tag{5.4.15}$$

Here, $\alpha = \pi l/2L$, $\alpha_1 = \pi(l+d)/2L$, $\ln F = \ln|F| + i \cdot \arg F$, and $-\pi < \arg F < \pi$. Function $\sqrt{\sin^2\alpha_1 - \zeta^2}$ is considered positive on the upper face of the slit $[-\sin(\pi l/2L), \sin(\pi l/2L)]$. The length of the plastic region d is related to length $2l$ of the crack and the acting load σ_y^∞ by

$$\sin\alpha/\sin\alpha_1 = \cos\beta, \qquad \beta = \pi\sigma_y^\infty/2\sigma_s. \tag{5.4.16}$$

The stresses and displacements in the plate are determined from formulas (5.4.4) and (5.4.15).

It is readily seen that the principal stresses, corresponding to solution (5.4.15), satisfy the Tresca–Saint Venant condition $\sigma_y^\infty \le \sigma_{1,2} \le \sigma_s$, where the equality takes place only when $y = 0$, $l < x - Ln < l + d$, and $-l - d < x - Ln < -l$.

5. Contact elastic1plastic problem for a plate

Let two straight and parallel cylinders of equal length be in contact so that the contact area is formed from a section of the cylindrical surface and bounded by the cylinder ends and some of their generators (Fig. 5.9). The cylinder ends are free from loads and the width of the contact area is small compared with the characteristic radius of curvature of the lateral surface. Under these conditions, the deformed bodies in the vicinity of the contact area will be in the plane stress state, and the boundary conditions in this vicinity

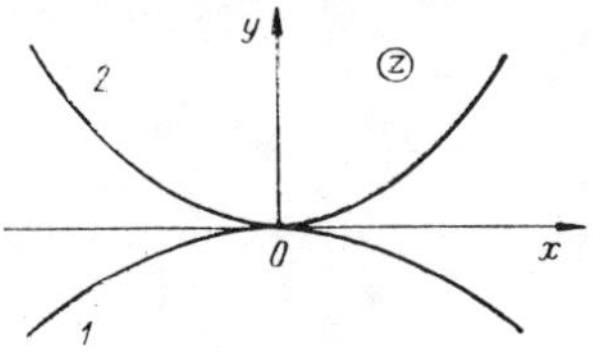

FIG. 5.9.

can be transferred to plane $y=0$. The material of the body satisfies the Tresca–Saint Venant plasticity condition. Friction on the contact area is neglected.

In the Cartesian system of coordinates xy, the deformed bodies can be regarded as half planes $y>0$ and $y<0$, respectively. Hereafter, all the values related to the lower and upper half planes will be denoted by subscripts 1 and 2, respectively.

We shall show that in this elastic–plastic problem the plastic regions are lines of zero thickness located on the x axis, and their length is determined in the process of solution. Let the elastic half plane be subjected to the action of normal loads $\sigma_y = p(x)$, $p(x)\le 0$. The shear stresses are equal to zero, $\tau_{xy}=0$, and stresses at infinity vanish. Then the principal stresses σ_1 and σ_2 at every point of the half plane will satisfy the conditions $0\le\sigma_{1,2}\le p_{\max}$, where $p_{\max}$ is the largest value of the boundary load $p(x)$ and the equality sign applies only on the boundary of the half plane. From these conditions and from the plasticity condition, which can be written as $|\sigma_{1,2}|\le\sigma_s$, where σ_s is the yield limit in compression, it follows that the solution with a plastic line on the boundary of the half plane is an exact solution of the corresponding elastic–plastic problem for $p(x)\le\sigma_s$, $-\infty<x<\infty$.

In this contact problem, the plastic line will always occur on the contact area in the body with a lower σ_s near the largest concentration of contact pressure. Due to the plastic flow, characteristic protrusions form at the ends of the cylinder in the vicinity of the plastic line. We shall limit this discussion to the case of simple loading and zero initial stresses. Stresses $\sigma_x, \sigma_y, \tau_{xy}$ and displacements u, v are determined through the Kolosov–Muskhelishvili potentials:

$$\begin{gathered}
\sigma_x+\sigma_y = 2\left[\Phi_j(z)+\overline{\Phi_j(z)}\right], \qquad j=1,2,\\
\sigma_y - i\tau_{xy} = \Phi_j(z) - \overline{\Phi_j(z)} + (z-\bar z)\overline{\Phi_j'(z)},\\
2G\left(\frac{\partial u}{\partial x}+i\frac{\partial v}{\partial x}\right) = \varkappa\Phi_j(z)+\Phi_j(\bar z)-(z-\bar z)\overline{\Phi_j'(z)},\\
z=x+iy, \qquad \varkappa=(3-\nu)/(1-\nu).
\end{gathered} \tag{5.4.17}$$

Functions $\Phi_1(z)$ and $\Phi_2(z)$ are analytic in the entire plane z with the exception of the slit along the x axis, where the contact area is located, and satisfy the conditions

$$\overline{\Phi_1(z)} = -\Phi_1(z), \qquad \overline{\Phi_2(z)} = -\Phi_2(z), \qquad \Phi_2(z) = -\Phi_1(z). \tag{5.4.18}$$

Boundary conditions of the problem are with $y=0$,

$$\sigma_y=\sigma_s \quad \text{on } L$$

with $y=0$,

$$\frac{\partial v_1}{\partial x}-\frac{\partial v_2}{\partial x}=f'(x) \quad \text{on } M. \tag{5.4.19}$$

Here, L is the plastic line and M is the remaining contact area: $f(x)=f_2(x)-f_1(x)$, where $y=f_1(x)$ and $y=f_2(x)$ are the equations of the lateral surfaces of the cylinders before deformation.

Using relations (5.4.17) and (5.4.18) we derive the following boundary value problem of the theory of analytical functions:

$$\begin{aligned}\Phi_1^- - \Phi_1^+ &= \sigma_s \quad \text{on } L,\\ \Phi_1^+ + \Phi_1^- &= \lambda i f'(x) \quad \text{on } M,\end{aligned} \tag{5.4.20}$$

where $\lambda=4G_1G_2/[G_2(\varkappa+1)+G_1(\varkappa+1)]$.

The boundary value problem (5.4.20) is easily solved by quadratures for any number of sections L and M along the real axis (with the help of a technique used, for example, in [25]). The solution of this problem is sought in a class of completely bounded functions. The linear dimensions of the plastic lines and of the contact area for the case of smooth surfaces are determined from the solvability conditions and the conditions at infinity. The location and number of plastic lines are easy to find from the physical considerations.

Let us present solutions for two special cases.

a. *Rectangular rigid punch* (Fig. 5.10):

$$\Phi_1(z)=\frac{\sigma_s}{2\pi i}\ln\frac{(z+l)\left[\sqrt{(l^2-a^2)(z^2-a^2)}+lz-a^2\right]}{(z-l)\left[\sqrt{(l^2-a^2)(z^2-a^2)}+lz+a^2\right]}. \tag{5.4.21}$$

Here, when $z\to\infty$, $\sqrt{z^2-a^2}=z+O(z^{-1})$. The size of the plastic zone is determined by

$$a=l\sqrt{1-P^2/4l^2\sigma_s^2}. \tag{5.4.22}$$

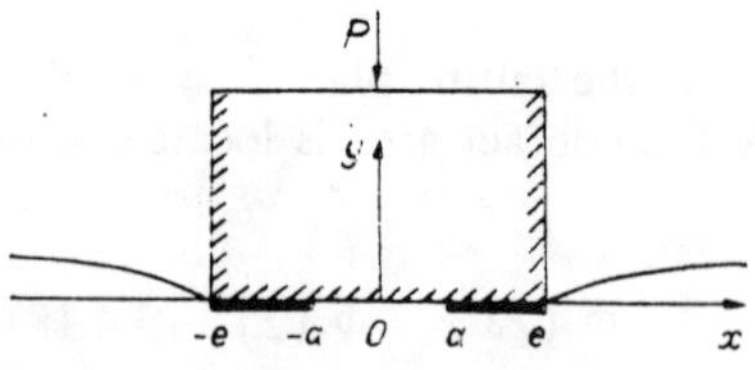

FIG. 5.10

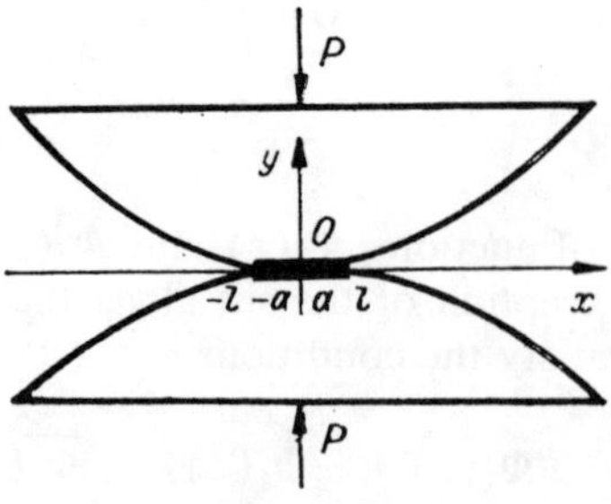

FIG. 5.11.

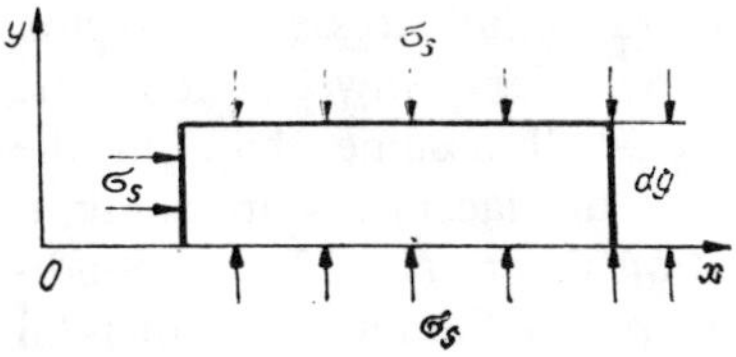

FIG. 5.12.

b. Two circular cylinders of radii R_1 and R_2 (Fig. 5.11):

$$\Phi_1(z) = 1/2\left(\frac{1}{R_1} + \frac{1}{R_2}\right)\lambda iz$$

$$-\frac{\sigma_s}{2\pi i}\sqrt{(z^2-a^2)(z^2-l^2)}\int_{-a}^{a}\frac{dt}{\sqrt{(l^2-t^2)(a^2-t^2)}\,(t-z)}. \quad (5.4.23)$$

Here, when $|z| \to \infty$, the root before the integral behaves as z^2 and the root under the integral is positive.

Parameters a and l, which determine the size of the contact area and plastic zone, are found from the system of equations

$$\frac{\pi\lambda}{\sigma_s} l\left(\frac{1}{R_1} + \frac{1}{R_2}\right) = 2\mathbf{K}\left(\frac{a}{l}\right),$$

$$\left(\frac{a^2}{l^2} - 1\right)\mathbf{K}\left(\frac{a}{l}\right) + 2\mathbf{E}\left(\frac{a}{l}\right) = \frac{P}{l\sigma_s}. \quad (5.4.24)$$

Here **K** and **E** are complete elliptic integrals.

The heavy lines in Figs. 5.10 and 5.11 represent the plastic zones. The physical significance of the solution becomes absolutely clear when the equilibrium and deformation of an infinitely small element of the body in the vicinity of the plastic line are considered (Fig. 5.12). When the plastic state is reached in this element, the material is extruded in the direction of the normal to plane xy due to plastic flow. Therefore, if the rate of change of the external loads is small in comparison with the characteristic flow velocity (which is not determined within the bounds of the theory of ideally plastic body), the plastic region will never be able to spread to a finite area. In real materials, the plastic line will have finite thickness as a result of hardening.

§ 5. CRACK

Let a thin plate with an arbitrary crack be subjected to tensile stresses normal to the crack. The plate material is assumed ideal elastic–plastic with Young's modulus E and yield point τ_s under simple shear and obeys the Guber–Mises plasticity condition.

We shall examine the vicinity of the crack tip, which is small in comparison with the characteristic linear dimension of the plate, but large compared with the characteristic linear dimension of the plastic region. The crack in plane xy is represented by a semiinfinite slit along the real x axis, which is free from external loads.

The problem is reduced to determination of strains and stresses in a region that comprises elastic and plastic regions. The boundary between the elastic and plastic regions is to be determined. We shall assume that, on the elastic–plastic boundary, the stresses, strains, and displacements are continuous. By assuming the relations of the deformation theory (1.2.12) and condition $T=\tau_s$ to be true, we shall utilize the principle of the minimum of total energy. The displacements u and v, which give the minimum of the functional

$$I(u,v)=\frac{1}{6}E\iint_{D_1}\Gamma^2\,dx\,dy+\iint_{D_2}\left(\tau_s\Gamma-\frac{3\tau_s^2}{2E}\right)dx\,dy, \tag{5.5.1}$$

$$\Gamma=2\left[\varepsilon_x^2+\varepsilon_y^2+\varepsilon_x\varepsilon_y+\tfrac{1}{4}\gamma_{xy}^2\right]^{1/2},\qquad \gamma_{xy}=2\varepsilon_{xy}.$$

correspond to the real equilibrium form. The boundary between the elastic D_1 and plastic D_2 regions is determined from condition $\Gamma=3\tau_s/E$.

At the nodal points on the boundary of region $D=D_1+D_2$, given were displacements from the solution of the problem of an elastic plate with a semiinfinite slit under tension at infinity. For numerical solution of the variational problem, the finite element method was used in conjunction with the method of local variations. The region, in this case, was broken into right-angle triangles. Region D was represented by a square with the center at the origin and had a slit along the real axis (Fig. 4.7).

The numerical solution of this elastic–plastic problem [13] showed that the equation of the elastic–plastic boundary can be written in polar coordinates in the form

$$r=\left[0.25(\pi-\theta)-0.15\sin\frac{3\theta}{2}\right]\frac{\pi K_{\mathrm{I}}^2}{8\sigma_s^2}. \tag{5.5.2}$$

Here, σ_s is the yield point in tension. The elastic–plastic boundary is shown in Fig. 5.13 in dimensionless coordinates

$$x_*=\frac{8\sigma_s^2}{\pi K_{\mathrm{I}}^2}x,\qquad y_*=\frac{8\pi}{\pi K_{\mathrm{I}}^2}y.$$

It is evident that the elastic–plastic boundary is quite similar to the Dugdale solution [29]. According to the Dugdale hypothesis, the plastic deformation is concentrated on the extension of the crack along a narrow layer of zero thickness (see § 4).

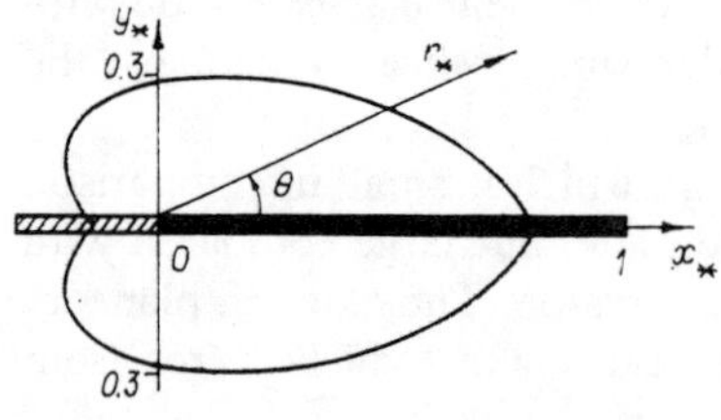

FIG. 5.13.

§ 6. NUMERICAL RESULTS

1. Plate with an elliptical hole

a. Let stresses σ_x^∞ and σ_y^∞ be applied at infinity. We assume that the plastic region entirely encloses the hole. This problem was solved by the Perlin method [20]. The equation of the ellipse is taken as $x^2 + 4y^2 - 4 = 0$. It is also assumed that the Tresca–Saint Venant condition is satisfied. The following results were obtained.

If the elastic–plastic boundary passes through points $A = 3$, $B = -1.5i$, then the mapping function is found in the form

$$z = \omega(\zeta) = 2.25(\zeta + 0.33/\zeta).$$

The locations of the characteristic points A and B of the elastic–plastic boundary and the corresponding values σ_x^∞ and σ_y^∞ are given in Table V.1. It is evident that the location of point B affects mainly stress σ_x^∞. It is not difficult to see that a change of the point A location results mainly in a change of stress σ_y^∞.

b. Let us consider an infinite plane with an elliptical hole under tension at infinity by mutually perpendicular forces d_2P_1 and d_2P_2 directed at angle θ_0 to the principal central axis of the ellipse (d_2 is a parameter).

We shall introduce the results obtained by the perturbation method and presented in reference [19] of Chapter IV. The equation of the elliptical hole is presented in the form

$$\rho = \alpha + \varepsilon\alpha d_1 \cos 2\theta - \varepsilon^2 \frac{3\alpha d_1^2}{4}(1 - \cos 4\theta) + \cdots .$$

Here, $\alpha = R/r_s^0$ (R is the radius of the initial circular hole; r_s^0 is the dimensional radius of the plastic zone in zero approximation), d_1 is a parameter, and $\varepsilon = (P_1 - P_2)/2$. If the contour of the hole is free from external forces, the equation of the plastic zone boundary L_s acquires the form

$$\begin{aligned}\rho_s = 1 &+ \varepsilon^*[4d_2 \cos 2(\theta - \theta_0) + 3\alpha d_1 \cos 2\theta] \\ &+ \varepsilon^{*2}\Bigg\{ d_1^2\left(\frac{\alpha^2}{4} - 8\alpha^4\right) - \left(18 d_1 d_2 \alpha \cos 2\theta_0 + 8d_2^2\right) \\ &\quad + \Bigg[-d_1^2\left(\frac{15}{4}\alpha^2 - 8\alpha^3 - \frac{3}{4}\alpha^4\right) + \big(18 d_1 d_2 \alpha \cos 2\theta_0 \\ &\qquad + 16 d_2^2 \cos 4\theta_0\big)\Bigg]\cos 4\theta \\ &\quad + \left[18 d_1 d_2 \alpha \sin 2\theta + 16 d_2^2 \sin 4\theta_0\right]\Bigg\} \sin 4\theta + \cdots .\end{aligned}$$

Here, $\varepsilon^* = \varepsilon/\alpha$.

TABLE V.1

Stress	$A=3$ $B=-1.25$	$A=3$ $B=-1.5$	$A+3$ $B=-2$
σ_x^∞	$0.794\sigma_s$	$0.808\sigma_s$	$0.854\sigma_s$
σ_y^∞	$0.594\sigma_s$	$0.600\sigma_s$	$0.626\sigma_s$

TABLE V.2

a_0	a_2	a_4	a_6	d_0	d_2	d_4	d_6
0.279	0.105	−0.05	—	0.105	0.220	0.01	—
0.268	0.085	−0.01	0.014	0.079	0.306	−0.082	0.035

When $d_1 = 0$, $\theta_0 = 0$, and $d_2 = 1$, we obtain a solution for biaxial tension of a thin plate with a circular hole; when $d_1 = 1$, we obtain the solution for uniform tension of a thin plate with an elliptical hole. In the latter case, the boundary of the plastic region is described by

$$\rho_s = 1 + 3\varepsilon\cos 2\theta + \varepsilon^2\left[\left(\frac{1}{4} - 8\alpha^2\right) - \left(\frac{15}{4} - 8\alpha - \frac{3}{2}\alpha^2\right)\cos 4\theta\right] + \cdots .$$

2. Plate with a circular hole

Let us consider the elastic–plastic problem for a thin plate with a circular hole of a unit radius partially enclosed by the plastic region. An approximate solution of the problem for ideal plasticity was obtained by Perlin [17]:

$$\Phi(\zeta) = a_0 + \frac{a_2}{\zeta^2} + \frac{a_4}{\zeta^4} + \frac{a_6}{\zeta^6} + \cdots ,$$

$$\Omega(\zeta) = z\Phi'(\zeta) + \Psi(\zeta) = d_0 + \frac{d_2}{\zeta^2} + \frac{d_4}{\zeta^4} + \frac{d_6}{\zeta^6} + \cdots ,$$

$$\sigma_x^\infty + \sigma_y^\infty = 4a_0, \qquad \sigma_y^\infty - \sigma_x^\infty = 2d_0 .$$

Ellipse $\omega(\zeta) = c(\zeta - c_1/\zeta^{-1})$ was taken as the contour to which the analytical extension was conducted. Coefficients c and c_1 were determined from the condition that the ellipse passes through points $A = 0.74 + i0.65$ and $B = 1.5$. Table V.2 gives the coefficients a_0, a_2, a_4, a_6, d_0, d_2, d_4, and d_6.

6. THREE-DIMENSIONAL PROBLEMS

§ 1. FUNDAMENTAL EQUATIONS. BRIEF REVIEW

The general case is a three-dimensional stress and strain state in which the equilibrium equations (1.2.2), the kinematic relation between displacements and strains (1.2.3), the plasticity condition (1.3.1), and the relationship

$$d\varepsilon_{ij} = d\varepsilon^e_{ij} + \lambda \frac{\partial f}{\partial \sigma_{ij}} \tag{6.1.1}$$

are used in solving elastic–plastic problems of ideal plasticity. The increments of elastic strains $d\varepsilon^e_{ij}$ are computed from Hooke's law (1.2.1). In the plastic regions, equation (6.1.1) applies; in the elastic regions, $\lambda = 0$ and equations (6.1.1) transform into Hooke's law. If the loads on the boundary of the body are given, we then have a complete system of equations for determination of stresses in the plastic region regardless of the strains, i.e., the problem is statically determinate. We assume that the stresses and displacements on the elastic–plastic boundary are continuous.

The spherical and axially symmetrical state is a special case of the general three-dimensional stress and strain state. The perturbation method proved useful in some three-dimensional problems: with its help, Ershov [11] and Semykina [18] solved interesting problems for near-spherical surfaces under stress state at infinity, which was close to hydrostatic compression. Some axisymmetric problems were also discussed [8–10].

Leonov and Panasyuk [15] analysed a penny-shaped crack with a plastic zone at its edge, which was assumed to be infinitely thin. The work of Kerimov [14] should also be noted.

The results obtained in the preceding papers essentially exhaust the analytical solutions of three-dimensional elastic–plastic problems. The number of analytical solutions of space problems in ideal plasticity that could be used in the solution of corresponding elastic–plastic problems is also limited. We should note the exact solutions of three-dimensional problems of ideal plasticity obtained by Hill (see [29], Chap. 1), Prager [17], Panarelli and Hodge [16], Ivlev (see [5], Chap. 1), Zadoyan [12], Annin [2, 3], and Senashov [19].

Numerical methods have not been widely used in three-dimensional elastic–plastic problems yet. In this regard, we should mention the work by Ayres [20], in which the method of finite differences was used to compute stresses and strains near a slit. A number of elastic–plastic problems for finite thin-wall bodies of revolution were solved numerically by Volchkov and Pavlov [7], Volchkov and Korobeinikov [5, 6], Burago and Kukudjanov [4]. The contents of §2 in this chapter are based on results obtained by Ershov [11] and Semykina [18]; §3 is based on results obtained by Leonov and Panasyuk [15]; §4 is based on papers by Ayres [20] and Volchkov and Korobeinikov [6].

§ 2. SPHERICAL CAVITY

1

Let the body have internal and external near-spherical boundaries. We assume that the boundary of the plastic zone completely encloses the internal surface.

For the elastic–plastic problem, the following fundamental equations apply:

Equilibrium equations:

$$\frac{\partial\sigma_\rho}{\partial\rho}+\frac{1}{\rho}\frac{\partial\tau_{\rho\theta}}{\partial\theta}+\frac{1}{\rho\sin\theta}\frac{\partial\tau_{\rho\varphi}}{\partial\varphi}+\frac{1}{\rho}(2\sigma_\rho-\sigma_\theta-\sigma_\varphi+\tau_{\rho\theta}\cot\theta)=0,$$

$$\frac{\partial\tau_{\rho\theta}}{\partial\rho}+\frac{1}{\rho}\frac{\partial\sigma_\theta}{\partial\theta}+\frac{1}{\rho\sin\theta}\frac{\partial\tau_{\theta\varphi}}{\partial\varphi}+\frac{1}{\rho}\left[(\sigma_\theta-\sigma_\varphi)\cot\theta+3\tau_{\rho\theta}\right]=0, \quad (6.2.1)$$

$$\frac{\partial\tau_{\rho\varphi}}{\partial\rho}+\frac{1}{\rho}\frac{\sigma\tau_{\theta\varphi}}{\sigma\theta}+\frac{1}{\rho\sin\theta}\frac{\partial\sigma_\varphi}{\partial\varphi}+\frac{1}{\rho}(3\tau_{\rho\varphi}+2\tau_{\theta\varphi}\cot\theta)=0.$$

Tresca–Saint Venant plasticity conditions:

$$\left(\sigma_\rho-\sigma+\tfrac{2}{3}k\right)\left(\sigma_\varphi-\sigma+\tfrac{2}{3}k\right)-\tau_{\rho\varphi}^2=0 \qquad \begin{matrix}\theta\\ \nearrow\ \searrow\\ \rho\leftarrow\varphi,\end{matrix} \quad (6.2.2)$$

Boundary conditions:

$$\sigma_\rho l+\tau_{\rho\theta}m+\tau_{\rho\varphi}n=Q_\rho \qquad \begin{matrix}\theta\\ \nearrow\ \searrow\\ \rho\leftarrow\varphi.\end{matrix} \quad (6.2.3)$$

The conditions of continuity of solutions on the elastic–plastic boundary apply as well. Here, as in Chapter I, the symbol $\begin{smallmatrix}\theta\\ \nearrow\ \searrow\\ \rho\leftarrow\varphi\end{smallmatrix}$ means that the remaining relationships are obtained by circular permutation of the subscripts; l, m, n are the direction cosines of the normal in system ρ, θ, φ.

We seek a solution in the form

$$\sigma_\rho=\sigma_\rho^0+\varepsilon\sigma_\rho^1, \qquad \tau_{\rho\theta}=\tau_{\rho\theta}^0+\varepsilon\tau_{\rho\theta}^1 \qquad \begin{matrix}\theta\\ \nearrow\ \searrow\\ \rho\leftarrow\varphi,\end{matrix} \quad (6.2.4)$$

assuming that the perturbation ε characterizes the deviation of the boundary of the body from the sphere, and also the boundary conditions. By substituting (6.2.4) into (6.2.2), we find the plasticity conditions for both approximations:

$$\left(\sigma_\rho^0-\sigma^0+\tfrac{2}{3}k\right)\left(\sigma_\varphi^0-\sigma^0+\tfrac{2}{3}k\right)-\tau_{\rho\varphi}^{02}=0 \qquad \begin{matrix}\theta\\ \nearrow\ \searrow\\ \rho\leftarrow\varphi,\end{matrix} \quad (6.2.5)$$

$$\left(\sigma_\rho^0-\sigma^0+\tfrac{2}{3}k\right)\left(\sigma_\varphi^1-\sigma^1\right)+\left(\sigma_\varphi^0-\sigma^0+\tfrac{2}{3}k\right)\left(\sigma_\rho^1-\sigma^1\right)-2\tau_{\rho\varphi}^0\tau_{\rho\varphi}^1=0. \quad (6.2.6)$$

We take the stresses for a hollow sphere of internal radius a and external radius b as a zero approximation.

Uniform pressures p and p_1 act on the internal and external contours of the sphere, respectively,

$$\sigma_\rho^{0p} = -4k\ln\frac{\rho}{a} - p, \qquad \sigma_\theta^{0p} = \sigma_\varphi^{0p} = -4k\ln\frac{\rho}{a} - 2k - p,$$

$$\tau_{\rho\theta}^{0p} = \tau_{\rho\varphi}^{0p} = \tau_{\theta\varphi}^{0p} = 0,$$

$$\sigma_\rho^{0e} = \frac{4k\beta_0^3}{3}\left(\frac{1}{\rho^3} - \frac{1}{b^3}\right) - p_1, \qquad \tau_{\rho\theta}^{0e} = \tau_{\rho\varphi}^{0e} = \tau_{\theta\varphi}^{0e} = 0, \tag{6.2.7}$$

$$\sigma_\theta^{0e} = \sigma_\varphi^{0e} = -\frac{2k\beta_0^3}{3}\left(\frac{1}{\rho^3} + \frac{2}{b^3}\right) - p_1.$$

Here, $\rho = \beta_0$ is the elastic–plastic boundary. By accounting for (6.2.7) and (6.2.6), we find $\sigma_\rho^{1p} = \sigma_\theta^{1p} = \sigma_\varphi^{1p} = \sigma^{1p}$, $\tau_{\theta\varphi}^{1p} = 0$. The equilbrium equations can now be written as follows:

$$\frac{2\sigma^{1p}}{\partial\rho} + \frac{1}{\rho}\frac{\partial\tau_{\rho\theta}^{1p}}{\partial\theta} + \frac{1}{\rho\sin\theta}\frac{\partial\tau_{\rho\varphi}^{1p}}{\partial\varphi} + \frac{1}{\rho}\tau_{\rho\theta}^{1p}\cot\theta = 0,$$

$$\frac{\partial\tau_{\rho\theta}^{1p}}{\partial\rho} + \frac{1}{\rho}\frac{\partial\sigma^{1p}}{\partial\theta} + \frac{3}{\rho}\tau_{\rho\theta}^{1p} = 0, \tag{6.2.8}$$

$$\frac{\partial\tau_{\rho\varphi}^{1p}}{\partial\rho} + \frac{1}{\rho\sin\theta}\frac{\partial\sigma^{1p}}{\partial\varphi} + \frac{3}{\rho}\tau_{\rho\varphi}^{1p} = 0.$$

We introduce function $U(\rho, \theta, \varphi)$ such that

$$\sigma^{1p} = -\frac{1}{\rho^2}\frac{\partial U}{\partial\rho}, \qquad \tau_{\rho\theta}^{1p} = \frac{1}{\rho^3}\frac{\partial U}{\partial\theta}, \qquad \tau_{\rho\varphi}^{1p} = \frac{1}{\rho^3\sin\theta}\frac{\partial U}{\partial\varphi}.$$

Then, the last two equations of (6.2.8) will be identically satisfied.

From the first equation of (6.2.8) we find function

$$U(\rho,\theta,\varphi) = \sum_{j=0}^{\infty}\sum_{i=0}^{j} C_{ij}\left(\rho^{3/2+\lambda} + D_j\rho^{3/2-\lambda}\right)\left(\cos i\varphi + E_{ij}\sin i\varphi\right)$$

$$\times P_j^{(i)}(\cos\theta), \qquad \lambda = \sqrt{\frac{9}{4} + j(j+1)}, \tag{6.2.9}$$

where C_{ij}, D_{ij}, and E_{ij} are constants determined by the boundary conditions and $P_j^{(i)}(x)$ is an associated Legendre polynomial. To linearize the boundary conditions, we transfer the boundary conditions to the spherical surface. Therefore, we present the boundary of the body in the form

$$\rho = \rho_0 - \varepsilon\rho_1(\theta,\varphi), \qquad \rho_0 = \text{const}.$$

As a first approximation, for $\rho = \rho_0$ we assume the following direction cosines

$$l \approx 1, \qquad m \approx -\varepsilon \frac{1}{\rho_0} \frac{\partial \rho_1}{\partial \theta}, \qquad n \approx -\varepsilon \frac{1}{\rho_0 \sin\theta} \frac{\partial \rho_1}{\partial \varphi}. \tag{6.2.10}$$

Further, by accounting for (6.2.4), (6.2.9), and (6.2.3), we obtain, as a first approximation, the following boundary conditions for $\rho = \rho_0$:

$$\begin{aligned}
&\sigma_\rho^1 + \frac{\partial \sigma_\rho^0}{\partial \rho}\rho_1 - \tau_{\rho\theta}^0 \frac{1}{\rho_0}\frac{\partial \rho_1}{\partial \theta} - \tau_{\rho\varphi}^0 \frac{1}{\rho_0 \sin\theta}\frac{\partial \rho_1}{\partial \varphi} = \frac{\partial Q_\rho}{\partial \rho}\rho_1,\\
&\tau_{\rho\theta}^1 + \frac{\partial \tau_{\rho\theta}^0}{\partial \rho}\rho_1 - \sigma_\theta^0 \frac{1}{\rho_0}\frac{\partial \rho_1}{\partial \theta} - \tau_{\theta\varphi}^0 \frac{1}{\rho_0 \sin\theta}\frac{\partial \rho_1}{\partial \varphi} = \frac{\partial Q_\theta}{\partial \rho}\rho_1,\\
&\tau_{\rho\varphi}^1 + \frac{\partial \tau_{\rho\varphi}^0}{\partial \rho}\rho_1 - \tau_{\theta\varphi}^0 \frac{1}{\rho}\frac{\partial \rho_1}{\partial \theta} - \sigma_\varphi^0 \frac{1}{\rho_0 \sin\theta}\frac{\partial \rho_1}{\partial \varphi} = \frac{\partial Q_\varphi}{\partial \rho}\rho_1.
\end{aligned} \tag{6.2.11}$$

By substituting (6.2.7) into (6.2.11), we obtain the boundary conditions for stresses in the plastic region when $\rho = a$:

$$\begin{aligned}
&\sigma_\rho^{1p} = 4k\rho_1(\theta, \varphi), \qquad \tau_{\rho\theta}^{1p} = \frac{2k}{a}\frac{\partial \rho_1(\theta,\varphi)}{\partial \theta},\\
&\tau_{\rho\varphi}^{1p} = \frac{4k}{a \sin\theta}\frac{\partial \rho_1(\theta,\varphi)}{\partial \varphi}.
\end{aligned} \tag{6.2.12}$$

The boundary conditions (6.2.12) are used to find the values of the arbitrary constants

$$\begin{aligned}
&C_{ij} = 4ka_{ij}\frac{\lambda + a - 3/2}{2\lambda}a^{1/2-\lambda},\\
&D_j = \frac{\lambda - a + 3/2}{\lambda + a - 3/2}a^{2\lambda}, \qquad E_{ij} = \frac{b_{ij}}{a_{ij}}.
\end{aligned} \tag{6.2.13}$$

Here, a_{ij}, b_{ij} are the coefficients of the expansion $\rho_1(\theta, \varphi)$ into series in terms of the spherical functions

$$a_{ij} = \frac{1}{N_{ij}} \int_0^{2\pi} \int_0^{\pi} \rho_1(\theta,\varphi) P_j^{(i)}(\cos\theta) \cos i\varphi \sin\theta \, d\theta \, d\varphi,$$

$$b_{ij} = \frac{1}{N_{ij}} \int_0^{2\pi} \int_0^{\pi} \rho_1(\theta,\varphi) P_j^{(i)}(\cos\theta) \sin i\varphi \sin\theta \, d\theta \, d\varphi,$$

$$N_{ij} = \frac{2\pi\delta_i (j+i)!}{(2j+1)(j-i)!}, \qquad \delta_i = \begin{cases} 2 & \text{when } i = 0,\\ 1 & \text{when } i > 0. \end{cases}$$

Let us represent the elastic–plastic boundary in the form

$$\rho_s = \beta_0 + \varepsilon\beta_1(\theta, \varphi), \qquad \beta_0 = \text{const.} \tag{6.2.14}$$

From the condition of coincidence of solutions for the elastic and plastic zones on the elastic–plastic boundary, we have

$$\begin{gathered}[\sigma_\rho] = 0, \qquad [\sigma_\theta] = 0, \qquad [\tau_{\rho\theta}] = 0, \\ [\tau_{\rho\varphi}] = 0 \quad \text{on } \rho = \rho_s.\end{gathered} \tag{6.2.15}$$

Here $[X]$ stands for the change in X associated with transition through the elastic–plastic boundary. In order to linearize conditions (6.2.15), they must be transferred to surface $\rho = \beta_0$. By omitting the intermediate operations, conditions (6.2.15) can finally be written as

$$\begin{gathered}\left[\sigma_\rho^1\right] = 0, \qquad \left[\tau_{\rho\theta}^1\right] = 0, \qquad \left[\tau_{\rho\varphi}^1\right] = 0, \\ \left[\sigma_\theta^1 + \frac{\partial \sigma_\theta^0}{\partial \rho}\beta_1\right] = 0.\end{gathered} \tag{6.2.16}$$

The first three conditions of (6.2.16), together with the boundary conditions on the external sphere $\rho_0 = b$, yield the elastic solution. The last condition of (6.2.16) allows to determine $\beta_1(\theta, \varphi)$.

It should be noted that the continuity of stresses σ_φ on ρ_s leads, generally speaking, to a different result for the elastic–plastic boundary. Therefore, the condition of continuity of all the stresses on the elastic–plastic boundary in linearized space elastic–plastic problems can result in overdetermination of the problem.

2

Let us present a solution of the elastic–plastic problem for a space with a spherical cavity of radius ρ_0. The stresses at infinity are

$$\begin{gathered}\sigma_x = \sigma_x^\infty, \qquad \sigma_y = \sigma_y^\infty, \qquad \sigma_z = \sigma_z^\infty, \\ \tau_{xy} = \tau_{xz} = \tau_{yz} = 0.\end{gathered} \tag{6.2.17}$$

As a zero approximation, we take the stresses

$$\sigma_x^\infty = \sigma_y^\infty = \sigma_z^\infty = \sigma^\infty. \tag{6.2.18}$$

We represent the external stresses as

$$\begin{gathered}\sigma_x^\infty = \sigma^\infty + \varepsilon\sigma_x^{1\infty}, \qquad \sigma_y^\infty = \sigma^\infty + \varepsilon\sigma_y^{1\infty}, \\ \sigma_z^\infty = \sigma^\infty + \varepsilon\sigma_z^{1\infty}, \qquad \varepsilon = \left(\sigma_x^\infty - \sigma_y^\infty\right) + \left(\sigma_z^\infty - \sigma_y^\infty\right).\end{gathered} \tag{6.2.19}$$

In the plastic region,

$$\sigma_\rho^{1p} = \sigma_\theta^{1p} = \sigma_\varphi^{1p} = \tau_{\rho\theta}^{1p} = \tau_{\theta\varphi}^{1p} = \tau_{\rho\varphi}^{1p} = 0. \tag{6.2.20}$$

In the elastic region,

$$\sigma_\rho^{1e} = \frac{\sigma_x^{1\infty} + \sigma_y^{1\infty} + \sigma_z^{1\infty}}{3} \frac{\rho^3 - \beta_0^3}{\rho^3} + \frac{2}{3}\left(1 - 5\frac{\beta_0^3}{\rho^3} + 4\frac{\beta_0^5}{\rho^5}\right)$$

$$+ \left[\sigma_x^{1\infty} P_2(\cos\theta_1) + \sigma_y^{1\infty} P_2(\cos\theta_2) + \sigma_z^{1\infty} P_2(\cos\theta_3)\right],$$

$$\sigma_\theta'^{e} = \sigma_x^{1\infty}\left\{\frac{2}{3}\cos^2\theta\cos^2\varphi - \frac{1}{6}\frac{\beta_0^3}{\rho^3}\cos 2\lambda_1 + \frac{2}{3}\frac{\beta_0^5}{\rho^5}\left[P_2(\cos\theta_1) - \cos^2\theta_1\cos 2\lambda_1\right]\right\}$$

$$+ \sigma_y^{1\infty}\left\{\cos^2\theta\sin^2\varphi + \frac{1}{6}\frac{\beta_0^3}{\rho^3} - \frac{2}{3}\frac{\beta_0^5}{\rho^5}\left[P_2(\cos\theta_2) + \cos^2\theta_2\cos 2\lambda_2\right]\right\}$$

$$+ \sigma_z^{1\infty}\left\{\frac{1}{6}\frac{\rho^3 - \beta_0^3}{\rho^3} - \frac{2}{3}\left(1 + \frac{\beta_0^5}{\rho^5}\right)P_2(\cos\theta) + \frac{2}{3}\frac{\beta_0^5}{\rho^5}\cos^2\theta\right\},$$

$$\sigma_\varphi^{1e} = \sigma_x^{1\infty}\left\{\frac{2}{3}\sin^2\varphi + \frac{1}{6}\frac{\beta_0^3}{\rho^3} - \frac{2}{3}\frac{\beta_0^5}{\rho^5}\left[P_2(\cos\theta_1) - \cos^2\theta_1\cos 2\lambda_1\right]\right\} \tag{6.2.21}$$

$$+ \sigma_y^{1\infty}\left\{\cos^2\varphi + \frac{1}{6}\frac{\beta_0^3}{\rho^3} - \frac{2}{3}\frac{\beta_0^5}{\rho^5}\left[P_2(\cos\theta_2) - \cos^2\theta_2\cos 2\lambda_2\right]\right\}$$

$$+ \sigma_z^{1\infty}\left\{\frac{1}{6}\frac{\beta_0^3}{\rho^3} - \frac{2}{3}\frac{\beta_0^5}{\rho^5}\left[P_2(\cos\theta) + \cos^2\theta\right]\right\},$$

$$\tau_{\theta\varphi}^{1e} = \sigma_x^{1\infty}\sin\lambda_1\cos\lambda_1\left[\sin^2\theta_1 + \frac{4}{3}\frac{\beta_0^5}{\rho^5}\cos^2\theta_1\right]$$

$$+ \sigma_y^{1\infty}\left\{\frac{1}{2}\sin 2\varphi\cos\theta - \frac{1}{3}\left[\frac{\beta_0^3}{\rho^3} - 2\frac{\beta_0^5}{\rho^5}P_2(\cos\theta_2)\right]\sin 2\lambda_2\right\},$$

$$\tau_{\rho\theta}^{1e} = \left(1 + \frac{5}{3}\frac{\beta_0^3}{\rho^3} - \frac{8}{5}\frac{\beta_0^5}{\rho^5}\right)\left(\sigma_x^{1\infty}\cos^2\varphi + \sigma_y^{1\infty}\sin^2\varphi - \sigma_z^{1\infty}\right)\sin\theta\cos\theta,$$

$$\tau_{\rho\varphi}^{1e} = \left(1 + \frac{5}{3}\frac{\beta_0^2}{\rho^3} - \frac{8}{3}\frac{\beta_0^5}{\rho^5}\right)\left(-\sigma_x^{1\infty}\sin\varphi + \sigma_y^{1\infty}\cos\theta\right)\sin\theta\cos\varphi.$$

Here $P_2(\cos\theta)$ is the second order Legendre polynomial

$$\cos\theta_1 = \sin\theta\cos\varphi, \qquad \cos\lambda_1 = -\frac{\cos\theta\cos\varphi}{\sin\theta_1},$$

$$\cos\theta_2 = \sin\theta\sin\varphi, \qquad \cos\lambda_2 = -\frac{\cos\theta\sin\varphi}{\sin\theta_2},$$

$$\sin\lambda_1 = \frac{\sin\varphi}{\sin\theta_1}, \qquad \sin\lambda_2 = \frac{\cos\varphi}{\sin\theta_2}.$$

For the elastic–plastic boundary, we have

$$\beta_1 = \frac{\beta_0}{6k}\left\{\sigma_x^{1\infty}\left[\frac{2}{3}\cos^2\theta\cos^2\varphi - \frac{1}{6}\left(\cos 2\lambda_1 + 4\cos^2\theta_1\right) + \frac{2}{3}P_2(\cos\theta_1)\right]\right.$$

$$+\sigma_y^{1\infty}\left[\cos^2\theta\sin^2\varphi + \frac{1}{6} - \frac{2}{3}\cos^2\theta_2\cos 2\lambda_2 - \frac{2}{3}P_2(\cos\theta_2)\right]$$

$$\left.+\sigma_z^{1\infty}\frac{2}{3}\left[\cos^2\theta_1 - 2P_2(\cos\theta)\right]\right\}. \tag{6.2.22}$$

If $\beta_1(\theta, \varphi)$ is determined from the continuity condition for stresses σ_φ, we obtain

$$\beta_1 = \frac{\beta_0}{6k}\left\{\sigma_x^{1\infty}\left[\frac{2}{3}\sin^2\varphi + \frac{1}{6} - \frac{2}{3}P_2(\cos\theta_1) + \frac{2}{3}\cos^2\theta_1\cos 2\lambda_1\right]\right.$$

$$+\sigma_y^{1\infty}\left[\cos^2\varphi + \frac{1}{6} - \frac{2}{3}P_2(\cos\theta_2) + \frac{2}{3}\cos^2\theta_2\cos 2\lambda_2\right]$$

$$\left.+\sigma_z^{1\infty}\left[\frac{1}{6} - \frac{2}{3}P_2(\cos\theta) - \frac{2}{3}\cos^2\theta\right]\right\}. \tag{6.2.23}$$

Figure 6.1 shows orthogonal projections of the elastic–plastic boundary. The sections bounded by the elastic–plastic boundary along the lines of action of stresses σ_x^∞, σ_y^∞, and σ_z^∞ are denoted by a, b, and c.

3

Let us consider an elastic–plastic stress state for a space with a spherical cavity surface ρ_0, when uniform pressure p is applied to the contour of the

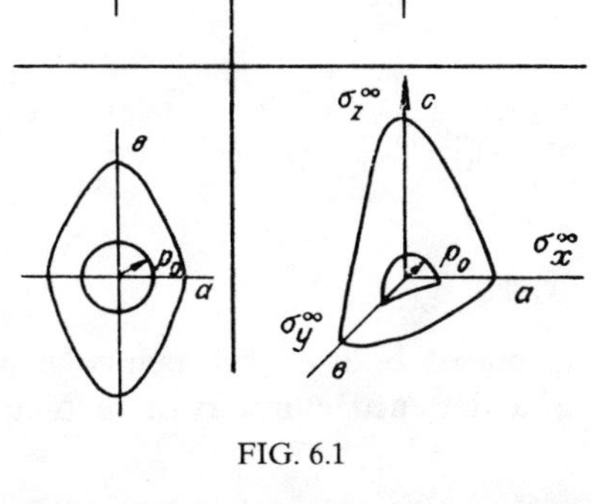

FIG. 6.1

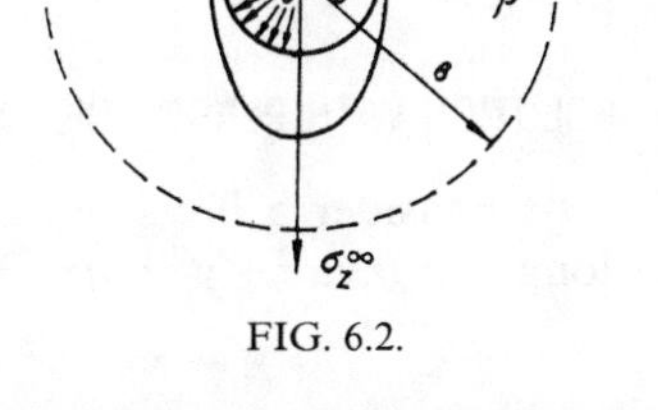

FIG. 6.2.

sphere. Constant stresses $\sigma_\rho = \sigma_\rho^\infty$ and $\sigma_z = \sigma_z^\infty$ are applied at infinity (Fig. 6.2).

We seek a solution in the form

$$\sigma_{ij} = \sigma_{ij}^0 + \sigma_{ij}^1. \tag{6.2.24}$$

The Tresca–Saint Venant condition is used here as the condition of plasticity

$$(\sigma_\rho - \sigma_\theta)^2 + 4\tau_{\rho\theta}^2 = 4, \qquad \sigma_\varphi = \tfrac{1}{2}(\sigma_\rho + \sigma_\theta) + 1. \tag{6.2.25}$$

As a zero approximation, we take the following stresses when $\sigma_\rho^\infty = \sigma_z^\infty$:

$$\begin{aligned} &\sigma_\rho^{0e} = q_0 + \tfrac{4}{3}\beta_0^3(1 - 1/\rho^3), && \sigma_\theta^{0e} = q_0 + \tfrac{4}{3}\beta_0^3(1 + \tfrac{1}{2}\rho^3), && \tau_{\rho\theta}^{0e} = 0, \\ &\sigma_\rho^{0p} = -p + 4\ln(\rho/\alpha), && \sigma_\theta^{0p} = -p + 2[1 + 2\ln(\rho/\alpha)], && \tau_{\rho\theta}^{0p} = 0, \end{aligned} \tag{6.2.26}$$

where $q_0 = \frac{1}{2}(\sigma_\rho^\infty + \sigma_z^\infty)$, $\rho = r/b$, $\alpha = a/b$, $\beta_0 = r_s^0/b$, a is the radius of the spherical cavity, b is the radius of a sphere surrounding the cavity, and r_s^0 is the radius of the plastic zone boundary.

As a first approximation, we have the following solution [11]:
In the plastic region,

$$\sigma_\rho^{1p} = \sigma_\theta^{1p} = \sigma_\varphi^{1p} = \tau_{\rho\theta}^{1p} = 0. \tag{6.2.27}$$

In the elastic region,

$$\begin{aligned} \sigma_\rho^{1e} &= \tfrac{4}{3}q'\left[1 - 5(\beta_0/\rho)^3 + 4(\beta_0/\rho)^5\right]P_2(\cos\theta) \\ &\quad - \tfrac{1}{3}q'\left[1 - (\beta_0/\rho)^3\right]P_0(\cos\theta), \\ \sigma_\theta^{1e} &= -\tfrac{4}{3}q'\left[1 + \tfrac{7}{3}(\beta_0/\rho)^5\right]P_2(\cos\theta) \\ &\quad + \tfrac{1}{3}q'\left[1 - \tfrac{1}{2}(\beta_0/\rho)^3 + \tfrac{4}{3}(\beta_0/\rho)^5\right]P_0(\cos\theta), \\ \tau_{\rho\theta}^{1e} &= \tfrac{2}{3}q'\left[1 + \tfrac{5}{3}(\beta_0/\rho)^3 - \tfrac{8}{3}(\beta_0/\rho)^5\right][dP_2(\cos\theta)]/d\theta. \end{aligned} \tag{6.2.28}$$

Here, $P_0(\cos\theta)$ is a zero order Legendre polynomial and $q' = \frac{1}{2}(\sigma_\rho^\infty - \sigma_z^\infty)$.

The elastic–plastic boundary has the form

$$\begin{aligned} \rho_s &= \beta_0 + \beta_1(\theta), \\ \beta_1 &= \tfrac{1}{54}\beta_0 q'\left[\tfrac{11}{2}P_0(\cos\theta) - 40P_2(\cos\theta)\right]. \end{aligned}$$

§ 3. SOLUTION WITH PLASTIC DISCONTINUITY LINES

Let us consider a homogeneous elastic–plastic body with a penny-shaped slit along $z = 0$, $x^2 + y^2 < a^2$. The slit is a flattened ellipsoid of revolution

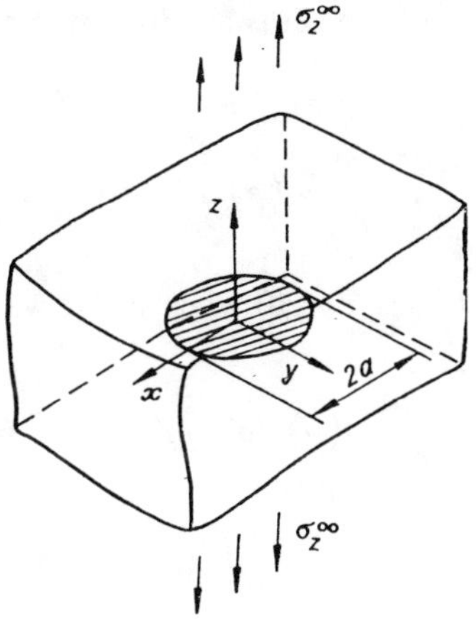

FIG. 6.3

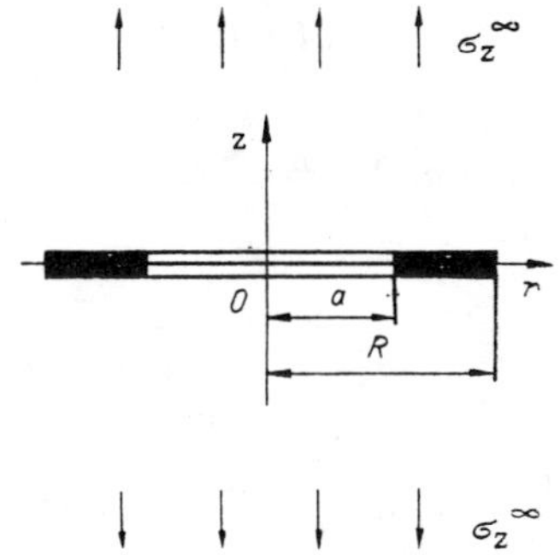

FIG. 6.4.

about axis Oz, and the plane of the slit coincides with plane xOy (Fig. 6.3). Stresses $\sigma_z = \sigma_z^\infty$ are applied at infinity. As discussed previously (see §4, Chap. 3, §7, Chap. 4, §4, Chap. 5), the development of plastic deformation at the initial stages is accompanied by the appearance of slip lines, which occupy insignificant volume of the body in comparison with the elastic part.

For the elastic–plastic problem, the region of plastic deformations will have the shape of a ring $a \leq r \leq R$ (Fig. 6.4) around the contour of the initial penny-shaped slit $r = a$. Here, r is a polar radius of the points, located in the plane of the slit; radius R, which separates the regions of elastic and plastic deformations, must be determined in the process of solving the elastic–plastic problem.

Let us write the boundary conditions of the problem. In plane $z = 0$,

$$\sigma_z(r,0) = \begin{cases} 0 & \text{when } r \leq a, \\ \sigma_s & \text{when } a < r \leq R; \end{cases} \tag{6.3.1}$$

at infinity

$$\sigma_z(r,\infty) = \sigma_z^\infty. \tag{6.3.2}$$

Stresses $\sigma_z(x, y, 0)$, which appear in the three-dimensional body with a penny-shaped slit, are determined by the formula [15]

$$\sigma_z(x, y, 0) = \frac{1}{\pi^2\sqrt{x^2+y^2-a^2}} \iint \frac{q(\xi,\eta)\sqrt{a^2-\xi^2-\eta^2}}{(x-\xi)^2+(y-\eta)^2}\, d\xi\, d\eta. \tag{6.3.3}$$

Here, $x^2+y^2 \geq a^2$, $\xi^2+\eta^2 < a^2$, and $q(x, y)$ is the normal pressure, applied to the faces of the slit.

If the normal pressure applied to the faces of the slit is axially symmetrical, formula (6.3.3) can be simplified to

$$\sigma_z(r,\beta,0) = \frac{2}{\pi\sqrt{r^2-a^2}} \int_0^a \frac{\rho\sqrt{a^2-\rho^2}}{r^2-\rho^2} q(\rho)\,d\rho, \tag{6.3.4}$$

where $r \geq a$, $0 \leq \beta \leq 2\pi$. In this case, for stresses $\sigma_z(r,0)$ when $r \geq R$ we have

$$\sigma_z(r,0) = \frac{2}{\pi\sqrt{r^2-R^2}} \int_0^R \frac{\rho\sqrt{R^2-\rho^2}}{r^2-\rho^2} q(\rho)\,d\rho + \sigma_z^\infty. \tag{6.3.5}$$

Here,

$$q(r) = \begin{cases} \sigma_x^\infty & \text{when } r \leq a, \\ \sigma_z^\infty - \sigma_s & \text{when } a < r \leq R. \end{cases} \tag{6.3.6}$$

By carrying out the integration in (6.3.5) and accounting for (6.3.6), we obtain

$$\sigma_z(r,0) = \frac{2}{\pi\sqrt{r^2-R^2}} \left\{ \sigma_z^\infty \left(R - \sqrt{r^2-R^2}\arctan\frac{R}{\sqrt{r^2-R^2}} \right) \right.$$
$$\left. - \sigma_s \left(\sqrt{R^2-a^2} - \sqrt{r^2-R^2}\arctan\frac{\sqrt{R^2-a^2}}{\sqrt{r^2-R^2}} \right) \right\} + \sigma_z^\infty. \tag{6.3.7}$$

The radius R of the elastic–plastic boundary is determined from the condition of stress boundedness, which leads in this case to the relationship

$$\sigma_z^\infty R - \sigma_s\sqrt{R^2-a^2} = 0. \tag{6.3.8}$$

From here

$$R = a\Big/\sqrt{1-(\sigma_z^\infty/\sigma_s)^2}.$$

In conclusion, we present a relation that defines the displacements $w(r,0)$ [15]:

$$w(r,0) = \frac{4(1-\nu^2)}{\pi E}\sigma_s \int_{\arcsin a/R}^{\varphi} \sqrt{a^2 - r^2\sin^2\alpha}\,d\alpha, \qquad r \leq R. \tag{6.3.9}$$

Here, $\varphi = \pi/2$ when $0 \leq r \leq a$, $\varphi = \arcsin(a/r)$ when $a \leq r \leq R$. From formula (6.3.9), we find the crack tip opening displacement (when $r = a$)

$$2w(a,0) = \frac{8(1-\nu^2)}{\pi E} a\sigma_s \left(1 - \frac{a}{R}\right). \tag{6.3.10}$$

§ 4. NUMERICAL RESULTS. METHOD OF FINITE DIFFERENCES

1

Due to recent advances in computational methods, space problems can be solved on the basis of a complete system of three-dimensional equations. We present here a solution scheme [20] based on the Prandtl–Reuss equations

$$\Delta\sigma_{ij} = 2G\left[\Delta\varepsilon_{ij}\delta_{ij} - \left(\Delta W/2\tau_s^2\right)\left(\sigma_{ij} - \sigma\delta_{ij}\right)\right] + E(1-2\nu)^{-1}\Delta e\delta_{ij},$$

$$G = E/2(1+\nu),$$

$$\Delta W = \begin{cases} \left(\sigma_{ij} - \sigma\delta_{ij}\right)\left(\varepsilon_{ij} - \Delta e\delta_{ij}\right) & \text{for } I = \tau_s^2, \\ 0 & \text{for } I < \tau_s^2, \end{cases} \tag{6.4.1}$$

$$I = \tfrac{1}{2}\left(\sigma_{ij} - \sigma\delta_{ij}\right)\left(\sigma_{ij} - \sigma\delta_{ij}\right), \qquad e = \varepsilon_{ij}\delta_{ij}/3, \qquad \sigma = \sigma_{ij}\delta_{ij}/3,$$

where Δ denotes an increment of the loading parameter. Assuming that the loading parameter is fixed and substituting (6.4.1) into the equilibrium equations written in terms of increments

$$\frac{\partial \Delta\sigma_{ij}}{\partial x_j} = 0, \qquad i, j = 1, 2, 3,$$

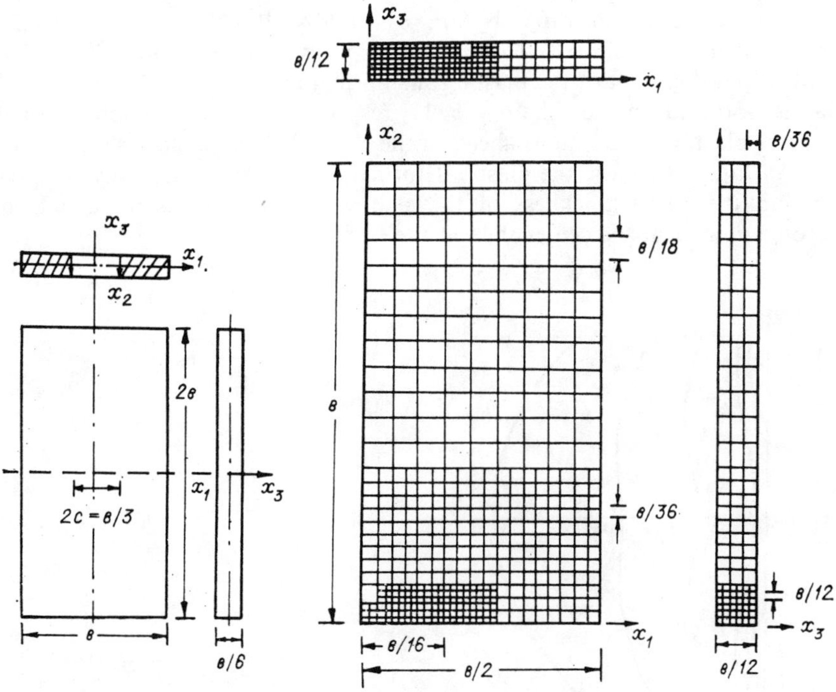

FIG. 6.5.

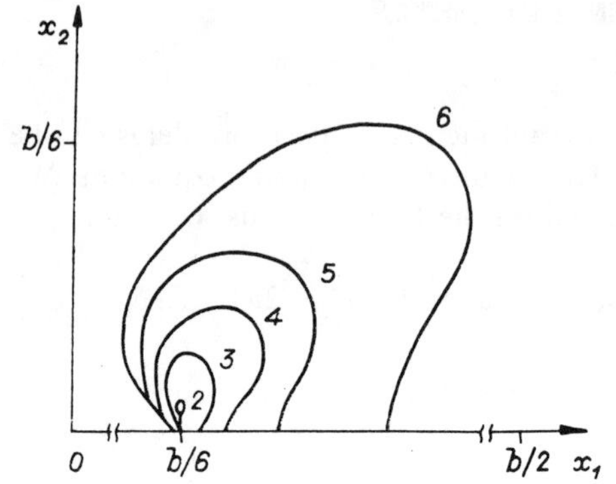

FIG. 6.6

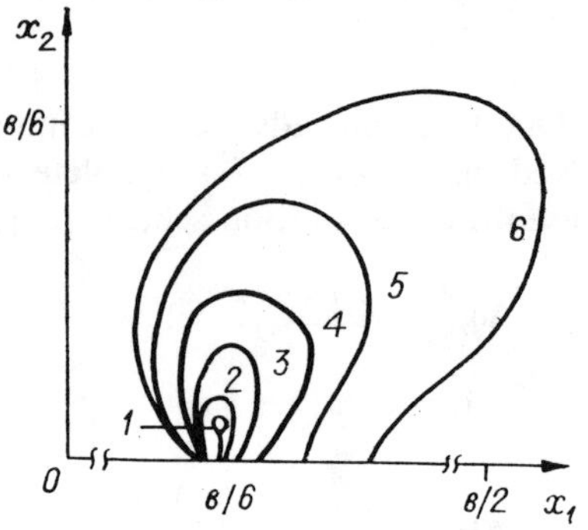

FIG. 6.7.

we obtain a system of three partial differential equations in terms of unknown components of the displacement-vector increments (Δu_1, Δu_2, Δu_3). To solve these equations, the method of finite differences is used. The system of finite-difference equations is solved by the relaxation method.

The results are summarized below:

1. The plastic zone at the tip of a through slit. A thick slab, having a rectangular through slit of length $b/3$, is loaded by a tensile stress σ_1 normal to the slit. The specimen is schematically shown in Fig. 6.5. The same figure shows three consecutive grids on one quarter of the symmetrical slab. The relaxation parameter is chosen equal to 1.4. The numerical solution of this elastic–plastic problem for $E = 2.06 \times 10^{11}$ N/m^2, $\sigma_s = 1.38 \times 10^9$ N/m^2, and $\nu = 0.3$ is presented in Figs. 6.6–6.8. The first two figures (Figs. 6.6 and 6.7) depict a gradual development of the plastic zone normal to the slit in the center of the slab $x_3 = 0$ and on the side $x_3 = b/12$, respectively. The development of the front of plastic deformations ahead of the slit and in its plane is shown in Fig. 6.8. As Fig. 6.8 shows, the plastic deformations have the tendency to spread through the entire thickness of the slab when the load is increased, and become greater at the center than at the sides.

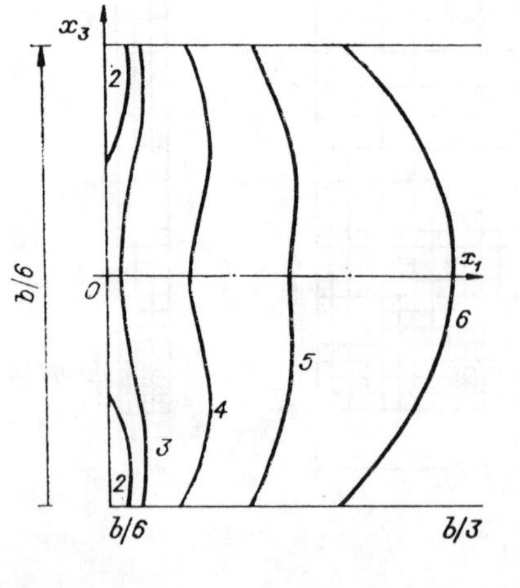

FIG. 6.8.

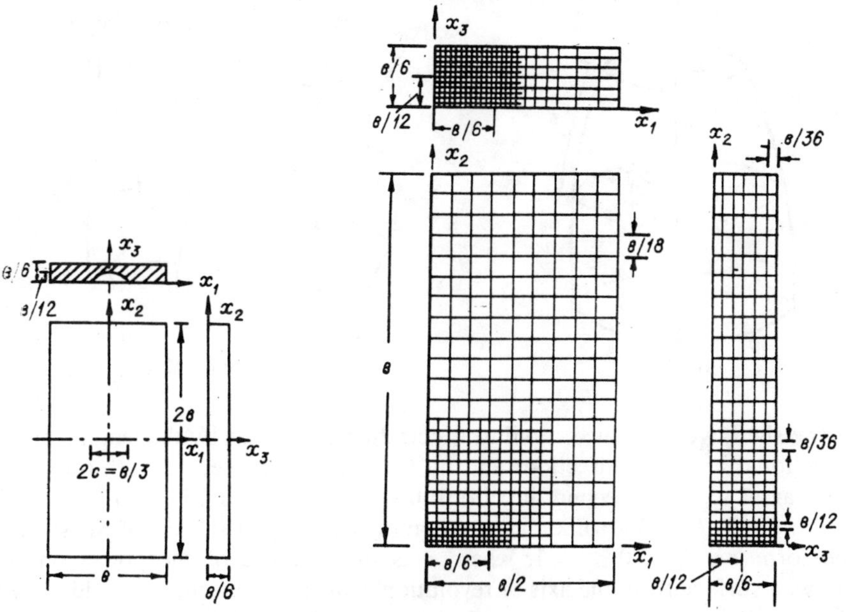

FIG. 6.9.

2. *Semielliptical slit*. A thick slab has a semielliptical slit with the major half-axis equal to $b/6$ and a minor half-axis equal to $b/12$ (Fig. 6.9). The slab is subjected to a uniform tensile stress normal to the slit. Figure 6.9 shows three consecutive grids on one quarter of the plate. The relaxation parameter is chosen equal to 1.4. The results of the numerical solution for the same values of parameters E, σ_s, and ν as before are presented in Figs. 6.10 and 6.11, which depict a gradual development of the plastic zone in a plane, perpendicular to the slit $x_3 = 0$ and in plane $x_1 = 0$, respectively.

2

Let us consider a thin-wall body of revolution, i.e., a shell of constant thickness h (Fig. 6.12) subjected to axisymmetric load. On the midwall surface

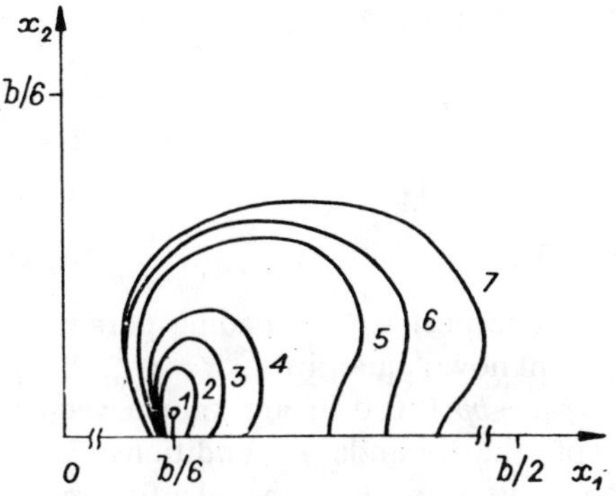

FIG. 6.10

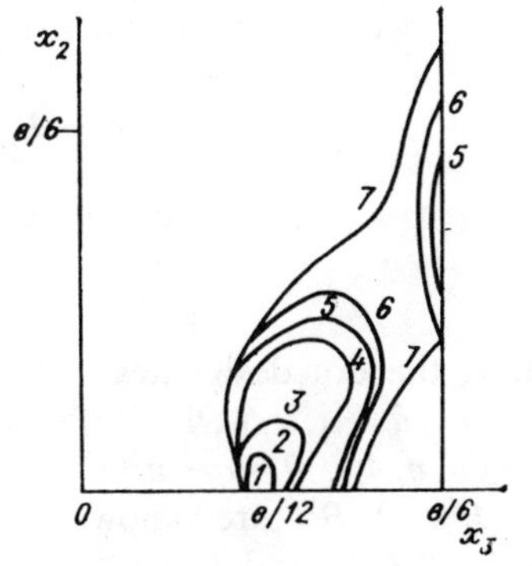

FIG. 6.11.

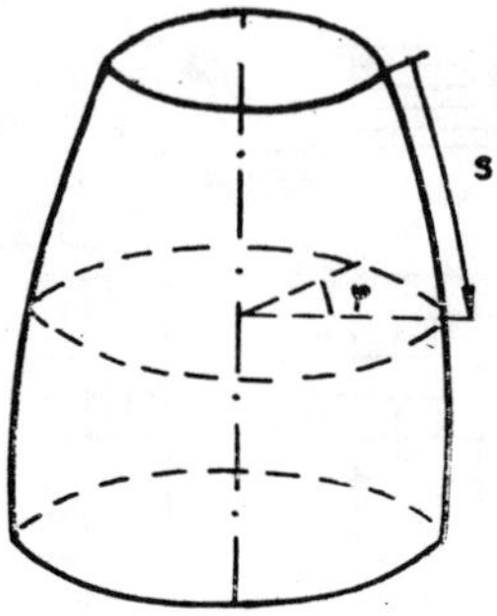

FIG. 6.12

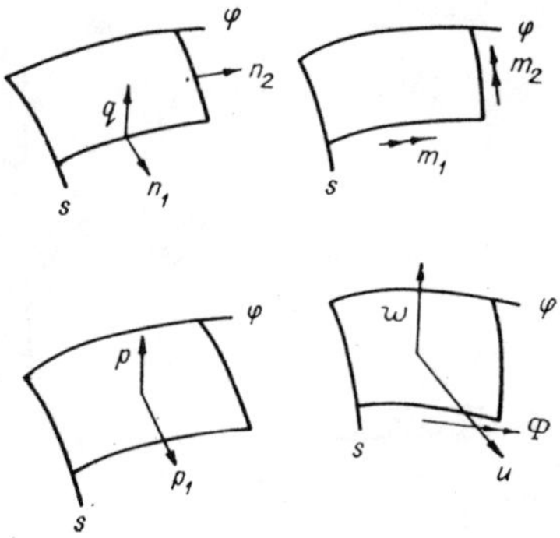

FIG. 6.13.

of the shell, we introduce curvilinear coordinates defined by the lengths of the meridional arc (first direction) and by the angle φ between an arbitrary and fixed axial sections (second direction): $0 \le s \le L$, $0 \le \varphi \le 2\pi$.

Let $\xi = s/L$ and assume the equation of the midwall surface of the shell in the form $r = r(\xi)$, $0 \le \xi \le 1$, where r is the distance of the points on the midwall surface from the axis of revolution, which is reduced, i.e., divided by a characteristic dimension such as L. Let σ_s be the yield point under uniaxial tension; $n_i, m_i (i = 1,2), q$ be dimensionless forces, bending moments, and shearing force, respectively (Fig. 6.13). In this case, we select $\sigma_s h^2/4$ as the characteristic moment and $\sigma_s h$ as the characteristic quantity for the shearing forces.

We use here the following designations: p_1, p_2 denote the components of an external load along direction 1 and along the normal divided by $b^2\sigma_s$, where $b = h/L$; u is the displacement of the points on the midwall surface along direction 1 divided by L; w is the displacement along the normal to the median surface divided by h; $\varepsilon_1, \varepsilon_2$ are the strains of the midwall surface along directions $1, 2$; χ_1, χ_2 are variations in principal curvatures R_1, R_2 divided by $4/h$; Φ is the angle of rotation of the normal around direction 2.

Based on the Sanders theory and the Kirchhoff–Love hypothesis, the equations of axisymmetric elastic–plastic deformation of a shell in case of small strains and moderate angles of rotation have the form

$$
\begin{gathered}
I_{ij}\, d\dot{z}_j/d\xi + H_{ij}\dot{z}_j = \dot{e}_i, \qquad i, j = 1, 2, \ldots, 6, \\
A^0_{lk}\dot{x}_k + B^0_{lk}\dot{y}_k = C^0_l \quad \text{for } \xi = 0, \\
A^1_{lk}\dot{x}_k + B^1_{lk}\dot{y}_k = C^1_l \quad \text{for } \xi = 1, \\
l, k = 1, 2, 3.
\end{gathered}
\tag{6.4.2}
$$

Here, the dot designates a derivative with respect to the loading parameter; $z = (n_1, \hat{q}, m_1, u, w, \Phi)$ is the vector of unknown quantities; $x = (n_1, \hat{q}, m_1)$, $y = (u, v, \Phi)$, $\hat{q} = q - n_1\Phi$, and $e = (-bp_1, -bp, 0, 0, 0, 0)$ are known vectors; A^0, B^0, A^1, B^1 are known matrices of the boundary conditions 3×3; (C^0_1, C^0_2, C^0_3) and (C^1_1, C^1_2, C^1_3) are known vectors, I is the identity matrix 6×6; and H is a 6×6 matrix, whose elements depend on vector z and on $\dot{z}$ (due to the physical nonlinearity).

The elements of matrix H are

$$
\begin{aligned}
&h_{11}=r_*(1-b_1), \quad h_{12}=a_1/b, \quad h_{13}=-r_*b_2, \quad h_{14}=-b_3r_*^2\\
&h_{15}=-r_*a_2b_3, \quad h_{16}=-br_*{}^2b_4/4+bp, \quad h_{21}=-(a_1+a_2b_1)/b,\\
&h_{22}=r_*, \quad h_{23}=-a_2b_2/b, \quad h_{24}=-a_2r_*b_3/b, \quad h_{25}=-a_2^2b_3/b,\\
&h_{26}=-a_2r_*b_4/4, \quad h_{32}=-4/b, \quad h_{31}=-(r_*b_5+4\Phi/b),\\
&h_{33}=r_*(1-b_6), \quad h_{34}=-r_*{}^2b_4,\\
&h_{35}=-r_*a_2b_4, \quad h_{36}=-\left(r_*{}^2bb_7/4+4n_1/b\right),\\
&h_{41}=-B_{33}/g, \quad h_{42}=0, \quad h_{43}=B_{13}/g,\\
&h_{44}=r_*c_1/g, \quad h_{45}=a_1+a_2c_1/g, \quad h_{46}=\Phi+c_2r_*b/4g,\\
&h_{51}=h_{52}=h_{53}=h_{55}=0, \quad h_{54}=-a_1/b^2, \quad h_{56}=1/b,\\
&h_{61}=4B_{13}/bg, \quad h_{62}=0, \quad h_{63}=-4B_{11}/bg,\\
&h_{64}=4r_*c_3/bg, \quad h_{65}=4c_3a_2/b_g, \quad h_{66}=r_*c_4/g.
\end{aligned}
\tag{6.4.3}
$$

Here,

$$
\begin{aligned}
&r_*=\frac{1}{r}\frac{dr}{d\xi}, \quad a_1=\frac{h}{R_1}, \quad a_2=\frac{h}{R_2},\\
&c_1=B_{12}B_{33}-B_{23}B_{13}, \quad c_2=B_{14}B_{33}-B_{34}B_{13},\\
&c_3=B_{11}B_{23}-B_{13}B_{12}, \quad c_4=B_{11}B_{34}-B_{13}B_{14},\\
&g=B_{11}B_{33}-B_{13}^2,\\
&b_1=(B_{33}B_{12}-B_{13}B_{23})/g, \quad b_2=(B_{11}B_{23}-B_{13}B_{12})/g,\\
&b_3=B_{22}+\left(2B_{12}B_{13}B_{23}-B_{12}^2B_{33}-B_{23}^2B_{11}\right)/g,\\
&b_4=B_{24}+(B_{12}B_{13}B_{34}+B_{13}B_{23}B_{14}-B_{12}B_{14}B_{33}-B_{11}B_{23}B_{34})/g,\\
&b_5=(B_{14}B_{33}-B_{13}B_{34})/g, \quad b_6=(B_{11}B_{34}-B_{13}B_{14})/g,\\
&b_7=B_{44}+\left(2B_{13}B_{14}B_{34}-B_{11}B_{34}^2-B_{33}B_{14}^2\right)/g.
\end{aligned}
$$

In this case, matrix $B=\|B_{ij}\|$, $i, j=1,2,3,4$, relates the rates of forces and rates of moments to strain rates and rates of curvatures

$$
\begin{pmatrix}\dot{n}_1\\ \dot{n}_2\\ \dot{m}_1\\ \dot{m}_2\end{pmatrix}=\begin{pmatrix}B_{11}B_{12}B_{13}B_{14}\\ B_{12}B_{22}B_{23}B_{24}\\ B_{13}B_{23}B_{33}B_{34}\\ B_{14}B_{24}B_{34}B_{44}\end{pmatrix}\begin{pmatrix}\dot{\varepsilon}_1\\ \dot{\varepsilon}_2\\ \dot{\varkappa}_1\\ \dot{\varkappa}_2\end{pmatrix}. \tag{6.4.4}
$$

The elements of matrix B were determined by two different methods [6].

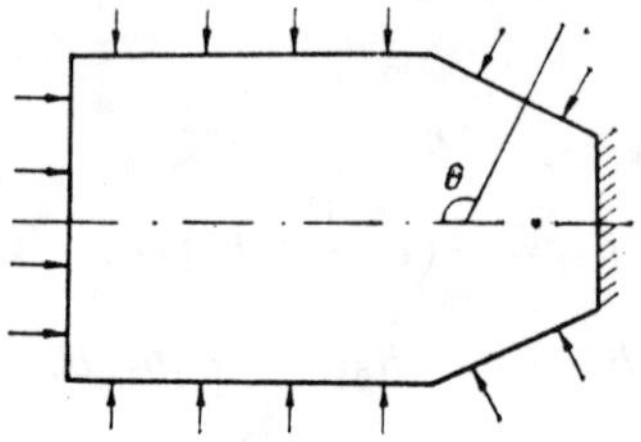

FIG. 6.14.

In the first method, the elements of matrix B were determined on the basis of the Prandtl–Reuss theory for the plane stress state. In the second method, the relationships of Ivanov [13] were used, which directly relate the rates of forces and rates of moments to the strain rates and rates of curvatures. It was assumed that the normal section of the shell is either completely plastic or completely elastic.

The results obtained by the first method [6] are given below. They show the development of plastic zones across the thickness of the shell. The shell was comprised of cylindrical and conical sections and subjected to external hydrostatic pressure (Fig. 6.14). Calculations were carried out for shells with parameters $R/h = 100$, $L_y/L_k = 1$, $L/R = 0.2$, $E/\sigma_s = 250$, and $\nu = 0.3$. The angle θ was varied within the range $90° \leq \theta \leq 130°$. Here, R is the radius of the midwall surface of the cylindrical section; L_y, L_k are the lengths of the meridians of the cylindrical and conical sections; θ is the angle between the

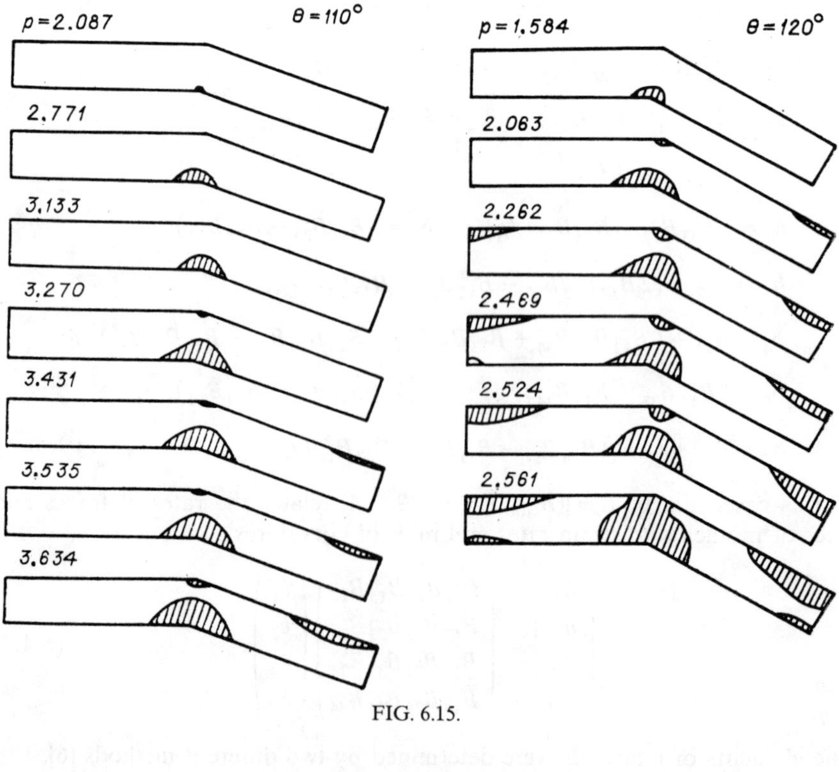

FIG. 6.15.

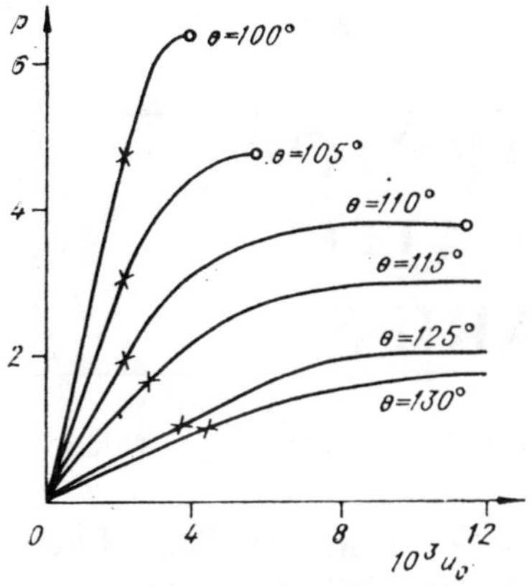

FIG. 6.16.

axis of rotation and the normal to the midwall surface of the conical section. The dimensionless axial displacement u_0 of the left end of the shell (axial displacement, ascribed to L) was taken as the loading parameter.

The following boundary conditions were used:

$$\dot{n}_1 = b^2 R\dot{p}/(2h), \qquad \dot{w} = \dot{\Phi} = 0 \quad \text{for } \xi = 0,$$
$$\dot{u} = \dot{w} = \dot{\Phi} = 0 \quad \text{for } \xi = 1. \tag{6.4.5}$$

These conditions mean that the left end of the shell is displaced in the axial direction under the action of force n_1, while the right end is stationary. At discontinuities of the normal to the midwall surface, conditions are used [5] that assure a smooth transition.

The derivatives with respect to the meridional coordinate were approximated by central differences of the second order of accuracy; the system of linear algebraic equations was solved by the Gauss method of elimination. The corresponding Cauchy problem with respect to the loading parameter was solved by the predictor–corrector method. The limit load was defined as the maximum on external force versus axial displacement ($p \sim u_0$) curve.

Figure 6.15 shows the development of plastic zones for shells with $\theta = 110°$ and 120°. When $p = 3.634$, a small plastic zone also appears on the cylindrical section. Figure 6.16 depicts the curves $p \sim u_0$ for various values of angle θ. The crosses indicate the value of the pressure at which the plastic zones first appear; the circles denote the limit (maximum) loads.

APPENDIX 1. A METHOD OF SOLVING PROBLEMS WITH UNKNOWN BOUNDARIES

Let us consider the following problem with an unknown boundary, which we shall call problem I.[1]

Problem I. In plane xy, we seek a singly connected region G bounded by a simple Jordan curve Γ and continuously differentiable function $z(x, y)$ in $G+\Gamma$ according to the following conditions:
a) In region G, function $z(x, y)$ is continuously differentiable twice and satisfies the equation

$$F(x, y, z, p, q, r, s, t)=0, \qquad p=\frac{\partial z}{\partial x}, \qquad q=\frac{\partial z}{\partial y},$$
$$r=\frac{\partial^2 z}{\partial x^2}, \qquad s=\frac{\partial^2 z}{\partial x\, \partial y}, \qquad t=\frac{\partial^2 z}{\partial y^2}. \tag{I.1}$$

Here, function F is defined and twice differentiable in the entire R^8 space $xyzpqrst$.

b) On contour Γ, which can be represented in the form (l is a parameter)

$$x=x(l), \qquad y=y(l), \qquad 0\le l<l^*,$$
$$[x(l_2)-x(l_1)]^2+[y(l_2)-y(l_1)]^2>0,$$
$$l_1, l_2\in[0, l^*),$$

function $z(x, y)$ satisfies the following conditions:

$$X(x, y, z, p, q)=\varphi(l), \qquad Y(x, y, z, p, q)=\psi(l),$$
$$Z(x, y, z, p, q)=f(l).$$

Here, functions X, Y, Z are independent, determinate, and continuous in the entire R^5 space $xyzpq$, and functions $x(l)$, $y(l)$, $\varphi(l)$, $\psi(l)$ and $f(l)$ are assumed continuous and periodic with a period l^*, so that

$$\varphi(0)=\varphi(l^*), \qquad \psi(0)=\psi(l^*), \qquad f(0)=f(l^*).$$

Note. The following problem,[2] for example, is reduced to Problem I. In plane xy, a singly connected region g is given, which is bounded by a sufficiently smooth curve γ. It is required to find a singly connected region $G\in g$ bounded by a Jordan curve Γ, and continuously differentiable function $z(x, y)$ in $g+\gamma$. In region G, function $z(x, y)$ is continuously differentiable twice and

[1]See reference [1] to Chap. 1.

[2]A special case of this problem was discussed in Chap. 3.

satisfies equation (I.1), while in region $g-(G+\Gamma)$ it satisfies the first order equation $\Omega(x, y, z, p, q)=0$. Function $z(x, y)$ on γ transforms into a given continuous function of arc α of contour γ:

$$z|_{\gamma}=\omega(\alpha). \tag{I.2}$$

Indeed, by solving the problem for equation $\Omega=0$ subject to condition (I.2) by the Cauchy method and then excluding the parameter along the characteristics from the obtained values x, y, z, p, q, we arrive, generally speaking, at Problem I for determination of region G and functions $z(x, y)$, $(x, y)\in G$. In this case, the length of arc α on contour γ is used as parameter l.

Let us introduce the following notation:

$$\frac{d}{dx}\equiv\frac{\partial}{\partial x}+p\frac{\partial}{\partial z}+r\frac{\partial}{\partial p}+s\frac{\partial}{\partial q}, \qquad \frac{D}{dx}\equiv\frac{\partial}{\partial x}+p\frac{\partial}{\partial z},$$

$$\frac{d}{dy}\equiv\frac{\partial}{\partial y}+q\frac{\partial}{\partial p}+s\frac{\partial}{\partial p}+t\frac{\partial}{\partial q}, \qquad \frac{D}{dy}\equiv\frac{\partial}{\partial y}+q\frac{\partial}{\partial z},$$

$$[AB]\equiv\frac{\partial A}{\partial p}\frac{DB}{dx}-\frac{\partial B}{\partial p}\frac{DA}{\partial x}+\frac{\partial A}{\partial q}\frac{DB}{\partial y}-\frac{\partial B}{\partial q}\frac{DA}{\partial y},$$

$$I(AB)=\frac{dA}{dx}\cdot\frac{dB}{dy}-\frac{dA}{dy}\cdot\frac{dB}{dx}.$$

Here, $A=A(x, y, z, p, q)$ and $B=B(x, y, z, p, q)$ are arbitrary continuous differentiable functions of the arguments.

Theorem. *Let functions* $X(x, y, z, p, q)$, $Y(x, y, z, p, q)$, *and* $Z(x, y, z, p, q)$ *be continuous and differentiable in* R^5: $xyzpq$ *and satisfy the relations*

$$[XY]=[XZ]=[YZ], \tag{I.3}$$

and let the following inequality exist for any $l_1, l_2\in[0, l^*]$:

$$[\varphi(l_2)-\varphi(l_1)]^2+[\psi(l_2)-\psi(l_1)]^2>0. \tag{I.4}$$

If[3] $J(XY)\neq 0$ *in* G, *then Problem I can be reduced to a boundary value problem.*

Proof. According to conditions (I.3), there exists [1,2] one and only one contact[4] transformation of space xyz:

$$x_1=X(x, y, z, p, q), \qquad y_1=Y(x, y, z, p, q),$$

$$z_1=Z(x, y, z, p, q), \qquad p_1=P(x, y, z, p, q), \qquad q_1=Q(x, y, z, p, q), \tag{I.5}$$

[3] Condition $I(X, Y)\neq 0$ in G can be satisfied as a result of either (I.1) or other conditions.

[4] Transformation (I.5) is called contact or tangential if relationship $dZ-PdX-QdY=\rho(dz-p\,dx-q\,dy)$ is satisfied identically and $\rho\neq 0$.

where functions P and Q are found from the simultaneous system

$$[XQ]=[YP]=[PQ]=0, \qquad [PZ]=\rho P, \qquad [QZ]=\rho Q,$$
$$[PX]=[QY]=\rho, \qquad \rho=\frac{\partial Z}{\partial z}-p\frac{\partial X}{\partial z}-q\frac{\partial Y}{\partial z}. \tag{I.6}$$

Let us assume that the solution of Problem I is such that $J(XY)\neq 0$ in G exists. Then [2], as a result of (I.4) the homeomorphic mapping of $G+\Gamma$ onto region $D+L$ can be performed[5] by formulas (I.5). Here, L is a boundary and D is determined from equation

$$x_1=\varphi(l), \qquad y_1=\psi(l), \qquad 0\leq l<l^*. \tag{I.7}$$

Let $z_1=z_1(x_1, y_1)$; $(x_1, y_1)\in D+L$ be the image of surface $Z(x, y)$, $(x, y)\in G+\Gamma$, for transformation (I.5), (I.6). Then

$$p_1=\frac{\partial z_1}{\partial x_1}=P(x, y, z, p, q), \qquad q_1\equiv\frac{\partial z_1}{\partial y_1}=Q(x, y, z, p, q),$$

$$r_1\equiv\frac{\partial^2 z_1}{\partial x_1^2}=\frac{R}{I(XY)}, \qquad s_1\equiv\frac{\partial^2 z_1}{\partial x_1\,\partial y_1}=\frac{S}{I(XY)}, \qquad t_1\equiv\frac{\partial^2 z_1}{\partial y_1^2}=\frac{T}{I(XY)}, \tag{I.8}$$

where R, S, T, $I(XY)$ are linear functions with respect to r, s, t, $(rs-t^2)$ with coefficients which depend on x, y, z, p, q. Surface $z_1=z_1(x_1, y_1)$, $(x_1, y_1)\in D+L$, is thus found from the conditions

$$F_1(x_1, y_1, z_1, p_1, q_1, r_1, s_1, t_1)=0 \quad \text{in } D, \tag{I.9}$$

$$z_1=f(l), \qquad 0\leq l<l^*, \text{ on } L, \tag{I.10}$$

where curve L is determined from equation (I.7) and equation (I.9) is derived from equation (I.1), taking into account (I.8). It should be noted that the boundary value problem (I.9), (I.10), and (I.7) for determination of function $z_1=z_1(x_1, y_1)$, $(x_1, y_1)\in D+L$, can be formulated without knowing an explicit form of function $z=z(x, y)$ which represents the solution of Problem I. If Problem (I.9), (I.10) has a solution, by performing a contact transformation which is inverse to the contact transformation determined by formulas (I.5), (I.6), we arrive, generally speaking, at the solution of Problem I.
Note. The proved theorem can easily be generalized for a number of independent variables $n>2$ if we use the properties of the contact transformation for n dimensions [4].

[5]The proved theorem in [2] goes as follows. Let ϕ be the differentiable mapping of the finite region G of n-dimensional Euclidean space onto itself, and the Jacobian of this mapping does not change the sign in G, such that the set where the Jacobian tends to zero has zero Lebesgue measure. Let the mapping ϕ map the boundary Γ of region G one-to-one onto the boundary L of some region D. Then, the mapping ϕ is a one-to-one mapping of $G+\Gamma$ onto $D+L$.

APPENDIX 2. NONLINEAR RIEMANN PROBLEM

1

Let it be required to determine a unique piecewise analytical function $\Phi(z)$ such that the boundary values on the contour satisfy condition

$$f[\Phi^+(t), \Phi^-(t)] = 0 \quad \text{on } L. \tag{II.1}$$

Here $f(\Phi^+, \Phi^-)$ is a given continuous function of two variables, where the plus and minus signs correspond to the left and right faces of the slit along the contour.

The boundary value problem (II.1) for the nonlinear function $f(\Phi^+, \Phi^-)$ is known as the nonlinear Riemann boundary value problem. Here, we shall seek to find all nonlinear Riemann boundary value problems, which can be reduced to the linear Riemann boundary value problem, and hence can be solved in closed form. More specifically, it is required to find such an operation that

$$F\{f[\Phi^+(t), \Phi^-(t)]\} = a_1(t)g^+[\Phi^+(t)] + a_2(t)g^-[\Phi^-(t)] + a_3(t). \tag{II.2}$$

Here, $g^+(z)$ and $g^-(z)$ are analytical functions, having only logarithmic singular points and isolated unique singular points. In the case of contour L, which divides the entire plane z into internal and external regions, functions g^+ and g^- can be different. In the case of an open contour L, $g^+(z) \equiv g^-(z)$, $a_1(t)$, $a_2(t)$, and $a_3(t)$ are almost everywhere continuous functions satisfying the Hölder condition on the continuity intervals.

It is easy to check that, for an arbitrary function $f(\phi^+, \phi^-)$, no function $F(f)$, satisfying condition (II.2), exists. The class of functions $f(\phi^+, \phi^-)$, for which $F(f)$ exists, is determined with the help of the following auxiliary lemma.

Lemma 1. *The functional equation*

$$F[f(x, y)] = a_1 g^+(x) + a_2 g^-(y) + a_3$$

is solvable when and only when function $f(x, y)$ can be represented in the form

$$f(x, y) = \varphi[a_1 g^+(x) + a_2 g^-(y) + a_3],$$

where $\varphi(z)$ has $\varphi^{-1}(z)$ as the inverse function and

$$F(z) = \varphi^{-1}(z).$$

The proof of the lemma is evident.

The following theorem, conforming to the initially stated question, is a direct consequence of the lemma.

Theorem. *The nonlinear Riemann boundary value problem (II.1) is reduced to a linear Riemann boundary value problem (and thus can be solved in a closed form) when and only when function $f(\Phi^+, \Phi^-)$ of the nonlinear Riemann boundary value problem can be represented in the form*

$$f[\Phi^-(t), \Phi^-(t)] = \varphi\{a_1(t)g^+[\Phi^+(t)] + a_2(t)g^-[\Phi^-(t)] + a_3(t)\}, \tag{II.3}$$

where $\varphi(z)$ has an inverse function $\varphi^{-1}(z)$, and g^+, g^-, a_1, a_2, and a_3 satisfy the above mentioned conditions.

The nonlinear Riemann boundary value problem (II.1) in this case is, evidently, solved by using operation φ^{-1}, introducing a new piecewise analytical function

$$\Psi(z) = \begin{cases} g^+[\Phi^+(z)] & \text{for } z \in D^+, \\ g^-[\Phi^-(z)] & \text{for } z \in D^-, \end{cases}$$

and subsequent solving of the obtained linear Riemann boundary value problem for $\Psi(z)$ in the class of analytical functions having logarithmic singularities and isolated unique singularities.

Using formula (II.3), it is not difficult to identify a large number of examples of nonlinear Riemann boundary value problems, which are solvable in a closed form. We shall discuss here one of the simplest boundary value problems, which has practical applications [3] (see Chapt. 2).

2

Let it be required to determine a unique piecewise analytical function $\Phi(z)$ such that the boundary values on contour L satisfy condition

$$\Phi^+(t)\Phi^-(t) = \beta t \quad \text{on } L. \tag{II.4}$$

It is evident that functions $\varphi(z)$, $g^{\pm}(z)$, $a_1(z)$, and $a_2(z)$ in representation (II.3) have, in this case, the form

$$\varphi(z) = e^z, \qquad g^+(z) = g^-(z) = \ln z,$$

$$a_1 = a_2 = 1, \qquad a_3(t) = \beta(t).$$

For the sake of simplicity, we assume that function $\beta(t)$ satisfies the Hölder condition everywhere on contour L and does not vanish.

Let us assume first that a simple smooth contour L divides the internal region D^+ and external region D^-. Let function $\Phi^+(z)$ have m zeros in D^+ at points $z = \mu_i$ and let $\Phi^-(z)$ have k zeros at points $z = \lambda_i$ ($m - k = \varkappa$, where $\varkappa = \operatorname{Ind}\beta(t) = [\arg\beta(t)]|_L/(2\pi)$ is the index of function $\beta(t)$). For the sake of definiteness, we assume that the origin of the coordinates lies in region D^+. Hence, the solution of problem (II.4) is determined with an accuracy of up to an arbitrary constant C by formulas

$$\Phi^{+}(z)=C\prod_{i=1}^{m}(z-\mu_i)\prod_{i=1}^{k}(z-\lambda_i)^{-1}e^{\Gamma^{+}(z)},$$

$$\Phi^{-}(z)=\frac{1}{C}z^{\varkappa}\prod_{i=1}^{m}(z-\mu_i)^{-1}\prod_{i=1}^{k}(z-\lambda_i)e^{-\Gamma^{-}(z)}, \tag{II.5}$$

$$\Gamma(z)=\frac{1}{2\pi i}\int_L\frac{\ln[\beta(t)t]^{-\varkappa}}{t-z}dt.$$

If additional requirements are not imposed on $\Phi^{+}(z)$ and $\Phi^{-}(z)$ regarding the number and location of zeros, the boundary value problem (II.4) will have an infinite number of solutions. The latter are determined by formulas (II.5) for arbitrary values of λ_i, μ_i, m, and k such that $\mu_i\in D^{+}$, $\lambda_i\in D^{-}$, and $m-k=\varkappa$. This is the most significant qualitative deviation of the boundary value problem (II.4) from the linear Riemann boundary value problem.

Let us assume now that the open smooth contour L comprises n arcs with ends a_k and b_k. It can be shown that the solution of the boundary value problem (II.4) either has, at the ends of sections a_k and b_k, a nonintegrable exponential singularity or is bounded so that its boundedness in the vicinity of these ends follows from the requirement of integrability of the solution at the ends of sections a_k and b_k. Let the analytical solution of problem (II.4) have m zeros at points $z=\nu_i$. We assume, for the sake of definiteness, that point $z=0$ belongs to contour L. It can be shown that the solution of problem (II.4) is determined by formulas

$$\Phi(z)=z^{-m}\prod_{i=1}^{m}(z-\nu_i)e^{\Gamma(z)},$$

$$\Gamma(z)=\frac{X_n(z)}{2\pi i}\int_L\frac{\ln\left[\beta(t)\tau^{2m}\prod_{i=1}^{m}(t-\nu_i)^{-2}\right]}{X_n^{+}(t)(t-z)}dt, \tag{II.6}$$

where the following $n-1$ conditions must be satisfied:

$$\int\frac{t^k}{X_n^{+}(t)}\ln\left[\beta(t)t^{2m}\prod_{i=1}^{m}(t-\nu_i)^{-2}\right]dt=0,\qquad k=0,1,2,\ldots,n-2. \tag{II.7}$$

Here,

$$X_n(z)=\prod_{i=1}^{n}(z-a_i)^{1/2}(z-b_i)^{1/2}.$$

Using the formula describing the behavior of the Cauchy-type integral near the point on the contour where its density has a logarithmic singularity, it is easy to find that solution (II.6) is bounded in the vicinity of point $z=0$.

Thus, for complete definiteness of the problem in the case of an open contour, it is necessary to specify the number and location of zeros of the sought solution. Otherwise, the number of solutions is infinite and the solutions are determined by formulas (II.6) and (II.7) for arbitrary m and ν. If the number of zeros, m, is specified, the solution is determined with an accuracy of up to m arbitrary constants.

The case of the coefficient $\beta(t)$, which either vanishes or has a finite number of the first order discontinuities, can be examined in a manner similar to the linear Riemann boundary value problem. Thus, at the points of the first order discontinuities in the coefficient $\beta(t)$, the solution of problem (II.4) has a power singularity.

3

Let it be required to determine a function $\omega(z)$ analytical in the upper half plane $\operatorname{Im} z > 0$ in accordance with the following nonlinear boundary condition on the real axis:

$$\begin{aligned} f\left[\omega(t), \overline{\omega(t)}\right] &= 0 \quad \text{on } L, \\ \operatorname{Re}\left[(a(t) - ib(t))\omega(t)\right] &= 0 \quad \text{on } M. \end{aligned} \tag{II.8}$$

Here, $a(t)$ and $b(t)$ are functions almost everywhere continuous and satisfying the Hölder condition on the intervals of continuity and at an infinitely distant point ($a + ib \neq 0$); $L = L_1 + L_2 + \cdots + L_p$, L_k is a section such that $-\infty < a_k \leq t \leq b_k < \infty$; M is a set of points on the real axis lying outside L; $f(\omega, \bar{\omega})$ is an arbitrary function. We seek the solution of the boundary problem (II.8) in a class of functions integrable everywhere in the upper half-plane including the boundary. For an arbitrary region, the boundary problem of this kind is reduced, in essence, to the problem stated with the help of conformal mapping.

When a function of two variables $f(x, y)$ can be represented in the form

$$\begin{aligned} f(x, y) &= -\alpha + f_0(x, y), \\ f_0(\lambda x, \lambda y) &= f_1(\lambda) f_2(x, y), \\ f_2(x, y) &= \varphi\left[a_1 g(x) + a_2 g(y) + a_3\right], \end{aligned} \tag{II.9}$$

where $\varphi(z)$, $g(z)$, a_1, a_2, and a_3 satisfy conditions of section 1 in this Appendix, the boundary problem (II.8) can be solved in a closed form.

Let us present a technique that can be used to solve problem (II.8) subject to conditions (II.9). The technique is based on reducing the problem to the nonlinear Riemann boundary value problem, which was discussed earlier.

Definition. A *canonical solution* of the nonlinear boundary value problem (II.8) is a piecewise analytical function $X(z)$, which is a canonical solution of the Riemann problem

$$\begin{aligned} X^+(t) &= G(t) X^-(t), \\ G(t) &= \begin{cases} -\dfrac{a(t) + ib(t)}{a(t) - ib(t)} & \text{on } M, \\ 1 & \text{on } L, \end{cases} \end{aligned} \tag{II.10}$$

where class $X(z)$ near the discontinuity points of coefficient $a+ib$ and at points a_k and b_k coincide with a given class of function $\omega(z)$.

The canonical solution of problem (II.8) thus defined is written as

$$X(z)=\prod_{k=1}^{n}(z-b_k)^{-\varkappa_k}e^{\Gamma(z)},$$
$$\Gamma(z)=\frac{1}{2\pi i}\int_{-\infty}^{\infty}\frac{\ln G(t)}{t-z}\,dt. \tag{II.11}$$

Here, $\varkappa=\sum\varkappa_k$ is the index of the Riemann problem (II.10), which is determined in the usual manner.

We introduce the following function $\Phi(z)$, which is analytic in the entire plane z with the exception, perhaps, of the real axis

$$\Phi(z)=\begin{cases}\omega(z)/X^{+}(z) & \text{when } \operatorname{Im} z>0,\\ \overline{\omega}(z)/X^{-}(z) & \text{when } \operatorname{Im} z<0.\end{cases} \tag{II.12}$$

Taking into account relation (II.9), the boundary condition (II.8) can be written with the help of formulas (II.12) and (II.10) as follows:

$$\begin{aligned}\varphi\{a_1(t)g[\Phi^{+}(t)]+a_2(t)g[\Phi^{-}(t)]-a_3(t)\}&\\ =\alpha(t)/f_1[X(t)]\quad &\text{on } L,\\ \Phi^{+}(t)=\Phi^{-}(t)\quad &\text{on } M.\end{aligned} \tag{II.13}$$

Thus, problem (II.8) has been reduced to the nonlinear Riemann boundary value problem, which was discussed earlier.

Let us introduce two simple examples of the nonlinear boundary problem of the theory of analytical functions, which have practical applications in the theory of elasticity and fluid mechanics [1,2] and belong to this class.

Let it be required to determine function $\omega(z)$, which is analytic in the upper half-plane $\operatorname{Im} z>0$, in accordance with the nonlinear boundary condition on the real axis

$$|\omega(t)|^2=\alpha(t),\qquad t\in L,$$
$$\operatorname{Re}[a(t)-ib(t)\omega(t)]=0,\qquad t\in M, \tag{II.14}$$

or according to the boundary condition

$$\arg\omega(t)=\alpha(t),\qquad t\in L,$$
$$\operatorname{Re}[(a(t)-ib(t))\omega(t)]=0,\qquad t\in M. \tag{II.15}$$

Functions φ, f_1, g, a_1, a_2, a_3, α [see (II.9)], have the form

$$\varphi(z)=e^{z},\qquad g(z)=\ln z,$$

$$f_1(z)=z^2,\qquad \alpha=\alpha(t),\qquad a_1=a_2=1,\qquad a_3=0$$

and

$$\varphi(z) = z, \qquad g(z) = \ln z, \qquad f_1(z) = 1,$$

$$\alpha = \alpha(t), \qquad a_1 = -\tfrac{1}{2}i, \qquad a_2 = \tfrac{1}{2}i, \qquad a_3 = 0.$$

The corresponding nonlinear Riemann problems on the basis of the general formula (II.13) acquire the form

$$\Phi^+(t)\Phi^-(t) = \alpha(t)/X^2(t), \qquad t \in L, \tag{II.16}$$

$$\Phi^+(t) = e^{2i\alpha(t)}\Phi^-(t), \qquad t \in L. \tag{II.17}$$

The Riemann boundary value problem (II.16) has already been discussed and the boundary value problem (II.17) is a well examined linear Riemann boundary value problem.

REFERENCES

CHAPTER 1

1. Annin, B. D. Two-Dimensional Elastic–Plastic Problems. Novosibirsk: Izd. Novosib. un-ta, 1968. 120 pp.
2. Annin, B. D. Experimental investigation of plastic properties of materials under complex loading. In Mechanics of a Solid Body. Warsaw: PAN, 1978, pp. 347–352.
3. Duvaut, G. and Lions, J.-L. Inequalities in Mechanics and Physics. Moscow: Nauka, 1980. 383 pp. (Translated from French; originally published by Dunod, Paris, 1972. Les Inequations en Mécanique et en Physique.)
4. Zienkiewicz, O. The Finite Element Method in Engineering Science. Moscow: Mir, 1975. 541 pp. (Translated from English; originally published by McGraw-Hill, London, 1971.)
5. Ivlev, D. D. Theory of Ideal Plasticity. Moscow: Nauka, 1966. 231 pp.
6. Ivlev, D. D. and Bykovtsev, G. I. Theory of Work-Hardening Plastic Body. Moscow: Nauka, 1971. 231 pp.
7. Ivlev, D. D. and Ershov, L. V. Perturbation Method in the Theory of Elastic-Plastic Body. Moscow: Nauka, 1978. 208 pp.
8. Ilyushin, A. A. Plasticity. Part I. Elastic–Plastic Deformations. Moscow/Leningrad: OGIZ, 1948. 367 pp.
9. Ilyushin, A. A. Plasticity. Principles of General Mathematical Theory. Moscow: Izd-vo AN SSSR, 1963. 271 pp.
10. Kachanov, L. M. Fundamentals of the Theory of Plasticity. Moscow: Nauka, 1969. 420 pp. (English edition by Mir Publishers, 1974, 445 pp.)
11. Klyushnikov, V. D. Mathematical Theory of Plasticity. Moscow: Izd-vo MGU, 1979. 208 pp.
12. Koiter, V. T. General theorems for elastic–plastic solids. Moscow: Izd-vo inostr. lit., 1961. 79 pp. (Translated from English; originally published in Progress in Solid Mechanics, vol. 1, I. N. Sneddon and R. Hill, eds. North-Holland 1960.)
13. Lavrentiev, M. A. and Shabat, B. V. Methods of the Theory of Complex-Variable Functions. Moscow: Fizmatgiz, 1958. 678 pp.
14. Morozov, E. M. and Nikishkov, G. P. The Method of Finite Elements in Fracture Mechanics. Moscow: Nauka, 1980. 250 pp.
15. Muskhelishvili, N. I. Some Basic Problems of the Mathematical Theory of Elasticity. Moscow: Nauka, 1966. 707 pp. (Translation of an earlier edition published by Noordhoff, Groningen, 1953.)
16. Muskhelishvili, N. I. Singular Integral Equations. Moscow: Nauka, 1968. 511 pp. (Translation of an earlier edition published by Noordhoff, Groningen, 1953.)
17. Nadai, A. Theory of Flow and Fracture of Solids, vol. 1. Moscow/Leningrad: Izd-vo inostr. lit., 1954. 647 pp. (translated from English; originally published by McGraw-Hill, New York, 1950.)
18. Nadai, A. Theory of Flow and Fracture of Solids, vol. 2. Moscow: Mir, 1969. 863 pp. (Translated from English; originally published by McGraw-Hill, New York, 1963.)
19. Novozhilov, V. V. Theory of Elasticity. Leningrad: Sudpromgiz, 1958. 370 pp. (English translation published by Pergamon Press, New York, 1961.)
20. Prager, W. and Hodge, F. G. Theory of Perfectly Plastic Solids. Moscow: Izd-vo inostr. lit., 1956. 398 pp. (Translated from English; originally published by Wiley, New York, 1951.)
21. Rabotnov, U. N. Creep Problems in Structural Members. Moscow: Nauka, 1966. 752 pp. (English translation published by North-Holland, Amsterdam, 1969.)
22. Rabotnov, U. N. Mechanics of Deformed Solid Bodies. Moscow: Nauka, 1979. 744 pp.
23. Savin G. N. Stress Distribution near Holes. Kiev: Naukova dumka, 1969. 888 pp.
24. Sedov L. I. Introduction to the Mechanics of a Continuous Medium. Moscow: Fizmatgiz, 1962. 284 pp. (English translation published by Addison-Wesley, Reading, Mass., 1965.)
25. Sedov, L. I. Continuum Mechanics, vols. I, II. Moscow: Nauka, 1973. 536 pp. 584 pp.
26. Sokolovsky, V. V. Theory of Plasticity. Moscow/Leningrad: Gostekhizdat, 1950. 396 pp. (In Russian with chapter summary in English.)
27. Theory of Plasticity. Collected Articles. Moscow: Izd-vo inostr. lit., 1948. 452 pp.
28. Freudenthal, A. and Geiringer, H. The Mathematical Theories of the Inelastic Continuum. Moscow: Fizmatgiz, 1962. 432 pp. [Translated from Encyclopedia of Physics (Handbuch der Physik), vol. VI, Elasticity and Plasticity, S. Flügge, ed. Springer-Verlag, Berlin, 1958, pp. 229–433.]
29. Hill, R. Mathematical Theory of Plasticity. Moscow: Gostekhizdat, 1956. 407 pp. (Translated from English; originally published by Clarendon Press, Oxford, 1950.)
30. Khristianovich, C. A. Continuum Mechanics. Moscow: Nauka, 1981. 483 pp.
31. Khristianovich, C. A. and Shemyakin, E. I. On the theory of ideal plasticity. Inzh. Zhurn.: Mekhanika Tverdovo Tela, 1967, no. 4, pp. 86–97.
32. Cherepanov, G. P. Mechanics of Brittle Fracture. Moscow: Nauka, 1974. 640 pp. (English translation published by McGraw-Hill, 1979, 939 pp.)
33. Cherepanov, G. P. Some problems of the theory of elasticity and plasticity with unknown boundaries.

In Application of the Theory of Functions in Continuum Mechanics, vol. I. Moscow: Nauka, 1965, pp. 135–150.
34. Cherepanov, G. P. and Ershov L. V. Fracture Mechanics. Moscow: Mashinostroenie, 1977. 224 pp.
35. Cherepanov, G. P. On the problem of nonsingularity in the theory of plasticity. Dokl. AN SSSR, 1974, vol. 218, no. 4, p. 1124.
36. Cherepanov, G. P. Mechanics of Brittle Fracture. New York: McGraw-Hill, 1979. 950 pp.
37. Zienkiewicz, O. C. and Cheung, Y. K. The Finite Element Method in Structural and Continuum Mechanics. New York: McGraw-Hill, 1967. 420 pp.

CHAPTER 2

1. Hoffman, O. and Zaks, G. Introduction to the Theory of Plasticity for Engineers. Moscow: Gosizdat. mash. lit., 1957. 279 pp. (Translated from English; originally published by McGraw-Hill, New York, 1953.)
2. Dobrovolsky, V. L. Solution of some problems of stress concentration in the theory of elasticity. Inzh. Zhurn. 1963, vol. 3, no. 4, pp. 732–736.
3. Kostrov, B. V. and Nikitin, L. V. Longitudinal shear crack with an infinitely narrow plastic zone. Prikl. Matematika i Mekhanika, 1967, vol. 32, no. 2, pp. 334–336 (J. Appl. Math. Mech., pp. 357–359).
4. McClintock, F. Ductile fracture instability in shear. Mekhanika. Sb. Perevodov i Obzorov Inostr. Periodich. Lit., 1959, no. 6, pp. 119–134 (J. Appl. Mech., 1958, vol. 25, pp. 581–588).
5. Neuber, G. Theory of stress concentration for shear-strained prismatical bodies with arbitrary non-linear stress-strain law. Mekhanika. Sb. Perevodov i Obzorov Inostr. Periodich. Lit., 1961, no. 4, pp. 116–130 (J. Appl. Mech., 1961, vol. 28, pp. 544–551).
6. Nekrasov, A. I. On plane parallel gas flow at subsonic speeds. Prikl. Matematika i Mekhanika, 1944, vol. 8, no. 2, pp. 249–266.
7. Nikolsky, A. A. and Taganov, G. I. Gas flow in local supersonic zone and some conditions of potential flow failure. Prikl. Matematika i Mekhanika, 1946, vol. 10, no. 4, pp. 481–502.
8. Applied Problems of Fracture Toughness. Collected translations. Moscow: Mir, 1968. 552 pp.
9. Segalov, A. E. On the shape of plastic zones due to antiplane deformation of half space weakened by a notch of finite width. Izv. AN SSSR. Mekhanika Tverdovo Tela, 1970, no. 5, pp. 139–45.
10. Slepyan, L. I. Mechanics of Cracks. Leningrad: Sudostroenie, 1981. 296 pp.
11. Sokolovsky, V. V. Concentration of shear stresses under nonlinear deformations. Inzh. Zhurn., 1962, vol. 2, no. 2, pp. 332–337.
12. Sukhikh, L. I. Elastic–plastic torsion of a bar with a longitudinal groove. Prikl. Mekhanika, 1968, vol. 4, no. 5. pp. 123–129 (Soviet Appl. Mech., 1968, vol. 4, no. 5, pp. 75–78).
13. Timoshenko, S. P. and Goodier, J. N. Theory of Elasticity. Moscow: Nauka, 1979. 560 pp. (Translated from English; originally published by McGraw-Hill, New York, 1970.)
14. Hult, J. Experimental study of fatigue crack propagation due to torsion. Mekhanika. Sb. Perevodov i Obzorov Inostr. Periodich. Lit., 1959, no. 6, pp. 81–110.
15. Hult, J. and McClintock, F. Elastic–plastic stress and strain distributions around sharp notches under repeated shear. Mekhanika. Sb. Perevodov i Obzorov Inostr. Periodich. Lit., 1959, no. 6, pp. 111–117. (Translated from the Proceedings of the Ninth International Congress for Applied Mechanics, Brussels, vol. 8, 1956, pp. 51–58.)
16. Cherepanov, G. P. Elastic–plastic problem under conditions of antiplane strain. Prikl. Matematica i Mekhanika, 1962, vol. 26, no. 4, pp. 697–708 (J. Appl. Math. Mech.).
17. Cherepanov, G. P. On a nonlinear boundary problem in the theory of analytical functions encountered in some elastic–plastic problems. Dokl. AN SSSR, 1962, vol. 147, no. 3, pp. 566–568.
18. Cherepanov, G. P. On the solution of statically indeterminate elastic–plastic problems under conditions of complex shear. Inzh. Zhurn., 1965, no. 6, pp. 1126–1127.
19. Cherepanov, G. P. The elastic–plastic problem. Izv. AN SSSR. Mekhanika Tverdovo Tela, 1969, no. 2, pp. 82–91 (Mechanics of Solids, pp. 73–80).
20. Shemyakin, E. I. Stress–strain state at the tip of a slit under antiplane deformation of rocks. Fiziko-tekhnicheskie problemi Razrabotki Poleznikh Iskopaemikh, 1973, no. 1, pp. 3–8.
21. Erdogan, F. Elastic–plastic antiplane problems for bonded dissimilar media containing cracks and cavities. Intern. J. Solids and Structures, 1966, vol. 2, no. 3, pp. 447–465.
22. Field, F. A. Yielding in a cracked plate under longitudinal shear. Trans. ASME, ser. E, 1963, vol. 30, no. 4.
23. Koskinen, M. F. Elastic–plastic deformation of a single grooved flat plate under longitudinal shear. Trans. ASME, ser. D, J. Basic Engng., 1963, vol. 85, no. 4.
24. Panc, V. Theory of elastic–plastic stress state for shear-strained prismatical bodies with symmetrical sharp notches. Acta Technica, 1965, vol. 10, no. 3.
25. Rice, J. R. Plastic yielding near a crack tip. Proc. 1st Intern. Conf. on Fracture. Sendai, Japan, 1965.
26. Rice, J. R. Contained plastic deformation near cracks and notches under longitudinal shear. Intern. J. Fracture Mech., 1966, vol. 2, no. 2.
27. Rice, J. R. Stresses due to a sharp notch in a work-hardening elastic–plastic material loaded by longitudinal shear. Trans. ASME, ser. E, J. Appl. Mech., 1967, vol. 34, no. 2.
28. Tuba, I. S. A perturbation theory of antiplane elastic–plastic deformations. Intern. J. Non-Linear Mech., 1969, vol. 4, no. 1, pp. 51–60.
29. Trefftz, E. Uber die Wirkung einer Abrundung auf die Torsionspannungen in der inneren Eckeeines Winkeleisens. Z. Angew. Math. und Mech., 1922, Bd. 2, H. 4.

30. Trefftz, E. Uber die Spannungsverteilung in tordierten Staben bei teilweiser Uberschreitung der Fliessgrenze. Z. Angew. Math. und Mech., 1925, Bd. 5, H. 1, S. 64–73.

CHAPTER 3

1. Annin, B. D. Elastic–plastic torsion of a cylindrical bar with oval cross section. Dokl. AN SSSR, 1963, vol. 149, no. 5, pp. 1043–1046.
2. Annin, B. D. Existence and uniqueness of the solution of the elastic–plastic torsion problem for a cylindrical bar of oval cross section. Prikl. Matematika i Mekhanika, 1965, vol. 29, pp. 777–887 (J. Appl. Math. Mech., pp. 1038–1047).
3. Annin, B. D. One property of solving equation $z_{xx}z_{yy} - z_{xy}^2 = 1$. Dokl. AN SSSR, 1966, vol. 168, no. 3, pp. 499–501.
4. Annin, B. D. Variational formulation of elastic–plastic problems. Dinamika Sploshnoy Sredi/ In-t Gidrodinamiki SO AN SSSR, Novosibirsk, 1979, no. 39, pp. 10–22.
5. Annin, B. D., Il'in, V. P., and Lebedev, C. V. Numerical solution of the elastic–plastic torsion problems. In Transactions of the Conference on Numerical Methods in Elasticity and Plasticity. Novosibirsk, 1969, pp. 37–40.
6. Annin, B. D. and Sadovsky, V. M. On a numerical solution of the problem of elastic–plastic torsion for a cylindrical bar with an oval cross section. Dinamika Sploshnoy Sredi / In-t Gidrodinamiki SO AN SSSR, Novosibirsk, 1978, no. 37, pp. 18–26.
7. Annin, B. D. and Sadovsky, V. M. Elastic–plastic torsion of a rectangular bar. Mekhanika Tverdovo Tela, 1981, no. 5, pp. 182–185.
8. Amosov, G. Y. On torsion of prismatic bars under elastic–plastic deformations. Vestnik MGU. Seria 1. Matematika, Mekhanika, 1966, no. 1, pp. 98–106.
9. Arutyunyan, N. Kh. and Abramyan, B. L. Torsion of elastic bodies. Moscow: Fizmatgiz, 1969, 686 pp.
10. Banichuk, N. V. Analysis of elastic–plastic torsion of bars by the method of local variations. Inzh. Zhurn.: Mekhanika Tverdovo Tela, 1967, no. 1, pp. 145–148.
11. Banichuk, N. V., Petrov, V. M., and Chernousko, F. L. Numerical solution of variational and boundary value problems by the method of local variations. Zhurn. Vychislit. Matematiki i Mat. Fiziki, 1966, vol. 6, no. 6, pp. 947–961.
12. Galin, L. A. Elastic–plastic torsion of prismatic bars with a polygonal cross section. Prikl. Matematika i Mekhanika, 1944, vol. 8, no. 4, pp. 307–322.
13. Galin, L. A. Elastic–plastic torsion of prismatic bars. Prikl. Matematika i Mekhanika, 1949, vol. 13, no. 3, pp. 285–296.
14. Galin, L. A. On the existence of solution for the problem of elastic–plastic torsion of prismatic bars. Prikl. Matematika i Mekhanika, 1949, vol. 13, no. 6, pp. 650–654.
15. Dembskaya, A. and Medukhovsky, A. Elastic–plastic torsion of composite bars. Izv. AN SSSR. Mekhanika Tverdovo Tela, 1969, no. 6, pp. 114–118 (Mechanics of Solids, pp. 101–105).
16. Kachanov, L. M. A solution of the elastic–plastic torsion problem using a variational method. In Investigations in Elasticity and Plasticity, vol. 1. Leningrad: Izd-vo LGU, 1961, pp. 151–161.
17. Koshelev, A. I. Existence of a generalized solution of the elastic–plastic torsion problem. Dokl. AN SSSR, 1954, vol. 99, no. 3, pp. 357–360.
18. Leonov, M. Y. and Shvayko, N. U. On discontinuous deformations in a solid body. Prikl. Mekhanika i Tekhn. Fizika, 1961, no. 2, pp. 96–103.
19. Leonov, M. Y. and Shvayko, N. U. Elastic–plastic deformation under torsion of a bar with a shallow semi-cylindrical groove on the surface. In Problems of Mechanics of a Real Solid Body. Kiev: Izd-vo AN UkSSR, 1962, no. 1, pp. 5–12.
20. Mikhlin, C. G. On an approximate solution of one-sided variational problems. Izv. Vuzov. Matematika, 1980, no. 2, pp. 45–58.
21. Perlin, P. I. Elastic–plastic torsion of oval bars. Inzh. Sb., 1961, vol. 31, pp. 202–205.
22. Rose, C. N. On the convergence of the L. Kachanov method. Vestn. Leningr. in-ta. Ser. Matematika, Mekhanika, Astronomia, 1961, vol. 19, no. 4, pp. 170–174.
23. Salvadori, M. G. Numerical Methods in Engineering. Moscow: Izd-vo inostr, lit., 1954. 247 pp. (Translated from the first Prentice-Hall edition.)
24. Sokolovsky, V. V. On a problem of elastic–plastic torsion. Prikl. Matematika i Mekhanika, 1942, vol. 6, nos. 2–3, pp. 241–246.
25. Christopherson, D. G. A theoretical investigation of plastic torsion in an I-beam. J. Appl. Mech., 1940, vol. 7, pp. A 1–4.
26. Davis, E. A. and Tuba, I. S. Elastic–plastic solutions for notched shafts in tor-sion. Trans. ASME, ser. E, J. Appl. Mech., 1966, vol. 88, no. 1.
27. Friedman, A. Free boundaries in elastic–plastic problems. Z. Angew. Math. und Mech., 1981, Bd. 61, H. 4, S. 2–8.
28. Heracovich, C. T. and Hodge, P. G. Jr. Elastic–plastic torsion of hollow bars by quadratic programming. Intern. J. Mech. Sci., 1969, vol. 11, no. 1, pp. 53–63.
29. Hodge, P. G. Jr., Herakovich, C. T., and Stout, R. B. On numerical comparisions in elastic–plastic torsion. Trans. ASME, ser. E, J. Appl. Mech., 1968 vol. 35, no. 3.
30. Lanchon, H. Torsion elastoplatique d'un arbre cylindrique de section simplement ou multiplement connexe. These a l'universite Paris VI, 1972. 169 pp.
31. Mises, R. Three remarks on the theory of the ideal plastic body. Reissner Anniversary Volume. Ann Arbor, Michigan: 1949, pp. 415–419.

32. Nadai, A. Der Begin des Fliessvorganges in einem tordirten Stab. Z. Angev. Math. und Mech., 1923, Bd. 3, H. 6, S. 442–454.
33. Okudo, H. The approximate decision of problem to torsion of a shaft with ring groove. The Reports Inst. High Speed Mech., 1953, pt. 3.
34. Onta, T. Fundamental research on the elastoplastic torsion of bar. Bull. Fac. Eng. Miyazaki. Univ., 1970, vol. 17, no. 12, pp. 41–52.
35. Shaw, F. S. The torsion of solid and hollow prism in the elastic and plastic range by relaxation methods. Australian Council for Aeronautics. Rep. ACA-11, 1944.
36. Southwell, R. V. Relaxation methods in theoretical physics, vols. 1, 2. Oxford University Press, 1946, 1956. 248 pp.; pp. 249–522.
37. Stout, Ray B. and Hodge, P. G. Jr. Elastic–plastic torsion of hollow cylinders. Intern. J. Mech. Sci., 1970, vol. 12, no. 1.
38. Takaci, N. On elastic–plastic torsion of a shaft with six semicircular longitudinal grooves, Trans. Japan Soc. Mech. Engrs., 1963, vol. 29, no. 197.
39. Ting, T. W. Elastic–plastic torsion of a square bar. Trans. Amer. Math. Soc., 1966, vol. 123, no. 2. pp. 369–401.
40. Ting, T. W. Elastic–plastic torsion. Archive for Ration Mech. and Analysis, 1967, vol. 25, no. 5, pp. 342–366.

CHAPTER 4

1. Aitaliev, Sh. M., Eskaliev, M. E., and Masanov, Zh. K. On elastic–plastic distribution of stresses and displacements in infinite anisotropic body with a round hole. Prikl. Problem Prochnosti i Plastichnosti, 1981, vol. 18. Statics and dynamics of deforming systems, pp. 129–136 (Vsesoyuzn. mezhvuz. sb., Gorky).
2. Alimzhanov, M. T. Elastic–plastic problem which accounts for inhomogeneity of mechanical properties of the material. Dokl. AN SSSR, 1978, vol. 242, no. 6, pp. 1281–1284.
3. Annin, B. D. A note on the paper by V. L. Dobrovolsky. "On the application of complex variables to the plane plastic strain." Prikl. Matematika i Mekhanika, 1965, no. 2, p. 400 (J. Appl. Math. Mech., pp. 455–456).
4. Annin, B. D. A plane elastic–plastic problem with an exponential yield condition. Inzh. Zhurn.: Mekhanika Tverdovo Tela, 1966, no. 3, pp. 122–123 (Mech. Solids, p. 83).
5. Annin, B. D. Elastic–plastic stress distribution in a plane with holes. Dokl. AN SSSR, 1969, vol. 184, no. 2, pp. 315–317.
6. Annin, B. D. Elastic–plastic stress distribution in a plane weakened by two circular holes. In Dynamics of Continuous Medium. Novosibirsk: Nauka, 1969, vol. 1, pp. 234–241.
7. Banichuk, N. V. Analysis of loading of elastoplastic body. Izv. AN SSSR. Mekhanika Tverdovo Tela, 1969, no. 1, pp. 128–135 (Mech. Solids, pp. 117–124).
8. Banichuk, N. V. Optimization of the shape of elastic bodies. Moscow: Nauka, 1980. 256 pp.
9. Batugin, C. A. and Pokhilko, V. A. Regions of existence and stability of solutions of elastic–plastic problems for the openings of circular shape. Fiz.-tekhn. Problemi Razrabotki Poleznikh Iskopaemikh, 1972, no. 3, pp. 10–16.
10. Bitsadze, A. V. Boundary value problems for elliptic equations of the second order. Moscow: Nauka, 1966. 203 pp.
11. Vorovich, I. I. and Kosmodamiansky, A. C. Elastic equilibrium of isotropic plate weakened by infinite row of single curvilinear holes. Izv. AN SSSR. Otd-niye Tekhn. Nauk, Mekhanika i Mashinostroenie, 1959, no. 4, pp. 69–76.
12. Galin, L. A. The plane elastic–plastic problem. Prikl. Matematika i Mekhanika, 1946, vol. 10, no. 3, pp. 367–386.
13. Galin, L. A. Analogy for the plane elastic–plastic problem. Prikl. Matematika i Mekhanika, 1948, vol. 12, no. 6, pp. 757–760.
14. Gakhov, F. D. Boundary Value Problems. Moscow: Fizmatgiz, 1963. 639 pp. (English translation published by Pergamon Press, New York, 1966.)
15. Grigolyuk, E. I. and Filshtinsky, L. A. Perforated plates and shells. Moscow: Nauka, 1970. 556 pp.
16. Ershov, L. V. and Ivlev, D. D. Elastic–plastic state of an elliptical tube under internal pressure. Izv. AN SSSR. Otd-nie Tekhn. Nauk, 1957, no. 9, pp. 130–134.
17. Ivlev, D. D. On determination of displacements in the Galin problem. Prikl. Matematika i Mekhanika, 1957, vol. 21, no. 5, pp. 716–717.
18. Ivlev, D. D. Approximate solution of elastic–plastic problems in the theory of ideal plasticity by the perturbation method. Dokl. AN SSSR, 1957, vol. 113, no. 2, pp. 294–296.
19. Ivlev, D. D. Approximate solution of plane elastic–plastic problems in the theory of ideal plasticity by the perturbation method. Vestnik Mosk. in-ta. Seria Matematika, Mekhanika, 1957, no. 5, pp. 17–26.
20. Karinsky, C. U. and Masanov, Zh. K. Determination of boundary state regions near a round opening in anisotropic mass. Izv. AN KazSSR. Ser. Fiz.–Mat., 1975, no. 3, pp. 42–48.
21. Kosmodamiansky, A. S. Elastic–plastic problem for an isotopic mass weakened by infinite row of single round openings. Izv. AN SSSR. Otd-nie Tekhn. Nauk, Mekhanika i Mashinostroenie, 1961, no. 4, pp. 187–188.
22. Koshelev, P. F. and Uzhik, G. V. Investigation of plastic deformation at stress raisers by the etching method. Izv. AN SSSR. Mekhanika i Mashinostroenie, 1959, no. 1, pp. 111–118.
23. Kudryavtsev, B. A., Parton, V. Z., Peskov, U. A., and Cherepanov, G. P. On the local plastic zone

near the tip of a crack (plane strain). Izv. AN SSSR. Mekhanika Tverdovo Tela, 1970, no. 5, pp. 132–138 (Mech. Solids, pp. 119–124).

24. Kuznetsov, A. I. Plane deformation of inhomogeneous plastic bodies. Vestnik Leningr. un-ta. Seria Matematika, Mekhanika, Astronomia, 1958, vol. 13, no. 3, pp. 112–131.
25. Kurshin, L. M. and Suzdalnitsky, I. D. Elastoplastic problem for a sheet weakened by a doubly periodic system of circular holes. Prikl. Matematika i Mekhanika, 1968, vol. 32, no. 3, pp. 463–467 (J. Appl. Math. Mech, pp. 467–471).
26. Leonov, M. Y. Fundamentals of Mechanics of Elastic Body, vol. 1. Frunze: Izd-vo AN KirgSSR, 1963. 328 pp.
27. Leonov, M. Y., Vitvitsky, P. M., and Yarema, C. Y. Plasticity bands due to tension of plates containing cracks. Dokl. AN SSSR, 1963, vol. 148, no. 3, pp. 541–544.
28. Mirsalimov, V. M. On the structure of plastic deformation at the tip of a crack. Izv. AN AzSSR. Ser. Fiz.–Tekhn. i Mat. Nauk. 1970, no. 6, pp. 24–29.
29. Mirsalimov, V. M. On biharmonic solutions of elastic–plastic problems in the presence of stress field discontinuity. Izv. AN AzSSR. Ser. Fiz.–Tekhn. i Mat. Nauk, 1972, no. 3, pp. 34–38.
30. Mirsalimov, V. M. Solution of elastic–plastic problems for a plane with a round hole in the presence of nonuniform temperature field. Dokl. AN AzSSR, 1973, vol. 29, no. 10, pp. 7–11.
31. Mirsalimov, V. M. On the solution of elastic–plastic problems for a plane with a single periodic system of round holes. Dokl. AN AzSSR, 1973, vol. 25, no. 5, pp. 11–15.
32. Mirsalimov, V. M. Some doubly periodic elastic–plastic problems under plane strain conditions. Izv. AN SSSR. Mekhanika Tverdovo Tela, 1975, no. 6, pp. 75–81 (Mech. Solids, pp. 63–69).
33. Mirsalimov, V. M. Solution of some periodic elastic–plastic problems. Zhurn. Prikl. Mekhaniki i Tekhn. Fiziki, 1975, no. 6, pp. 115–121.
34. Mirsalimov, V. M. Solution of some plane elastic–plastic problems. In Mechanics of Deformable Solid Bodies. Baku, 1975, pp. 118–130.
35. Mirsalimov, V. M. On an elastic–plastic problem for a mass weakened by two identical circular openings. Fiz.–Tekhn. Problemi Razrabotki Poleznikh Iskopaemikh, 1975, no. 5, pp. 142–146.
36. Mirsalimov, V. M. Some elastic–plastic problems for a plane weakened by a periodic system of circular holes. Prikl. Matemtika i Mekhanika, 1976, vol. 40, no. 1, pp. 152–158 (J. Appl. Math. Mech., pp. 133–139.)
37. Mirsalimov, V. M. Some elastic–plastic problems for a plane weakened by double periodic system of round holes. Zhurn. Prikl. Mekhaniki i Tekhn. Fiziki, 1974, no. 4, pp. 133–138.
38. Mirsalimov, V. M. Equal-strength opening in rock mass. Fiz.–Tekhn. Problemi Razrabotki Poleznikh Iskopaemikh, 1979, no. 4, pp. 24–29.
39. Mikhlin, C. G. Mathematical theory of plasticity. In Some New Problems of Continuum Mechanics, C. A. Khristianovich, C. G. Mikhlin, and B. B. Davison, eds. Moscow/Leningrad: Izd-vo AN SSSR, 1938, pp. 155–216.
40. Monakhov, V. N. Boundary value problems with free boundaries for elliptic systems of equations. Novosibirsk: Nauka, 1977. 424 pp.
41. Olszak, W., Rychlewski, J., and Urbanovski, V. Plasticity under nonhomogeneous conditions. Moscow: Mir, 1964. 156 pp. (Translated from English; originally published by Academic Press, New York/London, 1962.)
42. Ostrosablin, N. I. On an elastic–plastic problem. Fiz.–tekhn. Problemi Razrabotki Poleznikh Iskopaemikh, 1969, no. 4, pp. 118–121.
43. Ostrosablin, N. I. Elastic–plastic problem for a plane with two identical round holes. Dinamika Sploshnoy Sredi / In-t gidrodinamiki SO AN SSSR, Novosibirsk, 1970, no. 4, pp. 114–118.
44. Ostrosablin, N. I. Elastic–plastic stress distribution in a plane weakened by a finite number of circular apertures. Prikl. Mekhanika, 1973, vol. 9, no. 10, pp. 124–128 (Soviet Appl. Mech., May 15, 1975, pp. 124–127).
45. Ostrosablin, N. I. Determination of displacement in the Galin problem. Dinamika Sploshnoy Sredi / In-t Gidrodinamiki SO AN SSSR, Novosibirsk, 1973, no. 14, pp. 67–70.
46. Ostrosablin, N. I. Elastic–plastic stress distribution near round opening under exponential yield condition. Gornoe davlenie v kapitalnikh i podgotovietelnikh vyrabotkakh / In-t gornovo dela SO AN SSSR, Novosibirsk, 1973, pp. 32–39.
47. Ostrosablin, N. I. On an elastic–plastic problem for a plane with a hole. Dinamika Sploshnoy Sredi / In-t Gidrodinamiki SO AN SSSR, Novosibirsk, 1977, no. 28, pp. 81–88.
48. Ostrosablin, N. I. Equal strength hole in a plate under discontinuous stress state. Zhurn. Prikl. Mekhaniki i Tekhn. Fiziki, 1981, no. 2, pp. 155–63.
49. Parasyuk, O. C. Elastic–plastic problem with non-biharmonic plastic state. Dokl. AN SSSR, 1948, vol. 63, no. 4, pp. 367–370
50. Perlin, P. I. Elastic–plastic stress distribution around holes. In Investigations in Mechanics and Applied Mathematics. Moscow: Oborongiz, 1960, pp. 30–40. (Tr. Mosk. Fiz.–Tekhn. in-ta, no. 5).
51. Perlin, P. I. Approximate method of solution of elastic–plastic problems. Inzh. Sb., 1960, vol. 28, pp. 145–150.
52. Perlin, P. I. Solution of plane elastic–plastic problems for doubly connected regions. Inzh. Zhurn., 1961, vol. 1, no. 4, pp. 68–76.
53. Protodiyakonov, M. M. Generalized equation of envelopes to limit Mohr stress circles. In Investigation of Physico-Mechanical Properties of Rocks as Applied to the Problems of controlling Rock Pressures. Moscow: Izd-vo AN SSSR, 1962, pp. 27–38.
54. Rabotnov, U. N. and Stankevich, O. F. Experimental detection of plastic zones on models from titanium alloys. Izv. AN SSSR. Mekhanika, 1965, no. 2, pp. 108–109.
55. Reva, T. L. Biharmonic solutions of problems for elastoplastic bodies. Prikl. Mekhanika, 1971, vol.

7, no. 4, pp. 130–133 (Soviet Appl. Mech., Sept. 15, 1973, pp. 460–462).
56. Reva, T. L. Application of rational functions to the solution of elastoplastic problems. Prikl. Mekhanika, 1974, vol. 10, no. 11, pp. 119–124 (Soviet Appl. Mech., March 20, 1976, pp. 1252–1255).
57. Savin, G. M. and Parasyuk, O. C. Influence of nonuniform stress field on the plastic zone near a hole. Dopovidi AN UkRSR, 1948, no. 3, pp. 41–50.
58. Savin, G. M. and Parasyuk, O. C. Plastic zones around a round hole in plane nonuniform stress field. Naukovi Zapiski Lvovskovo Derzhavnovo Universiteta. Ser. Fiz–Mat., 1949, vol. 12, no. 3, pp. 5–20.
59. Sazhin, V. C. Elastic–plastic stress distribution around a near-square hole. Inzh. Zhurn., 1964, vol. 4, no. 2, pp. 364–368.
60. Sazhin, V. C. Elastic–plastic problem for infinite plane with a square hole. Prikl. Mekhanika, 1965, vol. 1, no. 11, pp. 134–137.
61. Sazhin, V. C. Elastic–plastic stress distribution around a square, oval, and vaulted openings. In Bases, Foundations, and Underground Structures. Moscow, Stroyizdat, 1967, pp. 84–91.
62. Sazhin, V. C. Determination of a non-elastic deformation region under variable rock cohesion. Fiz.–Tekhn. Problemi Razrabotki Polesnikh Iskopaemikh, 1967, no. 6, pp. 93–95.
63. Teplitsky, E. Boundary equilibrium plane problem of the theory of elasticity: contact of a punch with deformable half space. Prikl. Mekhanika, 1957, vol. 3, no. 3.
64. Fomin, V. L. Elastic–plastic equilibrium of a plane with a round notch under stationary temperature field. Uchen. Zap. Leningr. un-ta. Mekhanika, 1960, Vol. 280, no. 35, pp. 132–135.
65. Khoma, I. U. Plastic zones near a circular hole in a plane nonuniform stress field under normal and shear forces. Prikl. Mekhanika, 1963, vol. 9, no. 2, pp. 167–173.
66. Cherepanov, G. P. On a class of exact solutions of plane elastic–plastic problems. Izv AN SSSR. Otd-nie Tekhn. Nauk, Mekhanika i Mashinostroenie, 1963, no. 3, pp. 95–103.
67. Cherepanov, G. P. On a method of solving elastic–plastic problems. Prikl. Matematika i Mekhanika, 1963, vol. 27, no. 3, pp. 428–435 (J. Appl. Math. Mech.).
68. Cherepanov, G. P. Boundary value problems with analytical coefficients. Dokl. AN SSSR, 1965, vol. 161, no. 2, pp. 312–314.
69. Cherepanov, G. P. On quasibrittle fracture. Prikl. Matematika i Mekhanika, 1968, vol. 32, no. 6, pp. 1034–1042 (J. Appl. Math. Mech., pp. 1050–1058).
70. Cherepanov, G. P. Inverse problems of the plane theory of elasticity. Prikl. Matematika i Mekhanika, 1974, vol. 38, no. 6, pp. 963–974 (J. Appl. Math. Mech., pp. 915–930).
71. Cherepanov, G. P. An inverse problem of the theory of elasticity. Inzh. Zhurn. Mekhanika Tverdovo Tela, 1966, no. 3, pp. 119–130.
72. Chernousko, F. L. Method of local variations for a numerical solution of variational problems. Zhurn. Vychislit, Matematiki i Mat. Fiziki 1965, vol. 5, no. 4, pp. 749–754.
73. Shapiro, G. C. Elastic–plastic equilibrium of a wedge and discontinuous solutions in the theory of ideal plasticity. Prikl. Matematika i Mekhanika, 1952, vol. 16, no. 1, pp. 101–106.
74. Erlikhman, F. M. Determination of displacements in the Galin problem. Dinamika Sploshnoy Sredi / In-t Gidrodinamiki SO AN SSSR, Novosibirsk, 1970, no. 4, pp. 131–134.
75. Ault, R. T. and Spretnak, J. W. Initial yielding and fracture in notched molybdenum. Intern. J. Mech. Sci., 1965 vol. 7, no. 2, pp 87–102.
76. Dugdale, D. S. Yielding of steel sheets containing slits. J. Mech. Phys. Solids, 1960, vol. 8, no. 2, pp. 100–104.
77. Forman, R. G. Experimental program to determine effect of crack buckling and specimen dimensions on fracture toughness of thin sheet materials. Aerospace Res. Lab. Rep., N AFFDL-TR-65-146, Wright-Patterson Air Force Base, 1966.
78. Gerberich, W. W. Plastic strains and energy density in cracked plates. J. Experimental Techniques and Results Experimental Mech., 1964, pp. 335–344.
79. Hahn, G. T. and Rosenfield, A. R. Local yielding and extension of a crack under plane stress. Acta Metallurg, 1965, vol. 14, no. 3, pp. 293–306.
80. Hahn, G. T. and Rosenfield, A. R. Experimental determination of plastic constraint ahead of a sharp crack under plane strain conditions. Ship Structure Committee Report, SSC-180, 1966.
81. Hill, R. On discontinuous plastic states with special reference to a localized necking in thin sheets. J. Mech. and Phys. Solids, 1952, vol. 1, no. 1.
82. Lee, C. H. and Kobayashi, Sh. Elastoplastic analysis of plane-strain and axisymmetric flat punch indentation by the finite-element method. Intern. J. Mech. Sci., 1970. vol. 12, no. 4, pp. 349–370.
83. Levy, N., Mareal, P. V., Ostergren, W. J., and Rice, J. R. Small scale yielding near a crack in plane strain a finite element analysis. Div. Eng., Brown Univ. Tech. Report MGL40-002-080/1.
84. Nottrot, R. On the numerical solution of the plane elastoplastic problem. J. Engng. Math., 1967, vol. 1, no. 2.
85. Nottrot, R. and Timman, R. General method of solving the plane elastoplastic problem. J. Engng. Math., 1967, vol. 1, no. 1.
86. Rice, J. R. A path independent integral and the approximate analysis of strain concentration by notches and cracks. J. Appl. Mech., Trans. ASME, vol. 35, ser. E, 1968, no. 2, pp. 379–386.
87. Rosenfield A. R., Dai, P. K. and Hahn, G. T. Crack entension and propagation under plane stress. Proc. 1st Intern. Conf. on Fracture, Sendai, Japan, 1965.
88. Southwell, R. and de Allen, D. N. Relaxation methods applied to engineering problems. IV.- Philos. Trans. Roy Soc. London, ser. A, 1950, vol. 242.
89. Wesner, J. W. Jr. and Weinstein. Computerized relaxation applied to the plane-strain indenter. J. Appl. Mech., Trans. ASME ser. D, 1969, vol. 91, no. 4.

CHAPTER 5

1. Birger, I. I. Method of auxiliary deformations in problems of the theory of plasticity. Izv. AN SSSR. Otd-nie Tekhn. Nauk, Mekhanika i Mashinostroenie, 1963, no. 1, pp. 47–56.
2. Birger, I. I. Methods of elastic solutions in the theory of plastic flow. Izv. AN SSSR. Otd-nie Tekhn. Nauk, Mekhanika i Mashinostroenie, 1964, no. 2, pp. 116–118.
3. Vitvitsky, P. M. and Leonov, M. Y. On fracture of a plate with a crack. Prikl. Mekhanika, 1961, vol. 7, no. 5, pp. 516–520.
4. Vitvitsky, P. M. and Leonov, M. Y. Slip bands due to nonuniform deformation of a plate. In Problems of Mechanics of a Real Solid Body, vol. 1, Kiev: Izd-vo AN UkSSR, 1962, pp. 13–28.
5. Vitvitsky, P. M. and Leonov, M. Y. Tension of a plate with a circular hole beyond the elastic limit. Zhurn. Prikl. Mekhaniki i Tekhn. Fiziki, 1962, no. 1, pp. 109–117.
6. Vitvitsky, P. M. Slip bands in thin plates with rectilinear slits under tension. In Stress Concentration, vol. 1. Kiev: Naukova dumka, 1965, pp. 78–85.
7. Vitvitsky, P. M. On the development of plastic deformation at the crack tips in a thin plate. Dopovidi AN UkSSR, 1969, no. 4, pp. 316–320.
8. Galin, L. A. and Cherepanov, G. P. Contact elastic–plastic problem for plates. Dokl. AN SSSR, 1967, vol. 177, no. 1, pp. 56–58.
9. Zaslavsky, B. V. Plastic regions near round hole under biaxial tension of thin plate. Tr. Mosk. Aviatsionnovo in-ta, 1956, vol. 69, pp. 11–38.
10. Kerimov, R. U. Investigation of elastic–plastic stress state and plastic zones in a plate with round holes under tension. Izv. AN AzSSR. Ser. Fiziko-Tekhn. i Matematicheskikh Nauk, 1972, no. 7, pp. 130–134.
11. Korotkikh, U. G. Finite difference solutions of the plane problem in physically nonlinear bodies. Prikl. Mekhanika, 1966, vol. 3, pp. 50–57 (Soviet Appl. Mech., pp. 31–35).
12. Korotkikh, U. G. Computer-aided solution of a nonlinear elastic plane problem and problems of the theory of plasticity for compressible work-hardening material. In Construction Mechanics and Theory of Elasticity. Gorky, 1967. (Tr. Gorkovsk. Inzh.-Stroit. in-ta, vol. 50.)
13. Kudryavtsev, B. A., Parton, V. Z., Peskov, U. A., and Cherepanov, G. P. On a local plastic zone near a crack. Izv. AN SSSR. Mekhanika Tverdovo Tela, 1970, no. 1, pp. 61–64.
14. Kudryavtsev, B. A., Parton, V. Z., and Cherepanov, G. P. Elastic–plastic problem for a plane with rectilinear cracks. Izv. AN SSSR. Mekhanika Tverdovo Tela, 1969, no. 3, pp. 174–176.
15. Leonov, M. Y., Vitvitsky, P. M., and Yarema, C. Y. Plastic bands due to tension of a plate with a crack-like stress raiser. Dokl. AN SSSR, 1963, vol. 148, no. 3, pp. 541–544.
16. Panferov, V. M. Stress concentration under elastic–plastic deformations. Izv. AN SSSR. Otd-nie Tekhn. Nauk, 1954, no. 4, pp. 47–66.
17. Perlin, P. I. Elastic–plastic stress distribution around holes. In Investigations in Mechanics and Applied Mathematics. Moscow, 1960, pp. 30– 40. (Tr. Mosk, Fiziko-Tekhn. in-ta, no. 5.)
18. Rybakina, O. G. Tension of a strip with holes under large plastic deformations. Izv. AN SSSR, Mekhanika i Mashinostroenie, 1963, no. 4, pp. 144–146.
19. Sokolov, A. P. On elastic–plastic state of a plate. Dokl. AN SSSR, 1948, vol. 60, no. 1, pp. 33–36.
20. Solodilov, U. I. Elastic–plastic stress distribution in a plate with oval holes. Inzh. zhurn., 1961, vol. 1, no. 4, pp. 170–172.
21. Faerberg, I. I. Tension of a plate with a round hole beyond the elastic limit. Trudi TsAGI, 1947, no. 615, pp. 1–13.
22. Khoma, I. U. Stress concentration in a thin plate weakened by infinite number of round holes under elastic–plastic deformations. Transactions of IV All-Union Conference on the Theory of Shells and Plates, 1962, Erevan, 1964, pp. 959–964.
23. Cherepanov, G. P. On a method of solving elastic–plastic problems. Prikl. Matematika i Mekhanika, 1963, vol. 27, no. 3, pp. 428–435 (J. Appl. Math. Mech.).
24. Cherepanov, G. P. Solution of some problems of the theory of elasticity and plasticity with unknown boundaries. Prikl. Matematika i Mekhanika, 1964, vol. 28, no. 1, pp. 141–144 (J. Appl. Math. Mech.).
25. Cherepanov, G. P. The Riemann–Hilbert problem for the exterior of slits along a straight line or along a circle. Dokl. AN SSSR, 1964, vol. 156, no. 2, pp. 275–277.
26. Cherepanov, G. P. Boundary value problems with analytical coefficients. Dokl. AN SSSR, 1965, vol. 161, no. 2, pp. 312–314.
27. Cherepanov, G. P. On quasibrittle fracture. Prikl. Matematika i Mekhanika, 1968, vol. 33, no. 6, pp. 1034–1042 (J. Appl. Math. Mech., pp. 1050–1058).
28. Goodier, J. N. and Field, F. A. Plastic energy dissipation in crack. In Fracture of Solids, New York: Wiley Interscience, 1963.
29. Dugdale, D. S. Yielding of steel sheets containing slits. J. Mech. Phys. Solids, 1960, vol. 8, no. 2.
30. Southwell, R. and Allen, D. N. de G. Relaxation methods applied to engineering problems. Philos. Trans. Roy. Soc., London, ser. A, 1950, vol. 242, no. 850.

CHAPTER 6

1. Aleksandrov, A. Y. and Solovyov, U. M. Space problems of the theory of elasticity. Applications of the theory of complex variable functions. Moscow: Nauka, 1978. 462 pp.
2. Annin, B. D. Modern models of plastic bodies. Novosibirsk: Izd. Novosib. un-ta, 1975.
3. Annin, B. D. Group properties and exact solutions of plasticity equations of Mises and Tresca. In

Theoretical and Applied Mechanics. Trudi IV Kongress, vol. 1. Sofia: BAN, 1981. pp. 644–649.
4. Burago, N. G. and Kukugzhanov, V. N. Numerical method of solving geometrically nonlinear, axisymmetric problems for elastic–plastic shells of revolution. Stroitelnaya Mekhanika i Raschyot Sooruzheniy, 1976, no. 5, pp. 44–49.
5. Volchkov, U. M. and Korobeinikov, C. N. Numerical solution of elastic–plastic problems in the theory of shells. In Transactions of the V All-Union Conference on Numerical Methods of Solving Problems of the Theory of Elasticity and Plasticity, Part II. Novosibirsk: Izd. In-ta Teoret. i Prikl. Mekhaniki SO AN SSSR, 1978, pp. 40–47.
6. Volchkov, U. M. and Korobeinikov C. N. Estimation of limit loading for elastic–plastic shells of revolution. Zhurn. Prikl. Mekhaniki i Tekhn. Fiziki, 1981, no. 4, pp. 146–150.
7. Volchkov, U. M. and Pavlov A. V. Elastic–plastic deformation of a cylindrical shell with cross fins under hydrostatic pressure. In Transactions of VII All-Union Conference on the Theory of Shells and Plates. Moscow: Nauka, 1969, pp. 148–152.
8. Vulman, C. A. On the solution of axisymmetric elastic–plastic problems by the perturbation method. Izv. AN SSSR. Mekhanika Tverdogo Tela, 1969, no. 3, pp. 166–169.
9. Vulman, C. A. Approximate solution of elastic–plastic problems for hollow bodies with near-spherical surface. Izv. AN SSSR. Mekhanika Tverdogo Tela, 1971, no. 1, pp. 119–122.
10. Vulman, C. A. Solution of axisymmetric elastic–plastic problems for bodies of compressible material. Prikl. Mekhanika, 1971, vol. 7, no. 7, pp. 91–94.
11. Ershov L. V. Elastic–plastic state near a spherical cavity. Izv. AN SSSR. Otd-nie Tekhn. Nauk, Mekhanika i Mashinostroenie, 1960, no. 6, pp. 155–156.
12. Zadoian M. A. On a particular solution of the ideal plasticity equations. Dokl. AN SSSR, 1964, vol. 156, no. 1, pp. 38–49.
13. Ivanov, G. V. Elastic–plastic flow of shells under the Mises yield condition. Izv. AN SSSR. Mekhanika Tverdovo Tela, 1969, no. 3, pp. 85–90.
14. Kerimov, R. U. Stress distribution near a spherical cavity under elastic–plastic deformation. Dopovidi AN UkSSR, Ser. A, Fiz.-Mat. ta Tekhn. Nauka, 1976, no. 7, pp. 624–628.
15. Leonov, M. Y. and Panasyuk, V. V. Development of a crack having the shape of a circle in a plane. Dokl. AN UkSSR, 1961, no. 2, pp. 165–168.
16. Panarelli, J. E. and Hodge, P. G. Interaction of pressure, end load, and twisting moment for a rigid-plastic circular tube. Trans. ASME, ser. E, Applied Mechanics, vol. 30, no. 3, pp. 396–400.
17. Prager, W. Three-dimensional plastic flow under uniform stresses. Mekhanika. Sb. Perevodov i Obzorov Inostr. Lit., 1958, no. 3, pp. 23–27.
18. Semykina, T. D. On triaxial tension of elastic–plastic space weakened by a spherical cavity. Izv. AN SSSR. Otd-nie Tekhn. Nauk, Mekhanika i Mashinostroenie, 1963, no. 1, pp. 173–177.
19. Senashov, C. I. Invariant space solutions of the equations of ideal plasticity. Zhurn. Prikl. Mekhaniki i Tekhn, Fiziki, 1980, no. 3, pp. 159–163.
20. Ayres, D. J. A numerical procedure for calculating stress and deformation near a slit in three-dimensional elastic plastic solid. Eng. Fract., 1970, vol. 2, no. 2, pp. 143–156.
21. Francis, P. H. On the extent of plastic yielding surrounding a surface "inclusion" under remote tension, Int. J. Mech. Sci., 1971, vol. 13, pp. 461–469.

APPENDIX 1

1. Annin, B. D. On a problem with unkown boundaries for Poisson equation in space. In Partial Differential Equations and Problems with Free Boundaries. Kiev: Naukova dumka, 1983, pp. 12–15.
2. Kudryavtsev, L. D. On differentiable mappings. Dokl. AN SSSR, 1955, vol. 104, no. 1, pp. 9–11.
3. Ovsyannikov, L. V. Group analysis of differential equations. Moscow: Nauka, 1978, 399 pp.
4. Lie, S. Drei Kapitel aus dem unvollendeten zweite Bande der Geometrie der Berührungstransformationen. Math. Ann., 1907. Bd. 59, H. 1, 2.

APPENDIX 2

1. Arzhanov, G. V. On a nonlinear boundary value problem of the Riemann type. Sib. Mat. Zhurn., 1961, vol. 2, no. 4, pp. 481–504.
2. Solovyov, P. V. On a boundary problem in the theory of analytical functions. Dokl. AN SSSR, 1941, vol. 33, no. 3, pp. 190–192.
3. Cherepanov, G. P. On a nonlinear boundary problem in the theory of analytical functions encountered in some elastic–plastic problems. Dokl. AN SSSR, 1962, vol. 147, no. 3, pp. 566–568.
4. Cherepanov, G. P. Inverse elastic–plastic problem under conditions of antiplane deformation. Prikl. Matematika i Mekhanika, 1962, vol. 26, no. 6, pp. 1145–1147 (J. Appl. Math. Mech.).